Joachim W. Deitmer

# CALCIUM WAVES, GRADIENTS AND OSCILLATIONS

Ciba Foundation Symposium 188

# CALCIUM WAVES, GRADIENTS AND OSCILLATIONS

1995

JOHN WILEY & SONS

Chichester · New York · Brisbane · Toronto · Singapore

Published in 1995 by John Wiley & Sons Ltd
Baffins Lane, Chichester
West Sussex PO19 1UD, England
Telephone (+44) (243) 779777



*Other Wiley Editorial Offices*

John Wiley & Sons, Inc., 605 Third Avenue,
New York, NY 10158-0012, USA

Jacaranda Wiley Ltd, G.P.O. Box 859, Brisbane,
Queensland 4001, Australia

John Wiley & Sons (Canada) Ltd, 22 Worcester Road,
Rexdale, Ontario M9W 1L1, Canada

John Wiley & Sons (SEA) Pte Ltd, 37 Jalan Pemimpin #05-04,
Block B, Union Industrial Building, Singapore 2057

Suggested series entry for library catalogues:
Ciba Foundation Symposia

Ciba Foundation Symposium 188
ix+291 pages, 55 figures, 3 tables

*Library of Congress Cataloging-in-Publication Data*

Calcium waves, gradients and oscillations / [editors, Gregory R. Bock (organizer) and Kate Ackrill].
p. cm.—(Ciba Foundation symposium ; 188)
"Symposium on Calcium waves, gradients and oscillations, held at the Ciba Foundation, London, 26–28 April 1994."
Includes bibliographical references and index.
ISBN 0 471 95234 6
1. Calcium ions—Physiological effect—Congresses. 2. Cellular signal transduction—Congresses. 3. Calcium ions—Physiological transport—Congresses. 4. Calcium channels—Congresses. I. Bock, Gregory. II. Ackrill, Kate. III. Series.
QP535.C2C267 1995
574.87—dc20 94-42856
CIP

*British Library Cataloguing in Publication Data*

A catalogue record for this book is
available from the British Library

ISBN 0 471 95234 6

Phototypeset by Dobbie Typesetting Limited, Tavistock, Devon.
Printed and bound in Great Britain by Biddles Ltd, Guildford.

# Contents

*Symposium on Calcium waves, gradients and oscillations, held at the Ciba Foundation, London, 26–28 April 1994*

*This symposium is based on a proposal made by Lionel Jaffe*

*Editors: Gregory R. Bock (Organizer) and Kate Ackrill*

# Participants

**M. J. Berridge** *(Chairman)* The Babraham Institute Laboratory of Molecular Signalling, Department of Zoology, University of Cambridge, Downing Street, Cambridge CB2 3EJ, UK

**M. Cannell** Department of Pharmacology & Clinical Pharmacology, St George's Hospital Medical School, Cranmer Terrace, Tooting, London SW17 0PE, UK

**P. H. Cobbold** Department of Human Anatomy and Cell Biology, University of Liverpool, PO Box 147, Liverpool L69 3BX, UK

**F. S. Fay** Department of Physiology Program in Molecular Medicine, University of Massachusetts Medical School, 373 Plantation Street, Worcester, MA 01605, USA

**D. Friel** Department of Neurosciences, Case Western Reserve University, 10900 Euclid Avenue, Cleveland, OH 44106-4975, USA

**A. Galione** Department of Pharmacology, University of Oxford, Mansfield Road, Oxford OX1 3QT, UK

**A. Goldbeter** Service de Chemie Physique I, Faculty of Science, Université Libre de Bruxelles, Campus Plaine, CP 231, Boulevard du Triomphe, B-1050 Brussels, Belgium

**R. Irvine** Department of Development and Signalling, AFRC Babraham Institute, Babraham, Cambridge CB2 4AT, UK

**R. Jacob** Physiology Group, Biomedical Sciences Division, King's College London, Campden Hill, London W8 7AH, UK

**L. F. Jaffe** Marine Biological Laboratory, Woods Hole, MA 02543, USA

**H. Kasai** Department of Physiology, Faculty of Medicine, University of Tokyo, Hongo, Bunkyo-ku, Tokyo 113, Japan

**J. Lechleiter** Department of Neuroscience, University of Virginia Health Sciences Center, Charlottesville, VA 22908, USA

**T. Meyer** Department of Cell Biology, Duke University Medical Center, Box 3709, Durham, NC 27710, USA

**R. H. Michell** Centre for Clinical Research in Immunology and Signalling (CCRIS), The Medical School, University of Birmingham, Edgbaston, Birmingham B15 2TT, UK

**S. Miyazaki** Department of Physiology, Tokyo Women's Medical College, 8-1 Kawada-cho, Shinjuku-ku, Tokyo 162, Japan

**M. H. Nathanson** Department of Internal Medicine, Yale University School of Medicine, New Haven, CT 06510, USA

**I. Parker** Laboratory of Cellular and Molecular Neurobiology, Department of Psychobiology, University of California, Irvine, CA 92717-4550, USA

**O. H. Petersen** MRC Secretory Control Research Group, The Physiological Laboratory, University of Liverpool, Crown Street, PO Box 147, Liverpool L69 3BX, UK

**J. W. Putney Jr** NIEHS Cellular and Molecular Pharmacology, Calcium Regulation Section, NIEHS, PO Box 12233, Research Triangle Park, NC 27709, USA

**M. J. Sanderson** Department of Physiology, University of Massachusetts Medical Center, 55 Lake Avenue North, Worcester, MA 01655, USA

**S. J. Smith** Department of Molecular and Cellular Physiology, Beckman Center, Stanford University School of Medicine, Stanford, CA 94305, USA

**J. E. Speksnijder** Department of Genetics, Center for Biological Sciences, University of Groningen, Kerklaan 30, 9751 NN Haren, The Netherlands

**J. W. Stucki** Institute of Pharmacology, University of Berne, Friedbuhlstrasse 49, CH 3010 Berne, Switzerland

**A. P. Thomas** Department of Pathology and Cell Biology, Thomas Jefferson University, Room 271, 1020 Locust Street, Philadelphia, PA 19107, USA

**M. Whitaker** Department of Physiology, University College and Middlesex School of Medicine, University College London, Gower Street, London WC1E 6BT, UK

**W. G. Wier** Department of Physiology, University of Maryland School of Medicine, 655 West Baltimore Street, Baltimore, MD 21201, USA

**R. J. P. Williams** Inorganic Chemistry Laboratory, University of Oxford, South Parks Road, Oxford OX1 3QR, UK

**L. Tao** *(Ciba Foundation Bursar)* Department of Physiology, The Wellcome Research Laboratories, Langley Court, South Eden Park Road, Beckenham, Kent BR3 3BS, UK

# Introduction

Michael J. Berridge

*The Babraham Institute Laboratory of Molecular Signalling, Department of Zoology, University of Cambridge, Downing Street, Cambridge CB2 3EJ, UK*

I would like to welcome you to this meeting. The last Ciba Foundation symposium I attended was in 1985, Calcium and the cell (Ciba Foundation 1986), and quite a few of the people here were also at that meeting. The meeting was chaired by Peter Baker. His premature death was a great loss to the $Ca^{2+}$ field. He would have been fascinated by the developments that have taken place since that time. At that meeting there was almost no discussion of the spatiotemporal aspects of $Ca^{2+}$ signalling. It is really quite remarkable to think that most of what we know now has come about in less than a decade. It all began in 1986 with Peter Cobbold's lovely paper on $Ca^{2+}$ spikes in the hepatocyte (Woods et al 1986). That probably captured a lot of people's imagination. In much of what we're going to talk about in this meeting we shall be trying to understand the basic mechanisms responsible for such $Ca^{2+}$ spiking.

I'm not going to go into great detail on the progress that's been made, but I would like to summarize some of the key problems which I hope we'll deal with during the course of this meeting. One key aspect concerns the mechanisms that initiate the spikes and waves. In many cells there is a clear initiation site from which these $Ca^{2+}$ waves propagate throughout the cell. We shall also be interested in what happens when such waves impinge upon the nucleus. There's a general consensus now that in many cells these waves pass through the nucleus as if there were no barrier there, although this may not apply to all cell types. We shall also be interested in intercellular waves in those cases where there are cells connected together into epithelia or into sheets by gap junctions. One of the key questions here is the nature of the signal that's actually passing through the gap junction in order to perpetuate the wave in the next cell.

Another important aspect concerns the question of graded versus all-or-none $Ca^{2+}$ release. This problem arose first in the case of cardiac cells, in which, interestingly, the release, although it's all-or-none, is a regenerative process, yet can be graded with membrane potential. I raised this problem because there are indications that graded release is also found in non-muscle cells. The paradox we have to consider is how graded release can occur even though the release mechanism seems to depend upon an all-or-none regenerative process. A solution to this paradox may depend upon the role of luminal $Ca^{2+}$. One possible explanation for graded release is that cells have a population of stores with different sensitivities depending on the extent to which they are loaded with $Ca^{2+}$.

In many cells, the loading of the stores depends on $Ca^{2+}$ entry from the outside. David Friel has demonstrated that in sympathetic neurons there is a constant $Ca^{2+}$ influx from the outside responsible for recharging the internal stores, preparing them for $Ca^{2+}$ mobilization. But, of course, in non-excitable cells, there are various other mechanisms of $Ca^{2+}$ entry. Shun-ichi Miyazaki will be discussing the role of inositol 1,3,4,5-tetrakisphosphate in mammalian eggs. In the case of cardiac muscle, the entry of $Ca^{2+}$ through the voltage-operated gate actually provides the trigger for the initiation of $Ca^{2+}$ release. This entry of trigger $Ca^{2+}$ in cardiac cells may be somewhat analogous to what might be happening at the initiation site in non-muscle cells, where a small pulse of $Ca^{2+}$ is released spontaneously and then acts as a trigger to initiate a propagated $Ca^{2+}$ wave. This phenomenon has been described in *Xenopus* oocytes by Ian Parker, who has used the confocal microscope to study unitary events at the initiation site and how these then trigger a propagated wave.

In some cells, such as the mammalian egg, the propagation of $Ca^{2+}$ waves depends on the inositol 1,4,5-trisphosphate ($InsP_3$) receptor system, whereas in other cells, neurons for example, we are dealing with the ryanodine receptor. It has been interesting to find over the past few years that these two receptors seem to be functionally equivalent in terms of their capacity to generate spiking behaviour. In many cell types there are both $InsP_3$ receptors and ryanodine receptors, regulated by $InsP_3$ and $Ca^{2+}$, or, putatively, by $Ca^{2+}$ and cyclic ADP-ribose, respectively. When we started to think about this, in some modelling work we did with Albert Goldbeter, we had the idea that the $InsP_3$ system was a sort of priming system, with the ryanodine receptor acting as an amplification system. But the work that Peter Thorn and Ole Petersen have been doing is beginning to turn that idea on its head. They have suggested that the ryanodine receptor provides $Ca^{2+}$ which then recruits further release from the $InsP_3$ receptor system. The implication of all these studies is that we must begin to deal with the functional interactions between these two receptor types.

The way in which waves propagate through the cytoplasm is critically dependent on the organization of the $Ca^{2+}$ stores. We tend to draw these stores as neat little circular packets within the cell, but as Andy Thomas has pointed out, perhaps the system is functionally connected. From a functional point of view, it may not make a lot of difference whether the stores are separate or connected if the mode of propagation is from one intracellular receptor to another. We also have to remember that the nucleus is functionally connected to the endoplasmic reticulum and will also contribute to the movement of the wave. Andy Thomas and Tobias Meyer will be discussing the nature of the $Ca^{2+}$ signalling system within the nuclear compartment.

Finally, we need to consider in some detail the functional significance of oscillations and waves. Much attention has been focused on the mechanisms responsible for these spatiotemporal aspects and we haven't paid enough attention to their functional significance.

## References

Ciba Foundation 1986 Calcium and the cell. Wiley, Chichester (Ciba Found Symp 122)
Woods NM, Cuthbertson KSR, Cobbold PH 1986 Repetitive transient rises in cytoplasmic free calcium in hormone-stimulated hepatocytes. Nature 319:600–602

# Calcium waves and development

Lionel F. Jaffe

*Marine Biological Laboratory, Woods Hole, MA 02543, USA*

*Abstract.* Those calcium oscillations which go deep into cells take the form of 'fast' calcium waves. In fully active cells at room temperature, these move at 15–30 μm/s and are propagated by a reaction–diffusion mechanism governed by the Luther equation in which calcium ions are the only propagators and calcium-induced calcium release is the only reaction. However, they may be initiated by a second mode of $Ca^{2+}$-induced $Ca^{2+}$ release within the lumen of the endoplasmic reticulum (ER). In sea urchin fertilization, this second mode of $Ca^{2+}$-induced $Ca^{2+}$ release is in turn begun by calcium entering the sperm and thence the ER.

Subsurface calcium waves include an important class of surface contraction waves which move at 0.3–3 μm/s and are called 'slow' waves. Their prototype is the 0.5 μm/s wave which accompanies and controls cytokinesis in large eggs. Slow waves may be propagated by mechanical tension rather than by diffusion.

Recent work with *Dictyostelium* transfected with apoaequorin has provided the first views of free calcium patterns within a developing, multicellular organism. During most or all of development, those regions which will differentiate into stalk or stalk-like cells (as opposed to spores) exhibit frequent calcium pulses. These pulses are believed to be fast calcium waves and to feed back on these regions so as to favour non-spore differentiation.

*1995 Calcium waves, gradients and oscillations. Wiley, Chichester (Ciba Foundation Symposium 188) p 4–17*

## Fast calcium waves

We discovered fast calcium waves about 20 years ago in medaka fish eggs undergoing fertilization. First, we determined the course of light emission from an aequorin-loaded egg using a photomultiplier. Soon after insemination, light emission rises about 20 000-fold, indicating a 300-fold rise in free cytosolic calcium, which reaches a concentration of about 30 μM before slowly falling back to the resting level. The egg's cytosolic $Ca^{2+}$ concentration $[Ca^{2+}]_i$ rises exponentially, which indicates the operation of an autocatalytic mechanism—now called calcium-induced calcium release. Moreover, the exponential shape of this concentration increase indicates that a single calcium ion induces its own release (Ridgway et al 1977). When we observed such eggs with an image intensifier, we saw that this huge fertilization pulse takes the form of a wave, a moving band of high $[Ca^{2+}]_i$ which crosses the entire 1 mm wide egg at about 10 μm/s. We also saw that this wave fills the whole cytosol with high $[Ca^{2+}]_i$ (Gilkey et al 1978). Gilkey (1983) later confirmed that $Ca^{2+}$-induced $Ca^{2+}$ release propagates the wave, by injecting eggs with various calcium buffers using a method which rapidly raises subsurface $[Ca^{2+}]$ to a known level; in these conditions, activation waves could be regularly initiated at a trigger concentration of about 5 μM, a $[Ca^{2+}]$ well below the peak level (Gilkey 1983). Calcium waves moving at about the same speed are now known to initiate the development of a very wide variety of (and perhaps all) eggs (Jaffe 1991, 1993). Moreover, in the sea urchin egg, confocal microscopy has vividly confirmed that the fertilization wave goes deep into the egg so as to cross the whole cytosol as well as the nucleus (Stricker et al 1992).

Both ascidian and mammalian eggs also exhibit a series of periodic, post-fertilization waves which move two or three times faster than the fertilization wave which activates the eggs. Similar fast calcium waves underlie a wide variety of, or perhaps all, calcium signals that go deep inside cells and have been found recently to traverse certain multicellular tissues including brain slices and reconstituted airway epithelia (this volume: Dani & Smith 1995, Sanderson 1995). The waves in brain slices—which are propagated through the glial rather than the neural cells—are of particular medical interest because they may well underlie migraine attacks as well as other forms of spreading brain injury (Jaffe 1991, Leibowitz 1992).

The velocities of the fast, deep calcium waves which fully traverse active cells and tissues under more or less physiological circumstances are remarkably conserved, being about 15–30 μm/s at room temperature and increasing about 1.8-fold per 10 °C rise. By deep waves, I mean those travelling more than a few micrometres below the plasma membrane, including its extensions such as the transverse tubules of cardiac muscle cells. For reasons that are presented elsewhere (Jaffe 1993), I suspect that the somewhat faster (100 μm/s) subsurface waves seen in cultured ventriculocytes involve $Ca^{2+}$-induced $Ca^{2+}$ release speeded by calcium influx, whereas the somewhat slower (3–10 μm/s) subsurface(?) waves seen in short-term (but not long-term) cultures of vascular myocytes are propagated by $Ca^{2+}$-induced $Ca^{2+}$ entry rather than $Ca^{2+}$-induced $Ca^{2+}$ release. In any case, the conservation of velocity in deep calcium waves suggests that they are all propagated through cells by the same mechanism: but what is that general mechanism?

The mechanism is almost certainly among the reaction–diffusion mechanisms which are exemplified by the Belousov–Zhabotinsky reaction and governed by the Luther equation. Moreover, as stated above, the exponential rise in $[Ca^{2+}]_i$ seen in the fish egg on fertilization indicates that a single calcium ion induces its own release. For such reactions, the Luther equation is given by $v=2(Dk)^{0.5}$, where $v$ is the wave's speed, $D$ the propagator's diffusion constant and $k$ the autocatalytic reaction constant (given by the reciprocal of the time taken for the concentration of the autocatalyst or propagator to rise e-fold). Then, using the Luther equation, together with measurements of $v$ and $k$ from both fish eggs and hepatocytes, one learns (contrary to Albritton et al 1992) that the propagator's diffusion constant is comparable to that of the calcium ion in diverse cytosolic preparations at the relevant, high ($\approx 10\,\mu$M) calcium concentration. So, the data fit this quantitive test of the simple model of $Ca^{2+}$-induced $Ca^{2+}$ release. Furthermore, there is compelling quantitative evidence against inositol 1,4,5-trisphosphate ($InsP_3$) as a second propagator. This evidence is assembled in Jaffe (1993), but the strongest point of this article may be worth repeating here: "... fast waves travel deep inside of cells ... Hence the ... [$InsP_3$ propagator model] ... requires $Ca^{2+}$ to rapidly induce $IP_3$ [$InsP_3$] formation from inner membranes: yet no substantial evidence for such induction ... exists". Altogether, one can conclude that fast calcium waves are propagated by $Ca^{2+}$-induced $Ca^{2+}$ release, as originally proposed.

Although $InsP_3$ is not a propagator of fast calcium waves, it is surely a modulator. There is, for example, good evidence that its concentration rises greatly between fertilization and first cleavage in various eggs (Kamel et al 1985, Stith et al 1993) and that this rise helps account for the greater speed of calcium waves through fully activated cells than through eggs undergoing fertilization.

The mechanisms of initiation of fast waves by diverse stimuli must be more varied than those of propagation; nevertheless, I have proposed that they share a common pathway involving a second mode of $Ca^{2+}$-induced $Ca^{2+}$ release which occurs within the lumen of the endoplasmic reticulum (ER) (Jaffe 1991, 1993). According to this model, a fast wave is started by a conformational shift in a calsequestrin complex, itself triggered by a slow rise in free luminal [$Ca^{2+}$] to somewhere in the millimolar range. The long delays often observed between periodic calcium pulses or between sperm or agonist addition and wave intitiation can be attributed to the inherently slow process of pumping up luminal [$Ca^{2+}$] to the trigger level. The strongest evidence for this hypothesis lies in the biochemical observations of Ikemoto et al (1991) on calcium release from sarcoplasmic reticulum preparations. In the case of fertilization, current work indicates that the calcium which is slowly pumped into the ER flows through the fused sperm from the medium into the narrow space between the egg's plasma membrane and its subsurface ER. In short, this work supports the conduit model of egg activation.

## Slow calcium waves

Slow calcium waves were first seen a few years ago in cleaving fish eggs (Fluck et al 1991) and cleaving frog eggs (Miller et al 1991). In the (aequorin-loaded) fish egg, we saw two successive waves of luminescence moving along each of the first three cleavage furrows at a rate of about 0.5 $\mu$m/s. The first waves accompany furrow extension and the second ones accompany the subsequent apposition of the separating cells. If the first waves travel within the assembling contractile band, they would indicate local increases in free [$Ca^{2+}$] to about 5–8 $\mu$M. Similar, if less well studied, calcium waves have also been seen in *Xenopus* eggs. Injecting calcium shuttle buffers into *Xenopus* eggs showed that these slow calcium waves feed back to start, extend and maintain the eggs' furrows (Snow & Nuccitelli 1993, Miller et al 1993a).

Well before we had seen the first slow calcium waves, I had suggested that various, 0.3–3 $\mu$m/s, surface contractile waves which accompany development are driven by propagated increases in calcium (Jaffe 1985). Table 1 now updates that list. It also includes a recent observation by Brodland et al (1994) of a yet slower—0.05 $\mu$m/s—developmental wave. One wonders whether this will prove to be the first recognized member of another, ultraslow, class of developmental waves and whether it, too, is driven by calcium.

Two cogent arguments indicate that these slow, surface contractile waves are propagated by mechanical tension rather than molecular diffusion. First, such a mode of propagation would explain the fact that many of these waves grow out along a straight or curved line rather than radially. How could a diffusion-propagated wave spread in that way? Second, mechanical propagation could easily give such slow waves, whereas the Luther equation indicates that diffusion could not. Slow calcium waves are about 30 times slower than fast ones. This means that: (a) a slow wave's hypothetical propagator would have to diffuse about 100 times slower than calcium ions do in the cytosol; or (b) a slow wave's calcium release reaction would have to be about 100 times slower than that found in fast waves—both implausible possibilities—or (c) *both* would have to be about 10 times slower, which seems unlikely. Furthermore, cleavage furrows can elongate normally in both fish and frog eggs in a calcium-free medium. This indicates that the source of slow-wave calcium is internal, perhaps in the ER. This, in turn, suggests that the ER in these eggs contains stretch-sensitive calcium channels. A preliminary study has already shown that *Xenopus* egg extracts will repeatedly release large amounts of free calcium in response to controlled mechanical deformation (Miller et al 1993b). However, the deformations needed to release calcium greatly exceed those which accompany cytokinesis. We plan to pursue this question with whole eggs rather than extracts; in doing so we shall also keep in mind the interesting possibility that even slow calcium waves may sometimes go deep inside cells. Aspects of the suppression of cleavage by calcium shuttle buffers suggest that the furrowing signal is carried to the surface by a slowly growing structure called the diastema (Miller et al 1993a) or telophase disc (Margolis & Andreassen 1993). Because this structure seems to grow at slow-wave speed, and because internal tensions may be ubiquitous within animal cells (Ingber 1993), one wonders whether its growth is driven by an internal slow calcium wave.

**TABLE 1 Velocities of slow calcium waves accompanying development**

| *Organism* | *Stage* | *Speed ($\mu m/s$)* | *Reference* |
|---|---|---|---|
| Eggs | | | |
| Barnacle | Fertilization to cleavage | 0.3 | Lewis et al (1973) |
| Insect | Nuclear division | 0.4 | Wolf (1985) |
| Fish | Cytokinesis | 0.6 | Fluck et al (1991) |
| Amphibian | Post-fertilization | 1.1 | Hara et al (1977) |
| | Pre-cleavage | 0.8 | Yoneda et al (1982) |
| | Cytokinesis | 0.5 | Miller et al (1991) |
| Multicellular systems | | | |
| *Dictyostelium* | Close-packed aggregates | $0.7 \pm 0.1$[a] | Gerisch (1965) |
| | Mounds | 1.0 | F. Siegert & C. Weijer, personal communication (1994) |
| | Slugs | 0.8–1.2 | Durston & Vork (1979), Breen & Williams (1994) |
| Insect | Anatrepsis | 0.4–0.7 | Vollmar (1972) |
| Ascidian | Pre-metamorphosis | 1.0–3.0 | Torrence & Cloney (1981) |
| Chick | Stage 12, along axis | 3.0 | Robertson (1979) |
| | Stage 12, across axis | 1.0 | Robertson (1979) |
| Amphibian | Neural induction | 0.05[b] | Brodland et al (1994) |

[a]Waves through cell populations which are too sparsely plated to contact each other move more rapidly, e.g., 5 $\mu m/s$ (Devreotes 1982).
[b]This may be an ultraslow calcium wave, belonging to a different class (see text).

## Calcium patterns in developing *Dictyostelium*

A high local calcium concentration is essential for the initiation and maintenance of local growth in root hairs, in rhizoids, in pollen tubes and perhaps in all tip-growing plant cells. In fucoid eggs, for example, rhizoid initiation and the establishment of cell polarity can be suppressed by injecting a calcium shuttle buffer to suppress calcium gradients (Speksnijder et al 1989). One can also see high calcium zones at the poles of various animal eggs (Miller et al 1994) and at the rear end of polarizing eosinophils (Fay et al 1995, this volume). However, it seems best to focus here on our findings with *Dictyostelium*: not only are they new and exciting but they also support an idea put forward at an earlier Ciba Foundation symposium nearly a decade ago (Jaffe 1986).

At that symposium, it was speculated that the so-called antimeristemoid cells of Bünning would prove to be high calcium cells. If there ever were cells which fit this simple conception, it is the future stalk or stalk-like cells rather than the future spore cells of *Dictyostelium*. Moreover, Maeda (1970) long ago argued that a high free calcium concentration was a key to differentiation into stalk or stalk-like cells rather than spores. However, exploration of this idea has been frustrated by our inability to see patterns of free calcium within developing multicellular organisms. We have now overcome this difficulty by effectively transfecting *Dictyostelium* with apoaequorin, producing high concentrations of aequorin in such clones by exposing them to the cofactor coelenterazine, and visualizing the resultant patterns of luminescence with an ultra-low light-imaging system (A. B. Cubitt et al, unpublished results).

Our most striking finding is the consistent presence of a high calcium zone within the anterior quarter of the slug together with a secondary one which caps its posterior end. It is well known that these are the loci of future stalk or stalk-like cells. Moreover, a ring of high free calcium appears much earlier in development, at the base of the newly formed mound. There is good evidence that this early ring also contains cells destined to form stalk or stalk-like cells. Their is strong, direct evidence that future non-spore cells contain high concentrations of free calcium in both the mound and the slug stages (as well as in the final culminant stages) and thus throughout the multicellular development of *Dictyostelium*. There is also good indirect evidence that this association of high free calcium [$Ca^{2+}$] with a non-spore fate starts even earlier, at the very beginning of differentiation when the newly starved amoebae are still isolated and those cells which happen to be in the appropriate (S?) phase of the cell cycle start on the non-spore pathway. This evidence includes the finding that such cells soon develop high concentrations of *sequestered* calcium (Saran et al 1994) and the fact that regions of high sequestered calcium and high free calcium concentration appear to be colocalized in slugs. When one combines this evidence with the near certainty that the free calcium concentration within non-spore cells rises yet further as they die to complete their differentiation, one sees that a high calcium concentration does indeed characterize non-spore or antimeristemoid differentiation from start to finish.

There is also growing evidence that intervention to raise free calcium concentration favours non-spore differentiation in *Dictyostelium*, whereas lowering it favours differentiation to spore cells. For example, application of the calcium pump poison thapsigargin or of the calcium ionophore A23187 has been found recently to favour non-spore differentiation; so does 100 $\mu$M cyclic AMP (Chia 1975), a stimulus which induces the large, rapidly rising calcium pulses which are characteristic of fast calcium waves (Schlatterer et al 1994). This indicates that a high free calcium concentration is more than just a consequence of non-spore differentiation: it apparently feeds back to favour such differentiation.

It is of particular interest here that the zones of high free calcium seen in developing *Dictyostelium* consist largely of semiperiodic, rapidly rising calcium pulses which are almost certainly fast calcium waves. This raises the interesting question of what restricts fast calcium waves to future non-spore regions. One reason may be that the larger intercellular spaces found in pre-spore regions dilute the intercellular wave transmitter, cyclic AMP, to a concentration too low to relay fast waves. In any case, we now seem to have a considerable insight into the role of fast calcium waves in differentiation of *Dictyostelium*. As a spreading but self-controlled signal, they feed back to maintain the differentiation of whole, multicellular regions. By analogy with molecular phases (such as viral tactoids), the non-spore or antimeristemoid regions can be thought of as groups of cells which are kept separate from other groups by an interaction which supports fast calcium waves.

## *Acknowledgements*

The work on sea urchin fertilization was supported by a Universal Imaging Corporation Scholarship to Robbert Créton and NSF grant DCB-9103569 to L. F. J.; the work on *Dictyostelium* was supported by this same grant plus NIH grant GM37830 to Dr Richard A. Firtel.

## References

Albritton NL, Meyer T, Stryer L 1992 Range of messenger action of calcium ion and inositol 1,4,5-trisphosphate. Science 258:1812–1815

Breen EJ, Williams KL 1994 Optical flow analysis of the ventral cellular layer of the migrating *Dictyostelium discoideum* slug. Microbiology 140:1241–1252

Brodland GW, Gordon R, Scott MJ et al 1994 Furrowing surface contraction wave coincident with primary neural induction in amphibian embryos. J Morphol 219:131–142

Chia WK 1975 Induction of stalk cell differentiation by cyclic-AMP in a susceptible variant of *Dictyostelium discoideum*. Dev Biol 44:239–252

Cubitt AB, Firtel RA, Fischer G, Jaffe LF, Karplus E, Miller AL 1993 Patterns of free calcium in developing *Dictyostelium*. Cell, submitted

Dani JW, Smith SJ 1995 The triggering of astrocytic calcium waves by NMDA-induced neuronal activation. In: Calcium waves, gradients and oscillations. Wiley, Chichester (Ciba Found Symp 188) p 195–209

Devreotes PN 1982 Chemotaxis. In: Loomis WF (ed) The development of *Dictyostelium discoideum*. Academic Press, New York, p 117–169

Durston AJ, Vork F 1979 A cinematographic study of the development of vitally stained *Dictyostelium discoideum*. J Cell Sci 36:261–279

Fay FS, Gilbert SH, Brundage RA 1995 Calcium signalling during chemotaxis. In: Calcium waves, gradients and oscillations. Wiley, Chichester (Ciba Found Symp 188) p 121–140

Fluck RA, Miller AL, Jaffe LF 1991 Slow calcium waves accompany cytokinesis in medaka fish eggs. J Cell Biol 115:1259–1265

Gerisch G 1965 Stadienspezifische Aggregationsmuster bei *Dictyostelium discoideum*. Wilhelm Roux' Arch Entwicklungsmech Org 156:127–144

Gilkey JC 1983 The roles of calcium and pH in activation of eggs of the medaka fish, *Oryzias latipes*. J Cell Biol 97:669–678

Gilkey JC, Jaffe LF, Ridgway EB, Reynolds GT 1978 A free calcium wave traverses the activating egg of the medaka, *Oryzias latipes*. J Cell Biol 76:448–466

Hara K, Tydeman P, Hengst RTM 1977 Cinematographic observation of post-fertilization waves (PFW) on zygote of *Xenopus laevis*. Wilhelm Roux' Arch Dev Biol 181:189–192

Ikemoto N, Antoniu B, Kang JJ, Ronjat M, Mészáros LG 1991 Intravesicular calcium transient during calcium release from sarcoplasmic reticulum. Biochemistry 30:5230–5237

Ingber DE 1993 Cellular tensegrity: defining new rules of biological design that govern the cytoskeleton. J Cell Sci 104:613–627

Jaffe LF 1985 The role of calcium explosions, waves and pulses in activating eggs. In: Metz CB, Monroy A (eds) Biology of fertilization, vol 3: The fertilization response of the egg. Academic Press, Orlando, FL, p 144

Jaffe LF 1986 Calcium and morphogenetic fields. In: Calcium and the cell. Wiley, Chichester (Ciba Found Symp 122) p 271–288

Jaffe LF 1991 The path of calcium in cytosolic calcium oscillations: a unifying hypothesis. Proc Natl Acad Sci USA 88:9883–9887

Jaffe LF 1993 Classes and mechanisms of calcium waves. Cell Calcium 14:736–745

Kamel LC, Bailey J, Schoenbaum DL, Kinsey W 1985 Phosphatidylinositol metabolism during fertilization in the sea urchin egg. Lipids 20:350–356

Leibowitz D 1992 On an active role of neuroglia in spreading depression and migraine. Proc R Soc Lond B Biol Sci 250:287–295

Lewis CA, Chia F-S, Schroeder TE 1973 Peristaltic constrictions in fertilized barnacle eggs (*Pollicipes polymerus*). Experientia 29:1533–1535

Maeda Y 1970 Influence of ionic conditions on cell differentiation and morphogenesis of the cellular slime molds. Dev Growth & Differ 12:217–227

Margolis RL, Andreassen PR 1993 The telophase disc: its possible role in mammalian cell cleavage. Bioessays 15:201–207

Miller AL, McLaughlin JA, Jaffe LF 1991 Imaging free calcium in *Xenopus* eggs during polar pattern formation and cytokinesis. J Cell Biol 115:280(abstr)

Miller AL, Fluck RA, McLaughlin JA, Jaffe LF 1993a Calcium buffer injections inhibit cytokinesis in *Xenopus* eggs. J Cell Sci 106:523–534

Miller AL, Galione A, Karplus E, Jaffe LF 1993b Mechanically induced calcium release from *Xenopus* cell cycle extracts. Biol Bull 185:290–292

Miller AL, Karplus E, Jaffe LF 1994 The use of aequorin for $[Ca^{2+}]_i$ imaging. Methods Cell Biol 40:305–338

Ridgway EB, Gilkey JC, Jaffe LF 1977 Free calcium increases explosively in activating medaka eggs. Proc Natl Acad Sci USA 74:623–627

Robertson A 1979 Waves propagated during vertebrate development. J Embryol Exp Morphol 50:155–167

Sanderson MJ 1995 Intercellular calcium waves mediated by inositol trisphosphate. In: Calcium waves, gradients and oscillations. Wiley, Chichester (Ciba Found Symp 188) p 175–194

Saran S, Azhar M, Manogaran PS, Pande G, Nanjundiah V 1994 The level of sequestered calcium in vegetative amoebae of *Dictyostelium discoideum* can predict post-aggregative cell fate. Differentiation 57:163–169

Schlatterer C, Gollnick F, Schmidt E, Meyer R, Knoll G 1994 Challenge with high concentrations of cyclic AMP induces transient changes in the cytosolic free calcium concentration in *Dictyostelium discoideum*. J Cell Sci 107:2107–2115

Snow P, Nuccitelli R 1993 Calcium buffer injections delay cleavage in *Xenopus laevis* blastomeres by dissipating $Ca^{2+}$ gradients. J Cell Biol 122:387–394

Speksnijder JE, Miller AL, Weisenseel MH, Chen T-H, Jaffe LF 1989 Calcium buffer injections block fucoid egg development by facilitating calcium diffusion. Proc Natl Acad Sci USA 86:6607–6611

Stith BJ, Goalstone M, Silva S, Jaynes C 1993 Inositol 1,4,5-trisphosphate mass changes from fertilization through first cleavage in *Xenopus laevis*. Mol Cell Biol 4:435–443

Stricker SA, Centonze VE, Paddock SW, Schatten G 1992 Confocal microscopy of fertilization-induced calcium dynamics in sea urchin eggs. Dev Biol 149:370–380

Torrence SA, Cloney RA 1981 Rhythmic contractions of the ampullar epidermis during metamorphosis of the ascidian *Molgula occidentalis*. Cell Tissue Res 216:293–312

Vollmar H 1972 Die Einrollbewegung (Anatrepsis) des Keimstreifs im Ei von *Acheta domesticus* (Orthopteriodea, Gryllidae). Wilehelm Roux' Arch Entwicklungsmech Org 170:135–151

Wolf R 1985 Migration and division of cleavage nuclei in the gall midge, *Wachtliella persicariae*. 3. Pattern of anaphase-triggering waves altered by temperature gradients and local gas exchange. Wilhelm Roux' Arch Dev Biol 194:257–270

Yoneda M, Kobayakawa Y, Kubota HY, Sakai M 1982 Surface contraction waves in amphibian eggs. J Cell Sci 54:35–46

## DISCUSSION

*Galione:* Could you elaborate on your evidence that $Ca^{2+}$ influx is needed to activate the sea urchin egg and that it enters via the sperm?

*Jaffe:* The evidence comes from work by Robbert Créton (Créton & Jaffe 1995). If at zero time you add a high concentration of sperm to sea urchin eggs, a concentration known to initiate fertilization within one to two seconds in all of the eggs, then, up to about 10 s afterwards, long after the sperm have attached to the eggs, when you add lanthanum chloride to block the $Ca^{2+}$ channels, you completely block egg activation. This indicates that a continued entry of $Ca^{2+}$ from the outside is an essential part of the activation mechanism. The blockade becomes gradually less effective with time, with a half-time of 20 s; the kinetics is the same as in the famous Shen & Steinhardt (1984) experiment. The same results are obtained by simply removing external $Ca^{2+}$ with a buffer.

*Putney:* Human sperm, at least, show a capacitative $Ca^{2+}$ entry ($Ca^{2+}$ influx signalled by depletion of intracellular stores) which is probably triggered by $Ca^{2+}$ depletion from the nuclear envelope, since there's really no other endoplasmic reticulum. All that would be needed to start the entry process would be some inositol 1,4,5-trisphosphate ($InsP_3$) or some means of depleting $Ca^{2+}$ from the sperm after fusion.

*Jaffe:* The calculated amount of $Ca^{2+}$ inside the sperm is actually very small. Even though the $[Ca^{2+}]$ inside the sperm is high, the volume is so small that there's just too little total free $Ca^{2+}$ in the sperm to activate the egg.

*Putney:* That's not what I'm saying. Depletion of $Ca^{2+}$ in the nuclear envelope triggers the activation of $Ca^{2+}$ channels in the sperm membrane, which allows entry of $Ca^{2+}$ from the outside. That could be the initiating reaction.

*Jaffe:* Your proposition is that the egg jelly, which starts the acrosome reaction, releases $InsP_3$ inside the sperm, which in turn releases internal $Ca^{2+}$.

My main point is that the bulk of the $Ca^{2+}$ involved comes from the outside, not from the inside, because there's too little inside.

*Miyazaki:* Your idea is analogous to the microinjection of $Ca^{2+}$. Usually, fairly large amounts of $Ca^{2+}$ need to be injected into the egg to induce a propagating $Ca^{2+}$ wave. I wonder whether a sperm could provide such a large amount of $Ca^{2+}$.

*Jaffe:* The sperm, according to the sperm conduit model, is injecting $Ca^{2+}$ right at the surface of the egg, whereas microinjection with a pipette necessarily delivers the $Ca^{2+}$ rather deep in the cell. The best experimental simulation of what happens naturally is not injection in that manner but pricking of the egg. It's been known for many years that sea urchin, fish, frog and hamster eggs can be activated simply by pricking them with a needle. This activation by pricking works only if there is $Ca^{2+}$ in the medium (Jaffe 1985, p 128–129). It would be desirable, to explore this idea further, to re-do these old prick-activation experiments. You could also use electroporation to get a neatly controlled and population-effective introduction of $Ca^{2+}$ through a hole from the outside.

*Whitaker:* There's quite a lot of evidence against your idea. Almost 25 years ago, Chambers (1980) showed that sea urchin eggs could be fertilized in sea water containing high concentrations of $Ca^{2+}$ buffers, for example. The dogma is that extracellular $Ca^{2+}$ is not needed for fertilization. In some more recent experiments using confocal microscopy, which allows us to measure $Ca^{2+}$ entry through $Ca^{2+}$ channels in the egg membrane quite sensitively, we've found that when extracellular $Ca^{2+}$ is removed and there is no longer any detectable $Ca^{2+}$ influx, sea urchin eggs can still be fertilized (McDougall et al 1993). Your hypothesis cannot apply in the sea urchin, at least.

*Jaffe:* I entirely disagree. The old experiments you referred to, done in Chambers' and in Epel's laboratories, but actually most thoroughly published by Schmidt et al (1982), are worthy of careful study. If you look carefully at what they actually did, you discover that fertilization occurs in the absence of external $Ca^{2+}$ only if the sperm concentration is enormously higher than the usual concentration, 30 times higher, at least. You also discover, especially if you ask around, that these experiments worked only now and then. Also, Schmidt et al (1982) had to lower the temperature greatly for obscure reasons. I don't know why fertilization sometimes is possible under these conditions, but I would suggest that the mechanism is abnormal. In these experiments the sperm were, necessarily, pre-activated, so $Ca^{2+}$ had entered the sperm before the experiment. Calcium which enters sperm is known to accumulate in the mitochondria until it eventually kills the sperm. They transferred sperm from high $[Ca^{2+}]$ to zero $[Ca^{2+}]$ in a medium containing 500 mM $Na^+$. In the complete absence of $Ca^{2+}$, $Ca^{2+}$ channels are known to let $Na^+$ through. It's also known that high intracellular $[Na^+]$ releases $Ca^{2+}$ from the mitochondria. I would suggest that under the abnormal circumstances of those experiments,

$Na^+$ entry into the sperm releases $Ca^{2+}$ stored in the mitochondria, so that occasionally, probably when two sperm at these high concentrations cooperate to deliver $Ca^{2+}$ to a single region of the endoplasmic reticulum, there is release of enough pre-stored $Ca^{2+}$ from the mitochondria to bring about fertilization. In any case, it's hard to argue against Robbert Créton's experiments.

*Lu:* Does the membrane of the sperm which fuses with the egg membrane play a role in initiating the response?

*Jaffe:* The idea is that $Ca^{2+}$ which has been entering the sperm and going into the mitochondria is now diverted into the egg through the fused membrane. This particular patch of membrane serves for specific attachment. It has been argued by others, including my daughter (Jaffe 1993), that the mechanism of sperm action is analogous to that of hormone action, that G proteins and $InsP_3$ synthesis and so forth are turned on. However, when the sea urchin egg's receptor for sperm was isolated, it proved not to have the structure of a G protein-coupled receptor. This bit of egg membrane has some sort of glue but it doesn't have a mechanism for releasing $InsP_3$.

*Lu:* Does the sperm membrane bring extra channels to the egg?

*Jaffe:* In a sense it does, because the membrane of the egg becomes continuous with that of the sperm. It's possible that in the long times involved there is diffusion of channels within the lipid bilayer.

*Sanderson:* Pricking the egg would seem to provide a mechanical signal of short-lived duration, whereas your other points stressed that an extended period of $Ca^{2+}$ influx is required.

*Jaffe:* The hole is open long enough.

*Sanderson:* Do you think it doesn't heal up very quickly?

*Jaffe:* I haven't done the experiment with sea urchin eggs, but in *Drosophila* and medaka eggs you can see the wound for at least 20 min after you make a hole. The wound heals very slowly.

*Sanderson:* But I understood that even in the absence of extracellular $Ca^{2+}$ mechanical stimulation can still initiate a wave.

*Jaffe:* That's incorrect. Prick activation does not work in the absence of external $Ca^{2+}$.

*Cobbold:* How does this idea fit in with your $Na^+$ hypothesis? You said that in low $[Ca^{2+}]$ medium, $Na^+$ enters through $Ca^{2+}$ channels; if you are making a persistent hole with $Na^+$ outside, why don't you get a $Na^+$-mediated activation in the absence of $Ca^{2+}$?

*Jaffe:* In eggs, the mitochondria are well below the surface. Moreover, the surface:volume ratio of an egg is completely different from that of a sperm. I would think that $Na^+$ entering through a tiny hole in a giant egg would reach a far lower concentration in the vicinity of a mitochondrion than it would entering a sperm.

*Wier:* You offered an explanation as to why the waves in cardiac cells are anomalously fast. Your explanation involves a voltage-dependent mechanism, possibly $Ca^{2+}$ channels in the surface membrane. I take exception to that, on the basis of the observation that many of us have made that these $Ca^{2+}$ waves occur perfectly well at this velocity in voltage-clamped cells.

*Jaffe:* Clamped at what voltage?

*Wier:* At −80 mV, for example, at the negative membrane potential. In an unclamped cell the change in $[Ca^{2+}]$ would cause electrical depolarization through activation of $Na^{+}/Ca^{2+}$ exchange. There's little possibility that there could be activation of L-type $Ca^{2+}$ channels or any other voltage-dependent mechanism in clamped cells.

*Berridge:* Could you offer an explanation of why the rate of wave propagation is so fast in these cells?

*Wier:* I would imagine that this can be explained by the same reaction–diffusion mechanism that explains $Ca^{2+}$ waves in other cells.

*Cannell:* In fact, reaction–diffusion is all that is required to explain these propagation velocities. Backx et al (1989) showed clearly that you can get propagation velocities of 150 μm/s by diffusion plus $Ca^{2+}$-induced $Ca^{2+}$ release in a cardiac ventricular muscle cell model. The propagation velocity depends on the relationship between the cytosolic $Ca^{2+}$ and the rate of release of $Ca^{2+}$ from the sarcoplasmic reticulum. With a steep function for the dependence of the rate of release from the sarcoplasmic reticulum on cytosolic $Ca^{2+}$ you get more amplification and faster propagation velocities.

*Berridge:* What seems to be important is the relationship of the next store along the line.

*Cannell:* Absolutely. That's a factor Backx et al (1989) didn't consider, because they were dealing with an isotropic model. If the system is anisotropic with release sites spatially separated you get apparent saltation from one release site to the next. Of course, the trigger $Ca^{2+}$ still has to diffuse between release sites, but the propagation velocity is still not solely dependent on the rate of diffusion. The release function is still very important for determing propagation velocity, and low gain in that function results in lower propagation velocities.

*Goldbeter:* We have also done simulations of $Ca^{2+}$ waves in cardiac cells and have found (Dupont & Goldbeter 1992) that the propagation rate is much faster than the propagation rate in other types of cell. The main difference is that the kinetics of the $Ca^{2+}$ exchange processes, pumping and release, are much faster, by a factor of 60. This accounts for the difference in the propagation rate and also in the period of the oscillations, which is of the order of one second in cardiac cells and one minute in other cell types.

*Williams:* Dr Jaffe, what would you say if I talked about your first fast wave type as being anomalously slow, rather than about the waves in cardiac cells being anomalously fast? The waves you see as being very fast I would call 'slow'. The plot of velocity as a rate constant against temperature you have published (Jaffe 1991) has an activation energy of 20 kcal/mol. If that is true, if that is meaningful as an activation energy generally, it would mean that you are not following the free $Ca^{2+}$ diffusion but some process similar to the mechanical one you discussed in the second case. In other words, what you're looking at is the way in which the wave is being stopped and then started again at each of the little vesicles.

*Jaffe:* Take a look at the Luther equation, which really is correct. The Luther equations tells you that the velocity is a function not of the reaction constant but of its square root. The low $Q_{10}$ is a consequence of that fact. If you doubled that $Q_{10}$ you would get the $Q_{10}$ which really characterizes the reaction rate constant, and then you get a rather standard value.

*Williams:* You say that $K/k$ varies by a factor of two per 10 °C?

*Jaffe:* That's sufficient. With a twofold increase in $K/k$ you end up with $Q_{10}$ values of three or four, which are not unusual.

*Williams:* For free $Ca^{2+}$ ion diffusion?

*Jaffe:* No, a $Q_{10}$ for $Ca^{2+}$-induced $Ca^{2+}$ release.

*Williams:* That's what I mean. It doesn't matter whether the exact value is five or 20 kcal because that's not a characteristic of a process of free $Ca^{2+}$ ion diffusion but a process which is something to do with a mechanical device, which may be the vesicle release process.

*Meyer:* I would like to make a few remarks about $Ca^{2+}$ diffusion and $InsP_3$ diffusion. In the paper Dr Jaffe cited (Albritton et al 1992), we took a thin tube and filled it with cytosol from *Xenopus* oocytes that was diluted by less than 10% and then injected the tube with $^{45}Ca^{2+}$ or $[^3H]InsP_3$ under conditions in which $Ca^{2+}$ was not accumulated in the stores because uptake of $Ca^{2+}$ was blocked with thapsigargin. Degradation of $InsP_3$ was blocked by removing $Mg^{2+}$, which inhibits the phosphatase. We then measured how fast the two diffuse. It turned out that $InsP_3$ diffusion was very fast compared with $Ca^{2+}$. We think that even in a large cell such as an oocyte, $InsP_3$ can spread over quite a large distance in a short time. This rate of diffusion is about 300 $\mu m^2/s$. The situation with $Ca^{2+}$ is more complicated. The diffusion coefficient is less than 30 $\mu m^2/s$ at low $[Ca^{2+}]$ and becomes faster as $[Ca^{2+}]$ is increased. At a free $[Ca^{2+}]$ of around 3 $\mu M$, which could be reached in your system, the diffusion coefficient would be around 250 $\mu m^2/s$.

I went through different models trying to compare physiologically relevant amplifiers and how wave equations can describe them in simple terms, relating the spatial gradient at the front of a wave and the rise time that one would measure locally if one recorded from a point within the cell. Depending on the type of amplifier used, one can obtain differences in this simple relationship of factors of two to three. Although the Luther equation is not accurate for all types of amplifier, it is useful to obtain a rough estimate of how velocity, spatial graidents, local rise times of $Ca^{2+}$ and the diffusion coefficient are related. The equation is especially useful at the foot of the wave. If you have a $Ca^{2+}$ indicator with which you can measure $[Ca^{2+}]$ below the threshold at which amplification occurs, you should get an exponential profile at the front of the wave for any of the proposed models. If one measures either the length constant or the local time constant for the pre-threshold rise, one could determine the diffusion coefficient more accurately. Current measurements may not be sufficiently accurate to solve the problem of whether measured and predicted $Ca^{2+}$ diffusion coefficients for fertilization waves suggest $Ca^{2+}$ or $InsP_3$ is the diffusing messenger. From the current published results, the diffusion coefficient of $Ca^{2+}$ doesn't match the one calculated from the wave properties. It's possible to overcome this discrepancy by arguing that the $Ca^{2+}$ ions that are released from stores could directly bind to $InsP_3$ receptors before they equilibrate with the $Ca^{2+}$ buffers in the cell. This distance of free $Ca^{2+}$ diffusion might be around 0.1 $\mu$m and occurs over a time-scale of less than 1 ms, because most of the $Ca^{2+}$ buffers in the cell are expected to have equilibration kinetics of $10^7$–$10^{-8}\,M^{-1}\,s^{-1}$.

## References

Albritton NL, Meyer T, Stryer L 1992 Range of messenger action of calcium ion and inositol 1,4,5-trisphosphate. Science 258:1812–1815

Backx PH, De Tombe PP, Van Deen JHK, Mulder BMJ, Ter Keurs HEDJ 1989 A model of propagating calcium-induced calcium release mediated by diffusion. J Gen Physiol 93:963–977

Chambers EL 1980 Fertilization and cleavage of eggs of the sea urchin *Lytechninus variagatus* in $Ca^{2+}$-free sea water. Eur J Cell Biol 22:476(abstr)

Créton R, Jaffe LF 1995 The role of calcium influx during the latent period in sea urchin fertilization. Proc Natl Acad Sci USA, in press

Dupont G, Goldbeter A 1992 Oscillations and waves of cytosolic calcium: insights from theoretical models. Bioessays 14:485–493

Jaffe LA (ed) 1993 Fertilization. Dev Biol 158:9–78

Jaffe LF 1985 The role of calcium explosions, waves and pulses in activating eggs. In: Metz CB, Monray A (eds) Biology of fertilization, vol 3: The fertilization response of the egg. Academic Press, Orlando, FL

Jaffe LF 1991 The path of calcium in cytosolic calcium oscillations: a unifying hypothesis. Proc Natl Acad Sci USA 88:9883–9887

McDougall A, Gillot I, Whitaker MJ 1993 Thimerosal reveals calcium-induced calcium release in unfertilized sea urchin eggs. Zygote 1:35–42

Schmidt T, Patton C, Epel D 1982 Is there a role for the $Ca^{2+}$ influx during fertilization of the sea urchin egg? Dev Biol 90:284–290

Shen SS, Steinhardt RA 1984 Time and voltage windows for reversing the electrical block to fertilization. Proc Natl Acad Sci USA 81:1436–1439

# Subcellular organization of calcium signalling in hepatocytes and the intact liver

Andrew P. Thomas, Dominique C. Renard-Rooney, György Hajnóczky, Lawrence D. Robb-Gaspers, Chi Lin and Thomas A. Rooney

*Department of Pathology and Cell Biology, Thomas Jefferson University, Room 271, 1020 Locust Street, Philadelphia, PA 19107, USA*

*Abstract.* Hepatocytes respond to inositol 1,4,5-trisphosphate ($InsP_3$)-linked agonists with frequency-modulated oscillations in the intracellular free calcium concentration ($[Ca^{2+}]_i$), that occur as waves propagating from a specific origin within each cell. The subcellular distribution and functional organization of $InsP_3$-sensitive $Ca^{2+}$ pools has been investigated, in both intact and permeabilized cells, by fluorescence imaging of dyes which can be used to monitor luminal $Ca^{2+}$ content and $InsP_3$-activated ion permeability in a spatially resolved manner. The $Ca^{2+}$ stores behave as a luminally continuous system distributed throughout the cytoplasm. The structure of the stores, an important determinant of their function, is controlled by the cytoskeleton and can be modulated in a guanine nucleotide-dependent manner. The nuclear matrix is devoid of $Ca^{2+}$ stores, but $Ca^{2+}$ waves in the intact cell propagate through this compartment. The organization of $[Ca^{2+}]_i$ signals has also been investigated in the perfused liver. Frequency-modulated $[Ca^{2+}]_i$ oscillations are still observed at the single cell level, with similar properties to those in the isolated hepatocyte. The $[Ca^{2+}]_i$ oscillations propagate between cells in the intact liver, leading to the synchronization of $[Ca^{2+}]_i$ signals across part or all of each hepatic lobule.

*1995 Calcium waves, gradients and oscillations. Wiley, Chichester (Ciba Foundation Symposium 188) p 18–49*

Oscillations of cytosolic free $Ca^{2+}$ concentration ($[Ca^{2+}]_i$) in response to agonists acting through the inositol lipid-dependent signal transduction system have been observed in a wide variety of cell types (Berridge 1993). It is possible that all cells can demonstrate this behaviour when stimulated under appropriate conditions, which most often involve submaximal receptor activation. The phenomenon of hormone-induced $[Ca^{2+}]_i$ oscillations in non-excitable cells was first brought forward by the work of Cobbold and co-workers in isolated hepatocytes (Woods et al 1986, 1987). They found that sustained application of hormones known to act through an increase in the concentration of inositol 1,4,5-trisphosphate ($InsP_3$) and mobilization of intracellular $Ca^{2+}$ stores resulted in periodic spikes of increased $[Ca^{2+}]_i$. An important property of the $[Ca^{2+}]_i$ oscillations in hepatocytes was that the frequency increased with increasing hormone dose, whereas the amplitude of the $[Ca^{2+}]_i$ spikes was nearly constant at all effective levels of stimulation (Woods et al 1986, 1987, Rooney et al 1989). This led to the concept that $Ca^{2+}$-dependent processes within the cell might be regulated by the frequency of $[Ca^{2+}]_i$ spikes (frequency modulation) rather than by the amplitude of the $[Ca^{2+}]_i$ change (Berridge et al 1988, Goldbeter et al 1990). Thus, these initial findings in the hepatocyte brought about a major re-evaluation of the inositol lipid signalling system, which had at first appeared, on the basis of the well-correlated mono- or biphasic kinetics of inositol phosphate and $[Ca^{2+}]_i$ changes in cell population measurements, to act in a relatively simple manner. Clearly, bulk tissue studies are inadequate for a full understanding of this signal transduction pathway. This realization has spawned a great interest in single cell and imaging approaches to the study of $[Ca^{2+}]_i$ signalling. The use of imaging methods has become even more important with the discovery that $[Ca^{2+}]_i$ oscillations within individual cells are also organized spatially. The observations of $[Ca^{2+}]_i$ waves after fertilization of relatively large egg cells (Gilkey et al 1978, Jaffe 1991) have been followed by similar findings in small mammalian cells. Once again, the hepatocyte was in the vanguard of these advances in our understanding of the spatial organization of $[Ca^{2+}]_i$ oscillations (Rooney et al 1990, 1991). In the hepatocyte, each $[Ca^{2+}]_i$ oscillation is initiated at a specific subcellular locus and is propagated through the cell at a constant rate without loss of amplitude.

Several proposals have been put forward to explain how $[Ca^{2+}]_i$ oscillations and $[Ca^{2+}]_i$ waves might be generated (reviewed in Berridge 1993, Meyer & Stryer 1991, Cobbold et al 1991, Rooney & Thomas 1993). Initially, Cobbold and colleagues proposed that $[InsP_3]$ might oscillate as a result of the cyclical activation of diacylglycerol-dependent protein kinase C leading to feedback inhibition of $InsP_3$ formation via inactivation of the receptor/G protein-coupled phospholipase C (PLC). More recently, this model has been modified to incorporate activation of PLC by $Ca^{2+}$ feedback, to account for the propagation of $[Ca^{2+}]_i$ waves. Most of the other major oscillation models rely on feedback effects of $Ca^{2+}$, and each provides a potential mechanism for the propagation of $[Ca^{2+}]_i$ waves. Oscillations of $[Ca^{2+}]_i$ driven by $InsP_3$ oscillations have been proposed to result directly from $Ca^{2+}$ feedback activation of PLC, with the recovery phase between $[Ca^{2+}]_i$ spikes being a result of $Ca^{2+}$ uptake into $InsP_3$-insensitive stores and/or inactivation of the $InsP_3$ receptor. Two basic models have been proposed in which $[InsP_3]$ is not required to oscillate. The first of these is a two-pool model in which $Ca^{2+}$ released by $InsP_3$ triggers a secondary $Ca^{2+}$-induced $Ca^{2+}$ release from a distinct $InsP_3$-insensitive $Ca^{2+}$ pool, with the $[Ca^{2+}]_i$ spike being terminated through $Ca^{2+}$ resequestration. The discovery that the $InsP_3$ receptor is sensitive to both positive and negative modulation by $Ca^{2+}$ has led to the development of various one-pool models, in which the feedback effects of $Ca^{2+}$ on the $InsP_3$ receptor produce periodic surges of self-propagating $Ca^{2+}$ release and recovery. Although considerable evidence has been obtained in support of this one-pool model, it is possible that more than one of the mechanisms outlined here contribute to oscillatory $[Ca^{2+}]_i$ waves in a single cell type.

Here, we present some of our studies into the organization of $[Ca^{2+}]_i$ waves within individual hepatocytes and the relationship between subcellular structure and the path of $[Ca^{2+}]_i$ wave propagation. In particular, we have investigated the subcellular distribution of functional $Ca^{2+}$ stores and examined how different regions of these stores interact. In addition, we have extended our studies of $Ca^{2+}$ signalling from the isolated hepatocyte to investigate how $[Ca^{2+}]_i$ signals are organized at the level of individual cells in the intact liver.

## Calcium waves in isolated hepatocytes

Our intitial observations of $[Ca^{2+}]_i$ waves were made in hepatocytes maintained in primary culture for 24 h (Rooney et al 1990). Under these conditions the cells are somewhat flattened, which facilitates the detection of $[Ca^{2+}]_i$ waves. In addition, the culture conditions allow the regeneration of polarized membrane domains that are generally lost after dissociation of the cells from the liver. Global bath application of hormones to fura-2-loaded hepatocytes in these conditions stimulated $[Ca^{2+}]_i$ oscillations, which initiated at a discrete subcellular locus adjacent to a small region of the plasma membrane. Subsequently, the $[Ca^{2+}]_i$ increase propagated through the cell at 20–30 $\mu m\ s^{-1}$ until $[Ca^{2+}]_i$ was increased to a similar extent over the entire hepatocyte. The decay phase of each $[Ca^{2+}]_i$ spike was relatively homogeneous throughout the cell (Rooney et al 1991). The subcellular initiation site was the same for each $[Ca^{2+}]_i$ oscillation in a series, and was found to be identical when the same cell was sequentially stimulated with different agonists. Importantly, the rate of $[Ca^{2+}]_i$ wave propagation, like the amplitude, did not decay with increasing distance from the site of the wave origin. Moreover, the $[Ca^{2+}]_i$ wave propagation rates were not effected by agonist dose, even though the frequency increased in a dose-dependent manner. Table 1 summarizes $[Ca^{2+}]_i$ wave propagation kinetics under a variety of stimulation conditions in hepatocytes. More recently, similar $[Ca^{2+}]_i$ waves have been observed in freshly isolated hepatocytes with a confocal microscope in line-scan mode (Nathanson & Burgstahler 1992a,b). In these line-scan studies the rate of $[Ca^{2+}]_i$ wave propagation was considerably faster (80–110 $\mu m\ s^{-1}$) than that observed on our 24 h culture cells, but in our studies of fresh hepatocytes with full-frame fast imaging we find rates very similar to those

**TABLE 1 Characteristics of wave parameters in single hepatocytes**

| *Condition* | *Propagation rate* ($\mu m\ s^{-1}$) | *Frequency* ($min^{-1}$) | *Amplitude* ($nM\ Ca^{2+}$) |
|---|---|---|---|
| Cells in primary culture | | | |
| 2 $\mu$m Phenylephrine | $21 \pm 7$ | $0.59 \pm 0.08$ | $691 \pm 93$ |
| 10 $\mu$m Phenylephrine | $22 \pm 8$ | $0.93 \pm 0.12$ | $712 \pm 61$ |
| 10 $\mu$m Phenylephrine + 2.2 mM EGTA | $22 \pm 3$ | $0.39 \pm 0.06$ | $716 \pm 63$ |
| 50 mM NaF + 10 $\mu$M $AlCl_3$ | $22 \pm 4$ | $0.46 \pm 0.10$ | $634 \pm 38$ |
| 3nM Vasopressin | $23 \pm 4$ | $0.75 \pm 0.06$ | $741 \pm 52$ |
| 200 $\mu$m *tert*-butyl-hydroperoxide | $25 \pm 3$ | N.D. | $699 \pm 24$ |
| Fresh cells | | | |
| 0.5 $\mu$M Phenylephrine | $25 \pm 2$ | N.D. | N.D. |
| 50 $\mu$M Ryanodine | $29 \pm 5$ | N.D. | N.D. |

in cells maintained in culture for 24 h (Table 1). The reason for this discrepancy is unclear, although the path of $[Ca^{2+}]_i$ wave propagation in the freshly isolated cells is often more complex than in the cultured cells and in many cases more than one intitiation site is observed. In isolated hepatocyte couplets it is possible to observe the bile canalicular membrane domain and consequently to define the polarity of the cells. In this preparation hormone-induced $[Ca^{2+}]_i$ waves are generally, though not exclusively, initiated from the basal pole of the cell (Nathanson & Burgstahler 1992a).

In addition to receptor-linked agonists, several other interventions can initiate $[Ca^{2+}]_i$ waves in hepatocytes. Direct activation of the PLC-linked G protein by $AlF_4^-$, or sensitization of the intracellular $Ca^{2+}$ release channels to the basal $[InsP_3]$ following oxidation of cellular glutathione with *tert*-butyl hydroperoxide (Rooney et al 1990, 1991), causes $[Ca^{2+}]_i$ waves that propagate with similar kinetics and from the same origin as those induced by hormones (Table 1). In contrast, $Ca^{2+}$ mobilization with thapsigargin [an inhibitor of the $Ca^{2+}$ pumps in the endoplasmic reticulum (ER)] results in a relatively slow $[Ca^{2+}]_i$ increase over the whole cell that does not lead to any propagating $[Ca^{2+}]_i$ waves. Thus, it appears that $[Ca^{2+}]_i$ waves can be initiated in hepatocytes by any stimulus that activates $Ca^{2+}$ release via $InsP_3$-sensitive $Ca^{2+}$ channels. However, once initiated, the $[Ca^{2+}]_i$ waves propagate with a fixed velocity and amplitude regardless of the strength or nature of the initiating stimulus.

The stimulation of $[Ca^{2+}]_i$ waves by $AlF_4^-$ and *tert*-butyl hydroperoxide indicates that the $[Ca^{2+}]_i$ wave initiation domain in hepatocytes does not simply result from a polarized distribution of hormone receptors and suggests that some other element or elements of the signal transduction apparatus are also spatially organized within the hepatocyte. There may be a specialized subcellular region for $Ca^{2+}$ release that is more sensitive to $InsP_3$, as has been suggested for cells of the exocrine pancreas (Thorn et al 1993, Kasai et al 1993). The fact that the $[Ca^{2+}]_i$ wave kinetics are independent of the type and strength of the initiating stimulus suggests that the primary driving force for wave propagation is $Ca^{2+}$ itself, perhaps through feedback stimulation of PLC. However, the $Ca^{2+}$ requirement of this enzyme appears to be satisfied by the basal $[Ca^{2+}]_i$ in the hepatocyte because hormone-induced acitivation of PLC is unaffected when $[Ca^{2+}]_i$ is clamped at the basal level with a cytosolic $Ca^{2+}$ buffer (Renard et al 1987). Propagation of the $[Ca^{2+}]_i$ wave is more likely to occur either through $Ca^{2+}$-activated $Ca^{2+}$ release via $InsP_3$-sensitive channels or through another $Ca^{2+}$-activated channel in a more classical $Ca^{2+}$-induced $Ca^{2+}$ release process. In a recent study with aequorin-injected hepatocytes, Sanchez-Bueno & Cobbold (1993) reported that ryanodine reduces the frequency of $[Ca^{2+}]_i$ oscillations induced by vasopressin or angiotensin II, but not those induced by phenylephrine. Under the conditions we have used, ryanodine reduced the frequency of $[Ca^{2+}]_i$ oscillations in response to both vasopressin and phenylephrine in freshly isolated hepatocytes. However, addition of ryanodine alone generated oscillatory $[Ca^{2+}]_i$ waves with similar initiation sites and rates of propagation to those observed with hormones (Table 1). This direct $Ca^{2+}$-mobilizing effect of ryanodine makes it difficult to interpret $[Ca^{2+}]_i$ wave rates measured in the presence of both ryanodine and $InsP_3$-linked agonists. The similarity in the kinetics of ryanodine- and hormone-induced $[Ca^{2+}]_i$ waves suggests that similar mechanisms are involved in the actions of the two agents, but it remains to be determined whether ryanodine's $Ca^{2+}$-mobilizing effect is dependent on $InsP_3$-sensitive $Ca^{2+}$ channels.

## Spatial organization of calcium stores in single hepatocytes

The discrete origin and propagative nature of $[Ca^{2+}]_i$ waves in hepatocytes led us to investigate methods for obtaining spatially resolved functional measurements of the $Ca^{2+}$ stores in individual cells. One approach to this is to measure the release of $Ca^{2+}$ into the cytosolic compartment (Parker & Ivorra 1990, Thorn et al 1993, Kasai et al 1993), but the resolution can be limited by dissipation or self-propagation of the released $Ca^{2+}$, especially in small cells. Therefore, to improve the potential for spatial resolution, we developed techniques for imaging the functioning of the $InsP_3$-sensitive $Ca^{2+}$ stores using fluorescent dyes localized within the lumen of the store (Renard-Rooney et al 1993, Hajnóczky et al 1993). We did these experiments initially in permeabilized hepatocytes, where the level of messengers and other regulatory factors (e.g., $Ca^{2+}$) could be controlled. The $Ca^{2+}$ content of the intracellular stores was measured using chlortetracycline because this dye is membrane permeant and is sensitive to $[Ca^{2+}]$ over the millimolar range expected within the stores after ATP-dependent sequestration ($K_D$ for $Ca^{2+}$ of about 3 mM). Chlortetracycline undergoes a 20-fold enhancement of fluoresence on binding $Ca^{2+}$ in the membrane-delimited compartment. With this dye we could follow ATP-driven $Ca^{2+}$ uptake and $InsP_3$-induced $Ca^{2+}$ release from the luminal side of the $Ca^{2+}$ stores. In permeabilized hepatocyte suspensions, the luminal chlortetracycline signal could be shown to change inversely to the changes in $[Ca^{2+}]$ occurring in the cytosolic compartment measured simultaneously with quin-2 (Renard-Rooney et al 1993). We also made use of the fact that fura-2 can be extensively compartmentalized into the intracellular stores when intact cells are loaded with the acetoxymethyl ester (fura-2/AM) under appropriate conditions. This compartmentalized fura-2 is not suitable for monitoring $[Ca^{2+}]$ changes in the stores unless they are almost totally depleted of $Ca^{2+}$. However, as originally noted by Glennon et al (1992), activation of $InsP_3$-sensitive channels allows $Mn^{2+}$ to pass from the cytosolic compartment into the stores, where it quenches the fura-2 fluorescence. We have utilized this ability of $Mn^{2+}$ to act as a $Ca^{2+}$ surrogate, passing in a retrograde fashion through the $InsP_3$-activated channels, to obtain spatially-resolved information on the $Ca^{2+}$ stores. These experiments can be done in the presence of thapsigargin to eliminate indirect feedback effects of the released $Ca^{2+}$. In addition, the essentially stoichiometric nature of the quenching of fura-2 by $Mn^{2+}$ allows the quenching rate to be used as a relative indicator of the net permeability of the $InsP_3$-sensitive $Ca^{2+}$ channels (Hajnóczky et al 1993, Renard-Rooney et al 1993).

In imaging experiments using either chlortetracycline, to measure $Ca^{2+}$ disappearance from the stores directly, or $Mn^{2+}$ quenching of compartmentalized fura-2, to determine the distribution of $InsP_3$-activated channels, the $InsP_3$-sensitive $Ca^{2+}$ stores were found to be distributed over the entire cytoplasmic compartment of the hepatocyte (Renard-Rooney et al 1993, Hajnóczky et al 1993). The only region not containing these stores was the nuclear matrix. Even using submaximal $InsP_3$ doses, we were unable to observe any clear spatial heterogeneity in the responsiveness to $InsP_3$. In some experiments $[Ca^{2+}]_i$ waves in response to hormones were monitored in the intact cells before cell permeabilization, but no correlation was observed between the initiation site of the $[Ca^{2+}]_i$ waves and the subcellular distribution of the $InsP_3$-sensitive $Ca^{2+}$ stores. Thus, the pathway of $[Ca^{2+}]_i$ wave initiation and subsequent propagation within the hepatocyte does not appear to result from a co-localization of the $InsP_3$-sensitive $Ca^{2+}$ stores with the initiation site. However, it is possible that $InsP_3$ receptors in this subcellular domain are differentially regulated in the intact cell by factors that are lost after cell permeabilization. Nevertheless, it is clear that $InsP_3$-sensitive $Ca^{2+}$ stores are distributed throughout the cytoplasm in a manner that would be consistent with their playing a role in both initiation and propagation of $[Ca^{2+}]_i$ waves in the hepatocyte.

## Luminal continuity between calcium stores

$InsP_3$-induced $Ca^{2+}$ release in permeabilized cell populations has been reported to occur in a quantal, or incremental, manner, in which a submaximal $[InsP_3]$ releases only a fraction of the stored $Ca^{2+}$, with the residual $InsP_3$-sensitive $Ca^{2+}$ pool being unavailable until a higher $InsP_3$ dose is added. Quantal $Ca^{2+}$ release has been postulated to result from differences in sensitivity to $InsP_3$ between discrete $InsP_3$ receptor populations (Muallem et al 1989, Meyer & Stryer 1990, Oldershaw et al 1991, Ferris et al 1992). This phenomenon could be observed using both chlortetracycline and the $Mn^{2+}$ quenching technique in suspensions of hepatocytes (Renard-Rooney et al 1993, Hajnóczky et al 1993, 1994). Furthermore, imaging of individual cells permeabilized under the cell suspension conditions also revealed discrete steps of $Ca^{2+}$ release with increasing $InsP_3$ concentrations (Fig. 1A), indicating that quantal $Ca^{2+}$ release was not due to intercellular heterogeneity in $InsP_3$ sensitivity in the measurements made in cell populations. However, our imaging studies of $InsP_3$ action in hepatocytes which had been allowed to attach to a polylysine substratum for 2–24 h revealed a completely different response to graded doses of $InsP_3$. In these preparations the rate of $Ca^{2+}$ release was dependent on $[InsP_3]$, but the amplitude was essentially the same for all effective concentrations of $InsP_3$ (Fig. 1B). Moreover, the $InsP_3$-sensitive pool encompassed almost all of the ATP-dependent $Ca^{2+}$ stores in these attached cell preparations, whereas only about 50% of the intracellular $Ca^{2+}$ stores could be mobilized by a maximal $InsP_3$ dose when cells were permeabilized in suspension. Identical observations were made using the $Mn^{2+}$ quenching method to assess the accessibility of the $Ca^{2+}$ stores through the $InsP_3$-activated channels (Renard-Rooney et al 1993, Hajnóczky et al 1994).

The rates of $Ca^{2+}$ release and $Mn^{2+}$ entry indicated that activation of as few as 10% of the total $InsP_3$ receptor channels was sufficient to allow access to the entire $InsP_3$-sensitive store in the attached permeabilized cell preparation (Renard-Rooney et al 1993, Hajnóczky et al 1994). Because it is unlikely that the fundamental quantal activation properties of individual $InsP_3$ receptors are lost in these cells, we postulated that luminal continuity between $Ca^{2+}$ stores allowed $Ca^{2+}$ to redistribute between store domains with and without activated channels (Fig. 2). In contrast, the $Ca^{2+}$ stores in the cells permeabilized in suspension were considered to behave in a much more autonomous manner, allowing the quantal activation properties of individual $InsP_3$ receptors to be expressed as quantal steps of $Ca^{2+}$ release with increasing $[InsP_3]$ (compare left and right panels of Fig. 2). In support of this model, electron microscopy of the cells permeabilized in suspension revealed that the

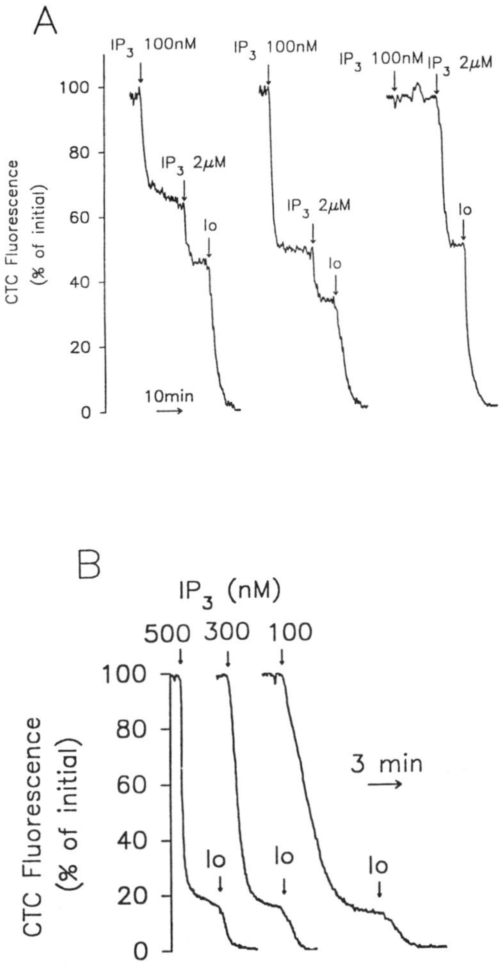


FIG. 1. Release of $Ca^{2+}$ from intracellular stores by inositol 1,4,5-trisphosphate ($IP_3$) in single permeabilized hepatocytes. Freshly isolated hepatocytes immobilized with Cell-Tak (A) or cells allowed to attach to polylysine-coated coverslips in primary culture for 24 h (B) were permeabilized with digitonin and loaded with $Ca^{2+}$ in the presence of ATP. Luminal $Ca^{2+}$ in individual cells was monitored by measuring chlortetracycline (CTC) fluorescence. The traces show $Ca^{2+}$ release in response to the indicated concentrations of $IP_3$ after completion of ATP-dependent $Ca^{2+}$ accumulation. The residual ($IP_3$-insensitive) stores were released with 4 μM ionomycin (Io). Adapted from Renard-Rooney et al (1993), where complete details of the methods may be found.

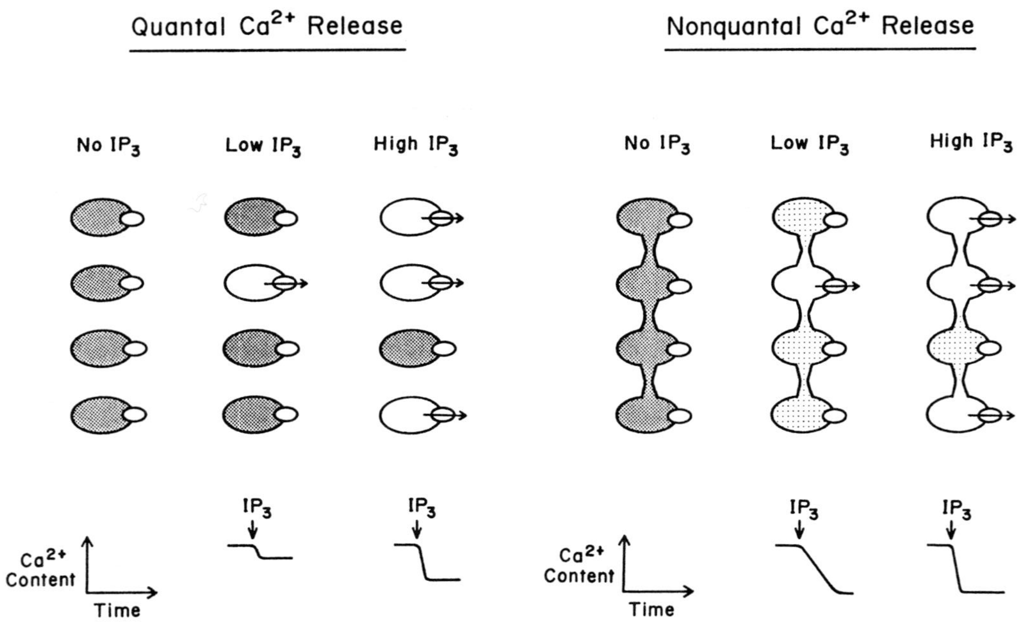


FIG. 2. Scheme showing the effect of luminal continuity between $Ca^{2+}$ stores on quantal release responses to inositol 1,4,5-trisphosphate ($IP_3$). Predicted responses to low and high [$IP_3$] are shown for a series of discrete $Ca^{2+}$ stores (*left*) or a luminally continuous store (*right*). In both cases the opening of individual $IP_3$-activated channels is assumed to occur in a quantal manner (specific populations of $IP_3$ receptors respond to each $IP_3$ dose). The predicted time course and magnitude of $Ca^{2+}$ release from the stores is indicated below each schematic representation of the final result of $IP_3$ action. Reproduced with the permission of the American Society for Biochemistry and Molecular Biology from Renard-Rooney et al (1993).

ER was highly fragmented, whereas the normal tubular ER structure was retained in attached cells that appeared to have continuous $Ca^{2+}$ stores. Luminal continuity could also provide an explanation for the much larger size of the $InsP_3$-sensitive pool in the attached permeabilized cells, because $Ca^{2+}$ in domains of the store completely devoid of $InsP_3$ receptors could be released via luminal $Ca^{2+}$ translocation.

The properties of the $InsP_3$-sensitive $Ca^{2+}$ stores were investigated in intact hepatocytes by preloading the cytosol with $Mn^{2+}$. $Mn^{2+}$ does not penetrate to the dye compartmentalized within intracellular stores under basal conditions; subsequent addition of an agonist that increases [$InsP_3$] results in activation of the intracellular channels and consequent quenching of the compartmentalized fura-2. Under conditions where the $Ca^{2+}$ stores were emptied and $[Ca^{2+}]_i$ changes eliminated by preincubation of the cells with thapsigargin in a $Ca^{2+}$-free medium, both low and high vasopressin concentrations induced a continuous and complete quenching of the entire store, with the quenching rate determined by the vasopressin concentration (Fig. 3A). These findings suggest that the luminal continuity evident in attached permeabilized cells is also a property of intact hepatocytes (Hajnóczky et al 1994). Nevertheless, under conditions where normal intracellular $Ca^{2+}$ fluxes occurred (no thapsigargin), the submaximal vasopressin dose induced discrete steps of fluorescence quenching by $Mn^{2+}$ at the level of the stores (Fig. 3B). These quenching steps, which do not reflect entry of $Mn^{2+}$ across the plasma membrane, were correlated with the frequency of $[Ca^{2+}]_i$ oscillations in parallel measurements of $[Ca^{2+}]_i$ (Fig. 3C). These steps presumably represent the opening of intracellular channels for short periods during each $[Ca^{2+}]_i$ oscillation. However, even under these conditions the entire store was accessible to the submaximal vasopressin dose, because the $Mn^{2+}$ quenching was complete after a small number of steps. Thus, there is probably luminal continuity between $Ca^{2+}$ stores in the intact cell, but discrete and transient focal $Ca^{2+}$ release can occur as a result of the feedback effects of $Ca^{2+}$ on the $InsP_3$ receptor.

## Interactions between inositol trisphosphate-sensitive calcium stores and the cytoskeleton

A possible explanation for the difference between the attached permeabilized cells and those in suspension is that the former may retain a more intact cytoskeleton, anchored at the attachment points of the substratum, which would protect the ER from fragmentation. We therefore examined the effect of disrupting the cytoskeleton on the behaviour of the $InsP_3$-sensitive $Ca^{2+}$ stores. Disruption of the cytoskeleton by cold treatment or cytoskeletal inhibitors reduced the size of the $InsP_3$-sensitive store measured as $InsP_3$-induced quenching by $Mn^{2+}$ in attached permeabilized cells (Hajnóczky et al 1994). In cells treated with a combination of cytoskeletal inhibitors

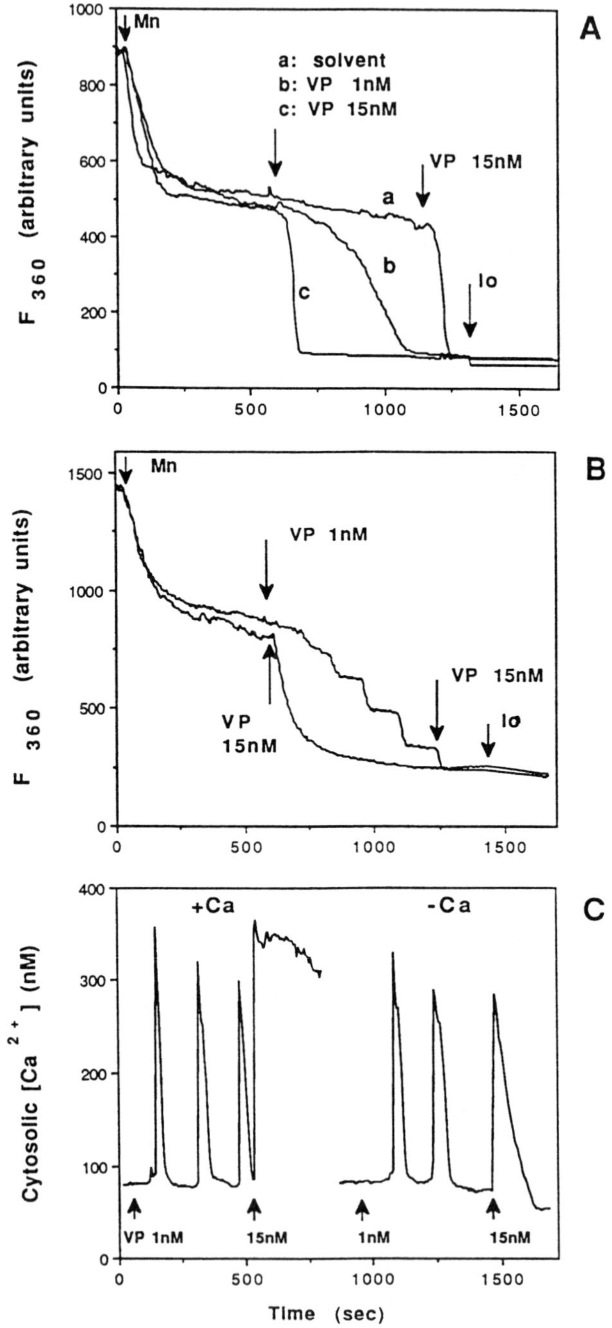

Mn
a: solvent
b: VP 1nM
c: VP 15nM
VP 15nM
a
b
c
Io
A
$F_{360}$ (arbitrary units)
B
VP 1nM
VP 15nM
Io
C
+Ca
-Ca
VP 1nM
15nM
1nM
15nM
Cytosolic $[Ca^{2+}]$ (nM)
Time (sec)

and cold, the $Mn^{2+}$ quenching response became quantal (Fig. 4), suggesting that cytoskeletal disruption leads to functional fragmentation of the $InsP_3$-sensitive $Ca^{2+}$ stores.

Previous studies have demonstrated that the size of the $InsP_3$-sensitive pool in permeabilized hepatocytes can be increased by incubation with GTP (Thomas 1988) in a manner that seems to involve the induction of luminal $Ca^{2+}$ conductance pathways between stores (Dawson & Commerford 1989, Ghosh et al 1989). A similar effect can be observed using the fluorescence quenching technique in suspensions of permeabilized cells. In this preparation, GTP increases the $Mn^{2+}$-quenchable pool size by adding a slowly accessible pool without increasing the initial rate of quenching (Hajnóczky et al 1994). Significantly, this effect is blocked by cytochalasin B, suggesting a role for the cytoskeleton in the process as it occurs in the permeabilized hepatocyte. Furthermore, addition of GTP can reverse the functional fragmentation of $Ca^{2+}$ stores that occurs when the cytoskeleton of attached permeabilized hepatocytes is disrupted by prolonged cold treatment.

## Functional implications of the structure of the calcium store

The results described above suggest that the structure of the $InsP_3$-sensitive $Ca^{2+}$ store is an important determinant of its functioning. The structure of the stores can be modulated in a manner that is controlled by GTP and is dependent on the integrity of the cytoskeleton. These control mechanisms provide a means by which intracellular $Ca^{2+}$ stores may be dynamically regulated through structural reorganization. For example, the distribution of $Ca^{2+}$ release sites or the size of the releasable pool could be modulated in this way. In addition, luminal continuity between $Ca^{2+}$ storage domains may provide a means to supply larger amounts of $Ca^{2+}$ at the required sites of release. Alternatively, the primary purpose of this organization may be to allow $Ca^{2+}$ to flow back to the release sites during $Ca^{2+}$ reuptake following a $Ca^{2+}$ spike, analogous to the partial segregation of $Ca^{2+}$ release and reuptake functions in striated

FIG. 3. Vasopressin-induced quenching of compartmentalized fura-2 by $Mn^{2+}$ in intact hepatocytes. Isolated hepatocytes were loaded with fura-2/AM under conditions designed to maximize compartmentalization of the dye, as described by Hajnóczky et al (1994). The cells were incubated in the absence of extracellular $Ca^{2+}$, except for the left trace of Panel C. For Panel A the cells were preincubated with 2 $\mu$M thapsigargin. Panels B and C show the quenching of fura-2 fluorescence (F) by $Mn^{2+}$ measured at the $Ca^{2+}$-insensitive wavelength of 360 nm, and Panel C shows $[Ca^{2+}]_i$ responses calculated from the 340:380 nm fluorescence ratio in parallel experiments. Additions were: Mn, 100 $\mu$M $MnCl_2$; VP, 1 nM or 15 nM vasopressin as indicated; Io, 5 $\mu$M ionomycin. Note that the 10 min preincubation with $Mn^{2+}$ was sufficient to saturate the cytosolic fura-2 fully such that subsequent fluorescence quenching responses on addition of vasopressin reflect stimulated entry of $Mn^{2+}$ into the intracellular stores. At 1 nM vasopressin there is a significant lag period in both the $[Ca^{2+}]_i$ and the $Mn^{2+}$ quenching responses. $Mn^{2+}$ quenching data adapted from Hajnóczky et al (1994).

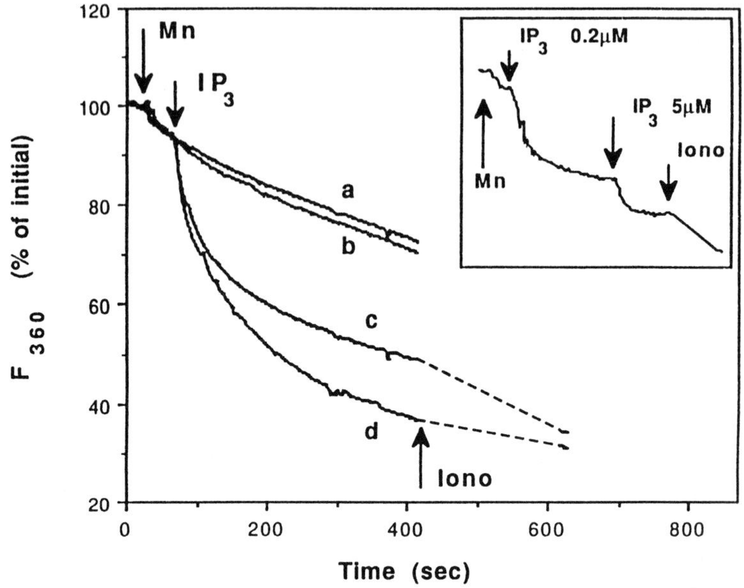


FIG. 4. Disruption of the cytoskeleton reduces inositol 1,4,5-trisphosphate ($IP_3$)-sensitive pool size and induces quantal responses in attached permeabilized cells. $IP_3$-induced quenching of compartmentalized fura-2 fluorescence (F) by $Mn^{2+}$ was measured in attached permeabilized hepatocytes in the presence of thapsigargin. The data represent the population averages of 30–60 cells from three independent experiments, except for the inset which is from a single cell. The hepatocytes were maintained in primary culture for 24 h before permeabilization. Traces b and d show control responses for $Mn^{2+}$ quenching with (b) or without (d) 5 μM $IP_3$. The cells for traces a and c were incubated with 4 μM cytochalasin B and 10 μM colcemid for the last 60 min of culture, followed by three hours' incubation on ice: trace a shows the $IP_3$-independent control quenching rate in these cells and trace c shows the response to 5 μM $IP_3$. The inset shows the quantal response to sequential doses of $IP_3$ in a single hepatocyte treated as for traces a and c. Iono, 5 μM ionomycin. Adapted from Hajnóczky et al (1994).

muscle sarcoplasmic reticulum. This would allow a more comprehensive distribution of $Ca^{2+}$ reuptake sites than release sites. In addition, separation of the release and reuptake sites into discrete but connected domains of the store could provide a way for $Ca^{2+}$ to be sequestered during the termination of a release cycle without its being immediately available for release. Thus, $Ca^{2+}$ redistribution within the lumen of the store might contribute to the refractory period observed with $[Ca^{2+}]_i$ oscillations.

## Organization of calcium signalling in the intact liver

Most studies of $Ca^{2+}$ signalling have been carried out in isolated cells or small cell clusters. Indeed, without these approaches the phenomena of $[Ca^{2+}]_i$

oscillations and waves would have been difficult to elucidate. Nevertheless, it is important to demonstrate that the mechanisms at work in the isolated cell preparation truly reflect the situation in the tissue or organ for which the cells are used as a model. One potentially important factor that can be lost or distorted in isolated cells is intercellular communication, through gap junctions, electrical continuity or paracrine factors. Several studies have demonstrated that $[Ca^{2+}]_i$ waves may propagate from cell to cell via gap junctions (Boitano et al 1992, Charles et al 1992; see also Sanderson 1995, this volume). $Ca^{2+}$ signals can also propagate between isolated hepatocyte couplets, with $InsP_3$ apparently being the messenger most likely to carry these intercellular signals (Saez et al 1989, Nathanson & Burgstahler 1992b). In addition to gap junctions, other factors such as cell polarity, blood flow and regional specialization may be important in determining how cells interact in the intact tissue. The liver is a good example of an organ where many of these factors come into play, including extensive gap junction connectivity, cell polarization and functional differences between cells in the periportal and perivenous regions of the hepatic lobule (the basic functional unit of the liver). We have therefore examined the $[Ca^{2+}]_i$ responses of individual hepatocytes *in situ* within the intact perfused rat liver, using conventional digital imaging fluorescence microscopy and confocal microscopy to monitor $[Ca^{2+}]_i$ in the first few cell layers near the liver surface.

The livers were loaded by perfusion with fura-2/AM on the stage of an inverted microscope. After washing out of the extracellular dye, the fluorescence of a small region (about 500 $\mu m^2$) was monitored with a cooled charge-coupled device camera using alternating excitation at 340 and 380 nm. Using this approach, we could observe $[Ca^{2+}]_i$ changes in individual hepatocytes following infusion of hormones via the hepatic portal vein. Figure 5 shows the individual fluorescence values and the calculated $[Ca^{2+}]_i$ changes shown by a single hepatocyte in the liver in response to sequential additions of 100 and 500 pM vasopressin. These data are qualitatively similar to those obtained in isolated hepatocytes (Rooney et al 1989). The lower concentration of vasopressin gave periodic $[Ca^{2+}]_i$ transients; the frequency of these transients increased with the higher hormone dose. However, there were also some differences from the isolated hepatocytes: the amplitude of the $Ca^{2+}$ spikes was not constant, increasing at the higher vasopressin concentration, and the kinetics of the rise and fall were slower in the intact liver than in isolated cells.

Comparison of the $[Ca^{2+}]_i$ oscillations in adjacent cells within the contiguous hepatocyte sheets of the intact liver revealed that the oscillations were synchronized through prolonged periods of constant vasopressin infusion. At low magnification, $[Ca^{2+}]_i$ oscillations were observed to follow very similar and closely coupled patterns over extensive regions that can be correlated with the lobular structure of the liver. In contrast, hepatocytes in separate lobules observed in the same image often differed markedly in their pattern of $[Ca^{2+}]_i$

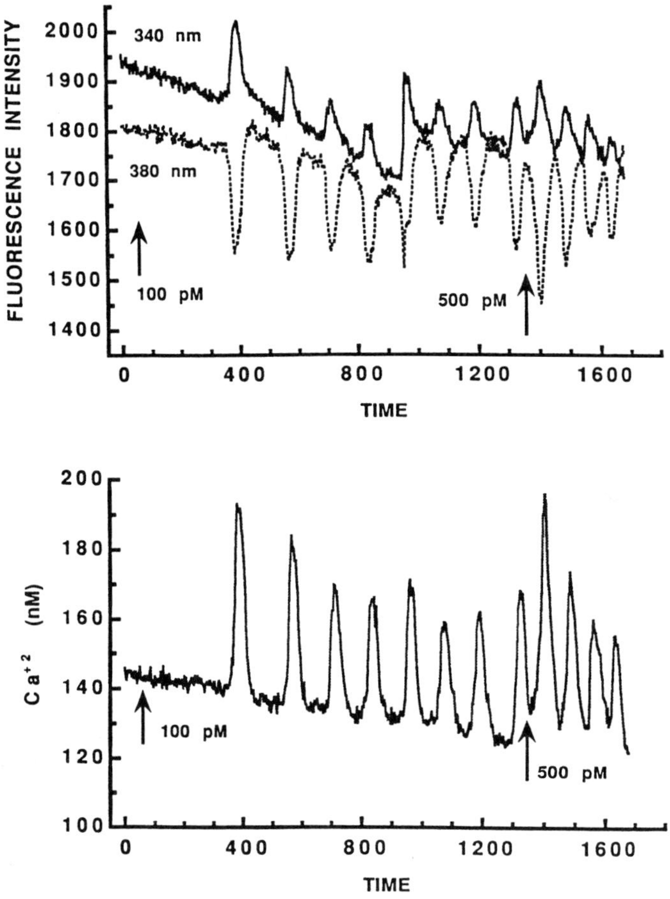


FIG. 5. Vasopressin-induced oscillations in cytosolic $Ca^{2+}$ concentration ($[Ca^{2+}]_i$) in a single hepatocyte within a perfused liver. Fluorescence images from a focal plane near to the surface of a fura-2/AM-loaded perfused rat liver were obtained using a 20× objective on an inverted epifluorescence microscope, essentially as described previously for isolated hepatocytes (Rooney et al 1989, 1991). The upper panel shows the fluorescence signals measured from a single cell with alternate excitation at 340 and 380 nm. The lower panel shows the $[Ca^{2+}]_i$ changes for this cell, calculated from the 340:380 nm ratio after correction for autofluorescence and non-cytosolic dye. The arrows indicate the times at which sustained infusions of vasopressin were initiated, first 100 pM and then 500 pM.

oscillations. The time course of $[Ca^{2+}]_i$ spikes within the coupled regions was not, however, exactly synchronized. Instead, $[Ca^{2+}]_i$ waves were first observed in individual cells and then propagated from cell to cell. The $[Ca^{2+}]_i$ waves occasionally skipped over cells, possibly reflecting propagation paths through the three-dimensional structure of the liver, out of the focal plane. On the basis of the closely parallel time courses of the $[Ca^{2+}]_i$ oscillations in several adjacent sheets of hepatocytes within a single lobule, we speculate that $[Ca^{2+}]_i$ wave propagation pathways exist to effectively synchronize the functioning of entire lobules, or at least the functional domains of the lobule. These propagating intercellular $[Ca^{2+}]_i$ waves most probably reflect communication via gap junctions, as shown previously for isolated hepatocytes (Saez et al 1989, Nathanson & Burgstahler 1992b). Extensive communication via a three-dimensional network could explain the differences in $[Ca^{2+}]_i$ oscillations observed between isolated cells and the intact liver. With isolated cells, each cell is effectively autonomous. However, with $[Ca^{2+}]_i$ waves in the intact liver propagating via many pathways, re-entrant circuits that prolong the responses of individual cells may come into play. Thus, it is possible that the $[Ca^{2+}]_i$ oscillations in the intact liver are a multicellular function, with significantly different properties from those in the isolated hepatocyte.

*Acknowledgements*

This work was supported by United States Public Health Service grants DK38422, AA07215 and AA07186.

## References

Berridge MJ 1993 Inositol trisphosphate and calcium signalling. Nature 361:315–325

Berridge MJ, Cobbold PH, Cuthbertson KSR 1988 Spatial and temporal aspects of cell signalling. Philos Trans R Soc Lond B Biol Sci 320:325–343

Boitano S, Dirksen ER, Sanderson MJ 1992 Intercellular propagation of calcium waves mediated by inositol trisphosphate. Science 258:292–295

Charles AC, Naus CCG, Zhu D, Kidder GM, Dirksen ER, Sanderson MJ 1992 Intercellular calcium signaling via gap junctions in glioma cells. J Cell Biol 118:195–201

Cobbold PH, Sanchez-Bueno A, Dixon CJ 1991 The hepatocyte calcium oscillator. Cell Calcium 12:87–95

Dawson AP, Commerford JG 1989 Effects of GTP on $Ca^{2+}$ movements across endoplasmic reticulum membranes. Cell Calcium 10:343–350

Ferris CD, Cameron AM, Huganir RL, Snyder SH 1992 Quantal calcium release by purified reconstituted inositol 1,4,5-trisphosphate receptors. Nature 356:350–352

Gilkey JC, Jaffe LF, Ridgway EB, Reynolds GT 1978 A free calcium wave traverses the activating egg of the medaka, *Oryzias latipes*. J Cell Biol 76:448–466

Ghosh TK, Mullaney JM, Tarazi FI, Gill DL 1989 GTP-activated communication between distinct inositol 1,4,5-trisphosphate-sensitive and -insensitive calcium pools. Nature 340:236–239

Glennon MC, Bird GSJ, Kwan C-Y, Putney JW Jr 1992 Actions of vasopressin and the $Ca^{2+}$-ATPase inhibitor, thapsigargin, on $Ca^{2+}$ signaling in hepatocytes. J Biol Chem 267:8230–8233

Goldbeter A, Dupont G, Berridge MJ 1990 Minimal model for signal-induced $Ca^{2+}$ oscillations and for their frequency encoding through protein phosphorylation.Proc Natl Acad Sci USA 87:1461–1465

Hajnóczky G, Gao E, Nomura T, Hoek JB, Thomas AP 1993 Protein kinase A modifies the permeability properties of the inositol 1,4,5-trisphosphate-sensitive $Ca^{2+}$ channel in a $Ca^{2+}$-independent manner in permeabilized hepatocytes. Biochem J 293:413–422
Hajnóczky G, Lin C, Thomas AP 1994 Luminal communication between intracellular calcium stores modulated by GTP and the cytoskeleton. J Biol Chem 269: 10280–10287
Jaffe LF 1991 The path of calcium in cytosolic calcium oscillations: a unifying hypothesis. Proc Natl Acad Sci USA 88:9883–9887
Kasai H, Li YX, Miyashita Y 1993 Subcellular distribution of $Ca^{2+}$ release channels underlying $Ca^{2+}$ waves and oscillations in exocrine pancreas. Cell 74:669–677
Meyer T, Stryer L 1990 Transient calcium release induced by successive increments of inositol 1,4,5-trisphosphate. Proc Natl Acad Sci USA 87:3841–3845
Meyer T, Stryer L 1991 Calcium spiking. Annu Rev Biophys Biophys Chem 20:153–174
Muallem S, Pandol SJ, Beeker TG 1989 Hormone-evoked calcium release from intracellular stores is a quantal process. J Biol Chem 264:205–212
Nathanson MH, Burgstahler AD 1992a Subcellular distribution of cytosolic $Ca^{2+}$ in isolated rat hepatocyte couplets: evaluation using confocal microscopy. Cell Calcium 13:89–98
Nathanson MH, Burgstahler AD 1992b Coordination of hormone-induced calcium signals in isolated rat hepatocyte couplets: demonstration with confocal microscopy. Mol Biol Cell 3:113–121
Oldershaw KA, Nunn DL, Taylor CW 1991 Quantal $Ca^{2+}$ mobilization stimulated by inositol 1,4,5-trisphosphate in permeabilized hepatocytes. Biochem J 278:705–708
Parker I, Ivorra I 1990 Localized all-or-none calcium liberation by inositol trisphosphate. Science 250:977–979
Renard D, Poggioli J, Berthon B, Claret M 1987 How far does phospholipase C activity depend on the cell calcium concentration? Biochem J 243:391–398
Renard-Rooney DC, Hajnóczky G, Seitz MB, Schneider TG, Thomas AP 1993 Imaging of inositol 1,4,5-trisphosphate-induced $Ca^{2+}$ fluxes in single permeabilized hepatocytes: demonstration of both quantal and nonquantal patterns of $Ca^{2+}$ release. J Biol Chem 268:23601–23610
Rooney TA, Thomas AP 1993 Intracellular calcium waves generated by $Ins(1,4,5)P_3$-dependent mechanisms. Cell Calcium 14:674–690
Rooney TA, Sass EJ, Thomas AP 1989 Characterization of cytosolic calcium oscillations induced by phenylephrine and vasopressin in single fura-2-loaded hepatocytes. J Biol Chem 264:17131–17141
Rooney TA, Sass EJ, Thomas AP 1990 Agonist-induced cytosolic calcium oscillations originate from a specific locus in single hepatocytes. J Biol Chem 265:10792–10796
Rooney TA, Renard DC, Sass E, Thomas AP 1991 Oscillatory cytosolic calcium waves independent of stimulated inositol 1,4,5-trisphosphate formation in hepatocytes. J Biol Chem 266:12272–12282
Sáez JC, Connor JA, Spray DC, Bennet MVL 1989 Hepatocyte gap junctions are permeable to the second messenger, inositol 1,4,5-trisphosphate, and to calcium ions. Proc Natl Acad Sci USA 86:2708–2712
Sanchez-Bueno A, Cobbold PH 1993 Agonist-specificity in the role of $Ca^{2+}$-induced $Ca^{2+}$ release in hepatocyte $Ca^{2+}$ oscillations. Biochem J 291:169–172
Sanderson MJ 1995 Intercellular calcium waves mediated by inositol trisphosphate. In: Calcium waves, gradients and oscillations. Wiley, Chichester (Ciba Found Symp 188) p 175–194
Thomas AP 1988 Enhancement of the inositol 1,4,5-trisphosphate-releasable $Ca^{2+}$ pool by GTP in permeabilized hepatocytes. J Biol Chem 263:2704–2711

Thorn P, Lawrie AM, Smith PM, Gallacher DV, Petersen OH 1993 Local and global cytosolic $Ca^{2+}$ oscillations in exocrine cells evoked by agonists and inositol trisphosphate. Cell 74:661–668
Woods NM, Cuthbertson KSR, Cobbold PH 1986 Repetitive transient rises in cytoplasmic free calcium in hormone-stimulated hepatocytes. Nature 319:600–602
Woods NM, Cuthbertson KSR, Cobbold PH 1987 Agonist-induced oscillations in cytoplasmic free calcium concentration in single rat hepatocytes. Cell Calcium 8:79–100

## DISCUSSION

*Meyer:* Your results in the liver are fascinating. Graf et al (1987) described pulsatile release of $^{45}Ca^{2+}$ from the entire liver.

*Thomas:* That experiment was done with a perfused liver, in the absence of extracellular $Ca^{2+}$, that was preloaded with $^{45}Ca^{2+}$ (Graf et al 1987). They found pulses of $Ca^{2+}$ release in the presence of a low-calcium perfusion medium with kinetics which were somewhat slower than the cellular $Ca^{2+}$ oscillations we have seen in the perfused liver. The liver is divided into lobules, functional units like pancreatic acini. It appears, although we need to analyse more data and get down to a lower magnification, that the entire lobule is oscillating—I would hesitate to say synchronously, because these are relatively slow waves—as a unit. Other lobules are oscillating with different patterns and with different frequencies, but within one lobule the patterns of $Ca^{2+}$ oscillations can be superimposed by a slight shift of time. That might explain the damped oscillations of $Ca^{2+}$ efflux from the whole liver, but of course we've looked at only a limited number of lobules. We looked at only a small region of the liver, whereas Graf et al (1987) had the whole liver pumping $Ca^{2+}$ out.

*Meyer:* I think they perfused vasopressin with a stepped gradient and found pulsatile release of $^{45}Ca^{2+}$.

*Thomas:* The $Ca^{2+}$ efflux oscillations ran down because there was no extracellular $Ca^{2+}$.

*Meyer:* Could it then be that the spikes in these experiments by Graf et al (1987) were synchronous only because they were triggered at the same time? Might spikes in the real liver be asynchronous because the vasopressin concentration would not be increased as rapidly?

*Thomas:* If you give a high dose, the whole liver responds at once with an increase in cytosolic $Ca^{2+}$ concentration ($[Ca^{2+}]_i$), and if you pick the right dose, you will get more than one spike, but you still get synchrony. The lower the dose, the lower is the probability that $Ca^{2+}$ oscillations will be synchronous, and the harder it will be to measure oscillations of $Ca^{2+}$ efflux.

*Fay:* Could you elaborate on your thoughts about the relationship between the cytoskeleton and the organization of the $Ca^{2+}$ stores? You have looked at both microfilaments and microtubules. The results of Dabora & Sheetz (1988) would suggest that microtubules might be the more important nucleation sites for assembly of endoplasmic reticulum (ER) membranes.

*Thomas:* It's well known that the ER essentially builds up along microtubules. It's easier to disrupt the actin cytoskeleton, and we can clearly do that, but because the microtubule cytoskeleton is in fact linked to the actin cytoskeleton we think we're disrupting it as well. The idea that we have (and it's really only an idea) is that the cytoskeleton is the scaffolding, that small G proteins direct the process, and that if you disrupt the cytoskeleton, you can make the ER fall apart. In a cell which is undergoing remodelling, perhaps during cell division or locomotion where the $Ca^{2+}$ stores may need to be moved, the cytoskeleton will be the scaffolding for the necessary trafficking in organelles, with G proteins directing the process. I would hypothesize that the $Ca^{2+}$ stores can be taken apart and put back together where they're needed.

*Whitaker:* Terasaki & Jaffe (1993) found that the ER fragmented during the $Ca^{2+}$ wave at fertilization in sea urchin eggs. When $[Ca^{2+}]_i$ reached about $2\,\mu M$, the ER fragmented into vesicles and became lamellar again when $[Ca^{2+}]_i$ came back down. Although what you're seeing with the permeabilized cells is presumably a pathological effect, there may actually be something physiological behind this lamellar–vesicular transformation.

*Thomas:* Their idea was that the ER was disassembled to allow the transit of the sperm nucleus through the egg. We've never seen anything to make us think that the hepatocytes have been disrupted, even when they've been exposed to high cytosolic $[Ca^{2+}]$ for a short time. However, hepatocytes don't like to sit in a dose of hormone which stimulates the maximal response for a long time, and they begin to lose their dye. Perhaps that has something to do with disruption or breakdown of the cytoskeleton.

*Whitaker:* I wondered if the fragmentation might be a limiting mechanism that would tend to reduce diffusion of $Ca^{2+}$ within the ER and, therefore, the size of the releasable pool and the degree of $Ca^{2+}$ release.

*Thomas:* One way to reduce the size of the releasable $Ca^{2+}$ pool is to break up the store so that you separate those regions which have no inositol 1,4,5-trisphosphate ($InsP_3$) receptors from the $InsP_3$-sensitive compartments. I have no evidence that that occurs under physiological conditions.

*Cobbold:* I don't think liver cells will lose fura-2 as rapidly as people think. We've been injecting fura-2–dextran and have found that there's something in liver cell cytoplasm, perhaps a heavy metal, that slowly quenches the signal over about half an hour. On the 360 nm signal the dye appears to be leaking out, but we're confident that it isn't actually leaking out in our cells.

*Thomas:* In our fluo-3-loaded liver cells the fluo-3 is definitely pumped out, because with the intact liver we see it arrive in the bile canaliculi. We have tried injecting the liver with dextran dyes, but got rather little signal, because of heavy metal quenching, we suspect. I don't know whether there was a progressive change because we never got a signal in the first place. This certainly is a problem, and the liver has a lot of heavy metals tied up with different enzymes.

*Cobbold:* We find we have about 20–30 min before the signal largely disappears.

*Thomas:* The fura-2 signal is stable over many hours in our hands and we can come back to cells again and again and get a good fluorescence response.

*Cobbold:* We use freshly isolated, non-attached hepatocytes, whereas you use cells maintained in culture overnight; that may be the difference.

*Petersen:* Dr Thomas, I would like to return to your original crucial observation that the $Ca^{2+}$ waves in liver cells always start from a particular location. Could you elaborate on what is special about that particular point?

*Thomas:* The distribution of $InsP_3$-sensitive pools does not seem to be the explanation. My hypothesis is that this site contains the $InsP_3$-generating machinery, the phospholipase C or the hormone receptor. The liver has a relatively high standing concentration of $InsP_3$ such that a small additional stimulation of the $InsP_3$ receptor is enough to initiate the $Ca^{2+}$ waves. My idea, and I think Tobias Meyer would probably disagree with this, is that there's an area of the cell which has all the signal transduction mechanisms and perhaps $Ca^{2+}$ pools that are more sensitive to $InsP_3$ and/or a slightly higher concentration of $InsP_3$. However, there's no direct evidence for this hypothesis.

*Nathanson:* Our immunohistochemical studies suggest that the $InsP_3$ receptor in hepatocytes within liver slices is localized to the canalicular pole (Nathanson et al 1994a). In the light of your results with chlortetracycline, might you alternatively hypothesize that $InsP_3$-mediated $Ca^{2+}$ release really occurs only at the apical pole, and that the other $Ca^{2+}$ stores from which you see release are not directly sensitive to $InsP_3$ but rather to $Ca^{2+}$ released from the initial $InsP_3$-mediated focus?

*Thomas:* I would be happy to accept that, but I don't think that our results agree with it. With both chlortetracycline and $Mn^{2+}$ quenching, the initial rate of the $InsP_3$-induced fluorescence change should be proportional to the density of $InsP_3$ receptors at the point at which you see fluorescence. In primary culture cells (we can't get good enough images in the freshly dispersed cells that you have used) we have yet to see any differences in rate over different areas of the cell, though we've looked carefully and repeatedly because this is an obvious question. $Ca^{2+}$ doesn't appear to be being released from a single subcellular region and flowing to the remainder of the cell.

When we do these $Mn^{2+}$-quenching experiments in non-permeabilized cells, we add hormone and look at the response of the compartmentalized dye to $Mn^{2+}$ entering the compartment, presumably through channels activated by $InsP_3$ but possibly by secondary $Ca^{2+}$-activated pathways. We find that the quenching begins in exactly the same place as the oscillatory $Ca^{2+}$ waves were observed to initiate in that same cell. The quenching then spreads through the cell as a wave following a similar path to the $Ca^{2+}$ wave. In the presence of thapsigargin, when there's no $Ca^{2+}$ feedback, we still see a wave of quenching although it's a slow wave. I would like to be able to tell you that this represents opening of $Ca^{2+}$ channels all the way across the cell, but the organization you suggest could also underly this response. It could be that $Ca^{2+}$ channels are activated in only one place and then the $Mn^{2+}$ enters there before being propagated through the lumen of the store. We will figure this out—we know how to.

*Berridge:* Can we take up the question of $InsP_3$ receptor localization? In pancreatic cells, a specific receptor type is thought to be located in the apical region.

*Nathanson:* The type 3 $InsP_3$ receptor is localized to the apical pole in the pancreas (Nathanson et al 1994b). Different $InsP_3$ receptor subtypes are expressed in the liver. In the liver, we've looked for types 1, 2 and 3, because there's some thought that more than one subtype is expressed. Our immunohistochemical evidence suggests the type 3 receptor is not present. In fact, the only antibody of ours that recognizes an $InsP_3$ receptor in liver recognizes a loop region that's common to the type 1 and type 2 receptor. With an antibody that recognizes only the type 1 receptor we get no labelling. There may be other subtypes that we haven't examined yet which are expressed in other parts of the hepatocyte.

In the pancreas we find immunohistochemical evidence for the type 3 receptor only, which is localized to the apical pole. We find no evidence for type 1 or type 2 receptor labelling.

*Jacob:* Dr Thomas, you mentioned that ryanodine stimulates waves. Do the properties of the ryanodine-stimulated waves match up with the properties of the agonist-stimulated wave that works through $InsP_3$?

*Thomas:* The difficulty here is that we see ryanodine-activated waves only in freshly isolated cells. In cells in culture, where resolution is much better because the cells are flatter and wider, we have not been able to induce waves with ryanodine. Sanchez-Bueno & Cobbold (1993) showed inhibition of some hormone-induced $Ca^{2+}$ oscillations in hepatocytes, and we can also see that in cells cultured for 24 h, but ryanodine itself causes waves only in freshly isolated cells. However, as Nathanson & Burgstahler (1992) showed, waves are propagated and they also have a site of origin—sometimes they have two sites of origin, which makes things more complex—but there are specific regions from which these waves propagate. In our experiments, ryanodine alone initiates $Ca^{2+}$ waves with the same origin as and similar amplitude and propagation rate to those induced by agonists. I have always thought that you have to open the $InsP_3$ receptors to get $Ca^{2+}$ waves in hepatocytes, but it seems that anything which will cause $Ca^{2+}$ release will do this. We agree with Peter Cobbold that ryanodine is inhibitory, albeit with less selectivity for different agonist types, but ryanodine doesn't seem to change the wave rate, although we are still analysing data on that.

*Cobbold:* This really concerns me. As well as differences in what ryanodine does, cyclic AMP (cAMP) does different things in our hands. In our system (freshly isolated, non-attached hepatocytes), cAMP has no effect on its own, yet you and Tom Rooney and also Thierry Capiod (unpublished work) find that cAMP induces $Ca^{2+}$ spikes in liver cells. My feeling is that we should all be working on the intact liver and forget about these cells in culture because the conditions of culture and isolation seem to cause large differences.

*Thomas:* One factor, I'm sure, is the basal [$InsP_3$], which declines after isolation.

*Cobbold:* Have you tested the effect of cAMP in your intact liver preparation?

*Thomas:* No.

*Speksnijder:* I would like to return to the localization of the initiation point of waves and the role of the cytoskeleton. Dr Jaffe described the oscillations that occur in ascidian eggs. We know that the initiation site of those waves becomes progressively localized at one side of the egg, the vegetal pole area. The localization of this initiation point is dependent on the cytoskeleton. If you treat the eggs with agents which disrupt actin, you don't see localization of the initiation point in the vegetal pole area. Using confocal microscopy, we have now seen that the ER is reorganized during this process, such that it becomes concentrated in the area from which the waves originate; this is also actin dependent. However, we have no information yet on the type of receptors present in this ER domain and their possible role in wave initiation.

*Thomas:* That's interesting. Phalloidin, which polymerizes actin, activates hepatocytes in culture. However, phalloidin also affects the basal [$InsP_3$] and the activity of phospholipase C, and it potentially changes the organization of the stores; I don't have a specific mechanism, but it appears that polymerizing actin is enough to trigger $Ca^{2+}$ spiking in our cells.

*Nathanson:* Rhodamine–phalloidin colocalizes with the $InsP_3$ receptor antibody in hepatocytes. Actin is known to be most concentrated at the apical pole of hepatocytes.

*Thomas:* Actin may be involved in bile canalicular contractions. We are working in a cell where we unfortunately are not able to see how functional polarization relates to the intact tissue, whereas you work with hepatocyte couplets, which retain a visible bile canalicular membrane domain.

*Sanderson:* It's certainly important to study $Ca^{2+}$ waves in intact tissue. However, with your intercellular waves, it doesn't seem to matter that the wave starts from a specific point, because the intracellular waves are moving along a cell in an opposite direction. Do the intracellular waves in single cells go in the opposite direction to the intercellular waves? In glial cells, an intercellular $Ca^{2+}$ wave can propagate out in one direction while the subsequent intracellular waves associated with oscillations can move in the opposite direction. Therefore, how important is it that a $Ca^{2+}$ wave is initiated from a certain point in liver cells?

*Thomas:* These $Ca^{2+}$ waves in hepatocytes are triggered by receptors in the plasma membrane. In the perfused liver, only a small, finite, domain of the plasma membrane of any given cell is seeing the hormone. I would say that one or more cells must have seen the initial signal at the appropriate subcellular domain. That raises some interesting questions. Did that cell respond with a $Ca^{2+}$ wave that propagated in a similar manner to those in isolated cells, with intercellular propagation predominating for the bulk of the tissue? Does the cell–cell propagation require the hormone to bathe all the cells, or is it enough for just one cell to see the hormone? We hope to address such questions by injecting cells in the liver. The waves do move, as far as we can tell from our confocal images, from one site of cell–cell contact to another, the site which, according to Dr Nathanson's data is in the region of the bile canalicular membrane. I assume the passage here is through gap junctions.

*Sanderson:* Do you have experimental evidence for the involvement of gap junctions?

*Thomas:* Dissociating the cells in the liver prevents the responses. However, gap junction inhibitors such as octanol perfused into the liver do not seem to get very far into the tissue.

*Sanderson:* Do hepatocytes in culture have gap junctions?

*Thomas:* The couplets that Mike Nathanson has looked at certainly seem to retain them. If the cells have gap junction connections when they are originally isolated, they may retain them in culture, but we use mostly dissociated cells, which never seem to re-establish gap junctions. So, once in a while, the waves propagate from cell to cell in culture.

*Fay:* I understand that you are investigating the linkage between $Ca^{2+}$ and mitochondrial function. Could you tell us about that?

*Thomas:* We have shown oscillations, in synchrony with but lagging slightly behind cytosolic $Ca^{2+}$ oscillations, of NADH and a flavoprotein which we've characterized as being associated with a mitochondrial dehydrogenase. Glennon et al (1992) and Rizzuto et al (1993) have actually measured mitochondrial $[Ca^{2+}]$ responses in intact cells, but not in the liver, unfortunately.

*Fay:* What comes first? Do you think there's increased electron transport and an increased driving force for $Ca^{2+}$ uptake, or is it the other way around?

*Thomas:* Our hypothesis is based on Jim McCormack and Dick Denton's work (Denton et al 1988). They have shown that the mitochondrial dehydrogenases, specifically pyruvate dehydrogenase and oxoglutarate dehydrogenase, are $Ca^{2+}$-activated enzymes in the region where we believe mitochondrial $[Ca^{2+}]$ is changing. Therefore, you are getting a push in the system. $Ca^{2+}$ activates the dehydrogenase, and NADH levels are increased, which will push the respiratory chain and more ATP will be made.

*Fay:* With L. Loew, we have measured mitochondrial membrane potential with dyes in cultured neurons and found that mitochondria become depolarized in response to an increase in cytosolic $[Ca^{2+}]$ (Loew et al 1994). The predominant effect in the cells is that $Ca^{2+}$ influx causes a fall in mitochondrial membrane potential rather than stimulation of the electron transport chain, which might be expected to increase membrane potential.

*Thomas:* If the membrane depolarization were the leading force, [NADH] should fall, whereas we see an increase. I presume the effect of $Ca^{2+}$ depends on the balance between the ability to drive energy metabolism, a major function of the liver, and the ability of the respiratory chain to compensate for the enhanced $Ca^{2+}$ cycling.

*Fay:* Have you or has anybody else measured mitochondrial membrane potential during $Ca^{2+}$ oscillations?

*Thomas:* Not during oscillations. It has been measured in the presence of hormone. In hormone-treated cells indirect measures suggest that there's a small increase in the mitochondrial membrane potential, which would be consistent again with the push hypothesis.

*Stucki:* We have measured mitochondrial membrane potential in hepatocytes during oscillations using rhodamine 123 (Emaus et al 1986) and JC-1 (Smiley et al 1991). There is no apparent difference from the mitochondrial membrane potential observed with conventional imaging methods. When you go down to confocal microscopy using JC-1, you find that the mitochondrial potentials are not homogeneous—there are regions with high potentials and others with lower potentials. You also find spontaneous changes in these potentials which are not necessarily related to hormone treatment.

*Thomas:* If the mitochondria have a lot of asynchrony, it may be difficult to resolve this question at the level of single mitochondria.

*Stucki:* Our thermodynamic calculations showed that the cellular phosphate potential, and therefore presumably also the mitochondrial membrane potential, is highly buffered (Stucki 1991). This is necessary because if it were not the case, any change in energy utilization would cause the whole system to break down.

*Thomas:* Nevertheless, in the intact perfused liver, it's possible to show enhanced flux through those pathways, through the citric acid cycle, through pyruvate dehydrogenase, that are activated when hormones which increase $[Ca^{2+}]_i$ are passed through the liver and which are $Ca^{2+}$ dependent.

*Stucki:* There is a redox feedback which presumably ensures that ATP production and the phosphate potential are always matched to cope with changing energy demands.

*Putney:* I would like to tell you about a paradox we have come across, and which has worried us somewhat, concerning oscillations in single hepatocytes. My colleague, Gary Bird, has shown that with caged $InsP_3$, and also with caged glycerophosphoinositol bisphosphate, increasing durations of flash photolysis produce a graded release of intracellular $Ca^{2+}$. However, we can never obtain a graded response to the peptide hormone vasopressin. For example, with 300 pM vasopressin, about half of the hepatocytes will respond and half will not. Those which do respond give a maximal, all-or-none response. When a massive flash of light is given on top of one of these all-or-none responses, liberating a supramaximal concentration of $InsP_3$, no additional release of $Ca^{2+}$ is observed. So the question is, why can we get a graded response to $InsP_3$ but not to a hormone which presumably acts through $InsP_3$?

I think that this has to do with an issue raised earlier by Dr Berridge, about the initiation of oscillations. Like Dr Thomas, we see all-or-none quenching of compartmentalized dye by $Mn^{2+}$ in cells which have no $Ca^{2+}$ outside and no $Ca^{2+}$ in their stores (because they have been previously depleted with thapsigargin). There is a variable delay after application of hormone, and then an abrupt, all-or-none quenching. This delay is inversely related to vasopressin concentration, as is the delay to the $[Ca^{2+}]_i$ spike, but it is somewhat longer. Our initial interpretation was that this was indicative of $Ca^{2+}$-induced $Ca^{2+}$ release in the spiking cells, but when we actually measured $[InsP_3]$, we found that the difference is more probably due to a faster rise in cellular $[InsP_3]$ in the spiking cells. Thus, the latency appears to be related to the time it takes for $[InsP_3]$ to reach some critical level.

Our conclusion is that a positively cooperative mechanism that is involved in the interaction of $InsP_3$ with its receptor and gating of the $InsP_3$-regulated channels is important in initiating spiking behaviour. Owing to inactivation mechanisms, this might not occur if $[InsP_3]$ is raised abruptly by photolysis, for example, or with very low affinity agonists. This might explain why it is difficult to get baseline $[Ca^{2+}]_i$ spikes with acetylcholine. The muscarinic acetylcholine receptor is of low affinity and thus shows a fast dissociation rate and rapid equilibration, such that subsequently the rise in $[InsP_3]$ would be rather abrupt. It is much easier to get baseline $[Ca^{2+}]_i$ spikes with peptide agonists, with which, because of the high receptor affinity, $[InsP_3]$ will increase as a function of hormone concentration more gradually.

*Thomas:* We have recently found with the $Mn^{2+}$ quenching technique that the $InsP_3$ receptor seems to become desensitized as a result of $InsP_3$ being bound to it (Hajnoczky & Thomas 1994). The effect occurs only at $Ca^{2+}$ concentrations above or in the region of the resting $[Ca^{2+}]$, but is clearly due to bound $InsP_3$ and not to other potential regulators that we have been able to eliminate. The desensitization appears to occur in a rather unusual way but one which is, perhaps, not too remarkable when you think about the $Ca^{2+}$ release phenomenon. The desensitization of the $InsP_3$ receptor is specific: with a low dose of $InsP_3$ only the population of channels that were activated becomes desensitized. A further step of $InsP_3$ will allow further quenching by $Mn^{2+}$, for example. The longer you pre-incubate with $InsP_3$ before adding $Mn^{2+}$, the lower the initial quenching rate is. This desensitization effect is temperature dependent, occurring at 37 and 30 °C but not at 4 °C.

*Putney:* We did our experiments at 20 °C.

*Thomas:* It would work there.

*Berridge:* This still doesn't help to explain why the $Ca^{2+}$ release process fails to become regenerative once it starts. Why is it graded?

*Thomas:* If all the receptors that can see the $InsP_3$ see it rapidly and immediately, they may become desensitized before complete $Ca^{2+}$ release and propagation of the release process can occur.

*Putney:* I'm saying that increasing [$InsP_3$] slowly is advantageous for propagation.

*Irvine:* Surely the difference here is the exact location at which you are actually starting the signal. When you flash the cell you raise [$InsP_3$] uniformly through the cell: when you put on a hormone you switch on this mysterious initial generation mechanism which is localized (near the hormone receptors), and which may elicit entirely different behaviour of $InsP_3$ receptors.

*Putney:* Are you suggesting there's a standing gradient of $InsP_3$?

*Irvine:* There might be. I would think more in terms of the fundamental difference in the setting up of the actual initiation of the wave at one point, as compared with simply raising [$InsP_3$] through the cell.

*Putney:* With caged glycerophosphoinositol bisphosphate in the cell, we have had some success in inducing baseline $[Ca^{2+}]_i$ spikes. That is also a caged compound distributed throughout the cell. I believe that to get the cells to produce baseline $[Ca^{2+}]_i$ spikes we must come as close as we can to mimicking the kinetics of what the peptide does when it raises [$InsP_3$].

*Cobbold:* We need to clear up the misconception that all the spikes should have the same peak free $[Ca^{2+}]$. With a virgin liver cell exposed to agonist, the first few spikes that you get are always higher than the subsequent stable peak height. Schofl et al (1993) showed that if you apply hormone in a pulsatile fashion, say for 10 s, and then leave the cell for progressively longer periods of time, you get back towards the peak height seen in a virgin cell. Indeed, if you give the hormone as a 10 s pulse, record a spike and then leave the cell without hormone for three minutes, when you next put the hormone on you will get a spike with a peak height corresponding to a free $[Ca^{2+}]$ of 2 $\mu$M rather than one of the standard 600 nM size. In an unstimulated hepatocyte there is actually continuous $Ca^{2+}$ entry across the plasmalemma. We have shown this using $Mn^{2+}$ quenching in fura-2–dextran-injected hepatocytes. When you first add $Mn^{2+}$ to a cell, whether or not there's hormone there, $Ca^{2+}$, as indicated by $Mn^{2+}$, is going in all the time, and we think that this $Ca^{2+}$ is available for refilling of the stores. The longer you leave the cell between pulses of hormone, or between pulses of caged $InsP_3$, the higher the peak you will get, because the stores are more loaded.

*Putney:* I'm not sure this is relevant to your point, but what we showed is that everything that's in the store is released. The peak height is related to the amount in the store, not how much gets released.

*Cobbold:* The stores cannot be emptied completely between spikes because you can get two to three spikes in zero $[Ca^{2+}]$.

*Putney:* That's because when the spike comes down the $Ca^{2+}$ gets taken back up. The store is empty at the top of the spike but not at the bottom of the spike.

We also find that the second spike is always smaller than the first, but if you flash caged $InsP_3$ at the top of the second spike you get no additional $Ca^{2+}$ release. The second spike is smaller than the first because the pool is not loaded to the same degree as it was in the resting state.

*Cobbold:* I would agree.

*Jacob:* We too are trying to work out what's happening with $Ca^{2+}$ release in endothelial cells. Our impression is that the spikes are not defined by the release of all the store, that only a small fraction of the store is released to form the spike, but we're not sure about that because there are still things that puzzle us.

*Putney:* You could say that $InsP_3$ induces no release because the receptor is completely desensitized, but *in vitro* the receptor never becomes desensitized to that degree, and you should always get a response to massive amounts of $InsP_3$. Perhaps we didn't release everything, and somehow in the intact cell $InsP_3$ is totally incapable of working.

*Meyer:* It was suggested that there might be a gradient of $InsP_3$ within cells. However, it is unlikely that a cell with dimensions of 10–30 $\mu$m has a gradient of $InsP_3$. The lifetime of $InsP_3$ that Dr Putney measured in hepatocytes and that other groups found in different cells is only a few seconds. Binding studies suggest that there aren't enough buffer sites for $InsP_3$ to reduce the free $[InsP_3]$ substantially. It's therefore unlikely that $InsP_3$ can be localized because it diffuses so rapidly, at 300 $\mu m^2/s$, which means there would be equilibration in one second even across a large cell.

*Putney:* We showed that after the blockade of muscarinic receptors with atropine, the half-life of the $InsP_3$ signal is of the order of a few seconds. However, that includes all the steps from putting on atropine to the decline in $[InsP_3]$, so we don't know what is rate limiting.

*Meyer:* Even if the lifetime were one second, which is probably the fastest estimate for most cells (calculated from phosphatase activity in a permeabilized cell assay), $InsP_3$ would be a general messenger, with about the same concentration everywhere in the cell.

*Putney:* If you inject fairly high concentrations of $InsP_3$ into *Xenopus* eggs, oscillatory activity hangs around for many seconds.

*Meyer:* In *Xenopus* eggs, the half-life is about 60 s. In most small mammalian cells, at least the cells we have studied, $Ca^{2+}$ transients can be triggered by UV uncaging of $InsP_3$ every 10 s. This suggests that $InsP_3$ lifetimes are much less than 10 s.

*Fay:* The difference between Dr Putney's caged $InsP_3$ experiment and the vasopressin experiment might also be related to the fact that in response to vasopressin not only is $InsP_3$ being generated but also diacylglycerol, and diacylglycerol will at least affect protein kinase C. In addition, other proteins may be activated via G protein pathways. Some of these proteins and second messengers may have profound influences on many steps involved in $Ca^{2+}$ signalling.

*Cobbold:* I was very keen on protein kinase C, because activators of protein kinase C inhibit phenylephrine-induced and vasopressin-induced spikes in hepatocytes at nanomolar concentrations, until Jane Dixon in my lab found that ATP-induced spikes will continue in 100 nM phorbol ester for half an hour.

*Putney:* In our experience, the influence of protein kinase C on this pathway is on $InsP_3$ production and before. Phorbol esters have no effect on responses to caged $InsP_3$ so there don't seem to be protein kinase C steps downstream.

*Fay:* There may be other targets for the G proteins.

*Petersen:* The most straightforward explanation for the discrepancy between the experiments with caged $InsP_3$ and vasopressin is that the kind of $InsP_3$ concentrations that you are generating by flash photolysis of caged $InsP_3$ may actually be higher than those that would be generated by hormone stimulation. In pancreatic acinar cells, the concentration of $InsP_3$ that you have to put in the cell in order to get $Ca^{2+}$ spiking is much higher than the level likely to be generated by hormones (Thorn & Petersen 1993). It now seems likely that hormone stimulation also produces an additional messenger, perhaps cADP-ribose (Thorn et al 1994).

*Putney:* You are saying that we aren't getting enough $Ca^{2+}$ because we're giving too much $InsP_3$.

*Petersen:* Yes. You will get much more desensitization with a high dose.

*Putney:* Yes, but we've tried a broad range of flash lengths, even with glycerophosphoinositol bisphosphate, and also a broad range of hormone concentrations, and you might expect there to be some overlap there.

*Irvine:* There is a small point you should remember when thinking about gradients. The $InsP_3$ phosphatase is mostly in the plasma membrane, right where the $InsP_3$ is generated, and although I accept that with low concentrations of $InsP_3$ the $InsP_3$ is metabolized through kinase rather than phosphatase, the kinase has also been reported to be localized to the plasma membrane in some cells. If you have removal right at the source of generation, or at a point remote from it, it may be possible to set up a gradient.

*Michell:* In cell fractionation studies, at least, although this hasn't been shown immunocytochemically, the plasma membrane $InsP_3$-5-phosphatase seems in the hepatocyte to be concentrated towards the bile canalicular pole rather than the pole at which $InsP_3$ is generated, suggesting that $InsP_3$ formation and destruction are spatially separated (Shears et al 1988).

*Meyer:* You can get a gradient only if the source of $InsP_3$ is at a site different from where it is degraded. The source and the sink have to be at opposite sites in a cell for a gradient to be established.

*Miyazaki:* Can you flash caged $InsP_3$ at the same small points at which the $Ca^{2+}$ gradient is detected?

*Putney:* That could be done with a confocal apparatus, but we haven't done that yet.

*Jaffe:* As I understood it, you said that when [$InsP_3$] went up, the latency went down?

*Putney:* Well, when the hormone concentration goes up the latency is reduced.

*Jaffe:* Would that suggest a sort of two-pool model in which the $InsP_3$ releases $Ca^{2+}$, with the time it takes for uptake into the ultimate release site determining the latency?

*Putney:* Perhaps. My simple interpretation, from the limited $InsP_3$ measurements we've made, is that if the hormone concentration is increased, the time required to reach some triggering level of $InsP_3$ is reduced.

*Parker:* I agree with your interpretation regarding the latency. In *Xenopus* oocytes we can step up the [$InsP_3$] but see no detectable $Ca^{2+}$ release whatsoever during the latent period. I think two processes are contributing. One is, as you say, a build up of $InsP_3$ to a threshold concentration, but beyond that if we give a very rapid step of $InsP_3$, a very short flash, there's an additional latency of only 200 ms, which is not due to a rise in [$InsP_3$] but suggests the involvement of some additional process in triggering the $InsP_3$ receptor.

*Putney:* But in *Xenopus* oocytes, the slow build up doesn't apply, because you can clearly flash $InsP_3$ and get instant wave propagation, whereas when we flash in hepatocytes we never seem to get a propagated response.

*Lechleiter:* In oocytes, when we flash caged $InsP_3$ we don't get wave propagation. The $InsP_3$ concentration must be at a certain level. We have to first inject $InsP_3(S)_3$ to stimulate $Ca^{2+}$ wave activity. Then when we give additional UV flashes, new waves are stimulated.

*Putney:* It's not so different, then.

*Cobbold:* Is it possible that $InsP_3$, through its receptor, is promoting loading of the $Ca^{2+}$ store between spikes? We don't really know what the free [$Ca^{2+}$] is in the lumen. We are not too sure if there's any voltage across the membrane at all. In one of Dr Thomas's traces showing $Mn^{2+}$ quenching of fura-2 sequestered in the ER (Fig. 3) there seemed to be a pacemaker-like quench before the spike itself lifted up. If $InsP_3$ were allowing $Ca^{2+}$ entry, and then somehow loading the ER, that might explain why you need continuous low concentrations of $InsP_3$ rather than a sudden burst.

*Putney:* But if loading were increasing, the spike height would increase, but it doesn't.

*Thomas:* As Dr Putney said, I think that the pre-spike latency and the accelerating phase reflect the initial time needed to build up [$InsP_3$], and then $InsP_3$ may continue to accumulate during part of the time course. But it's hard to get direct evidence for that; we need single cell [$InsP_3$] measurements.

*Putney:* I don't mean to say that there's no $Ca^{2+}$-induced $Ca^{2+}$ release in the initiation of the spike. One difference between $Mn^{2+}$ quenching of sequestered dye and the spike is that the rate of $Mn^{2+}$ quenching depends on vasopressin concentration, but if you allow the cell to spike, the rate of rise of the spike does not. $Ca^{2+}$-induced $Ca^{2+}$ release isn't involved in initiating the spike, but once it begins, the spike goes to an all-or-none release through a $Ca^{2+}$-induced $Ca^{2+}$ release mechanism because the process becomes $InsP_3$ independent.

*Berridge:* Dr Friel, you have been studying this with regard to ryanodine receptors. What's your impression of the initiation phase? Is it focal in your system?

*Friel:* There is no indication that initiation occurs focally, at least on the basis of non-confocal $[Ca^{2+}]_i$ imaging. In sympathetic neurons, $[Ca^{2+}]_i$ gradients dissipate in several hundred milliseconds, e.g., following a voltage step (Hernandez-Cruz et al 1990, Hua et al 1993), which is fast compared with the changes in $[Ca^{2+}]_i$ that occur during the oscillatory cycle. Therefore, significant $[Ca^{2+}]_i$ gradients may not develop. Theoretically, $[Ca^{2+}]_i$ oscillations could occur in the absence of spatial gradients. However, $[Ca^{2+}]_i$ measurements with high spatial and temporal resolution will be needed to determine whether or not this actually occurs.

*Berridge:* So what would your interpretation be of this latency preceeding the spike?

*Friel:* Several things could contribute to the accelerated rise in $[Ca^{2+}]_i$ that leads to the first spike after exposure to agonists. As Dr Thomas pointed out, it could reflect the rise in $InsP_3$ concentration, but time-dependent gating of $InsP_3$-sensitive channels could also play a role. In addition, the rate at which $[Ca^{2+}]_i$ approaches the threshold for spike generation may depend on the relative rates of $InsP_3$-induced $Ca^{2+}$ release and of $Ca^{2+}$ extrusion and buffering. Each of these factors needs to be examined in detail.

*Thomas:* In our $Mn^{2+}$ quenching experiments where we had physiological $[Ca^{2+}]_i$ and spiking going on, the rate at which the steps of quenching occurred, which are the steps that we think correlate with the opening of $InsP_3$ receptor channels during a spike, were substantially faster than the rate in the absence of $Ca^{2+}$ feedback with a low dose of hormone, and actually looked about as fast at the rate with the high vasopressin dose. This does suggest that there's some $Ca^{2+}$ feedback to enhance the opening.

*Petersen:* One of the most dramatic ways in which you can change latency is by changing, active or passive buffering. To me, that's a good indication that the initiation phase actually does involve $Ca^{2+}$-induced $Ca^{2+}$ release.

*Putney:* That's exactly what I did. If I hadn't measured $[InsP_3]$, I would have interpreted the longer latency seen when we moved the $Ca^{2+}$ signal as $Ca^{2+}$-induced $Ca^{2+}$ release, but unless you measure $[InsP_3]$ you can't be sure that that manipulation isn't changing the delay in the rise in $[InsP_3]$.

*Berridge:* Is this anything like the $Ca^{2+}$-induced formation of $InsP_3$ proposed by Tobias Meyer?

*Putney:* I don't think so. We were surprised, because all our evidence suggests that there's no regulation by $Ca^{2+}$ of the steady state, and I think the steady-state propagation and spiking is strictly driven by $Ca^{2+}$-induced $Ca^{2+}$ release. $Ca^{2+}$ does influence the initial time course, both in the absence of external $Ca^{2+}$ and in the absence of stored $Ca^{2+}$, and the only difference is the internal spike; there's no external $Ca^{2+}$ under either condition and yet it's clear that the release of stored $Ca^{2+}$ changes the time to the [$InsP_3$] plateau but not the height of the plateau. There are earlier observations in the literature that $Ca^{2+}$ outside the cell influences the latency, and that's been interpreted as $Ca^{2+}$-induced $Ca^{2+}$ release, or a luminal effect or whatever, but unless [$InsP_3$] is measured, you don't know that the explanation is not an effect on the kinetics of $InsP_3$ accumulation.

*Williams:* It is important to remember that the $Ca^{2+}$ rates on and off proteins are completely different in different cell types. There will be extensive buffering in a cell loaded with a protein such as calbindin or other $Ca^{2+}$-binding proteins, whereas other proteins don't buffer effectively against time. Some have slow time courses and some have fast time courses, according to the cell. You may find yourself confused by the rise times of various effects which depend on $Ca^{2+}$ binding to proteins of different kinds.

I have heard today that not only is there $Ca^{2+}$ spiking, but there is also possible oscillation of $InsP_3$ and spikes of $Ca^{2+}$ into the mitochondria which affect NADH reactions. What other cooperative phenomena occur? Does oscillating contraction occur in these cells? Are all the filaments responding to $Ca^{2+}$ as soon as you admit it? If so, there is a complete change in the cell as $Ca^{2+}$ is admitted so that the $Ca^{2+}$ spike is connected to the membrane. What happens to the cell shape? Do these cells breathe as the $Ca^{2+}$ spikes occur?

*Putney:* In secretory cells, where you can measure the response of a single cell in real time, the response of the cell will often oscillate with the same frequency as the $Ca^{2+}$ spikes. We can't measure glucose release from a single hepatocyte in real time.

*Thomas:* The whole liver will contract when you perfuse high concentrations of vasopressin through it, which makes imaging rather difficult.

## References

Dabora SL, Sheetz MP 1988 The microtubule-dependent formation of a tubulovesicular network with characteristics of the ER from cultured cell extracts. Cell 54:27–35

Denton RM, McCormack JG, Midgley PJW, Rutter GA, Thomas AP 1988 The role of $Ca^{2+}$ in the hormonal control of intramitochondrial metabolism in heart, liver and adipose tissue. Adv Second Messenger Phosphoprotein Res 21:157–164

Emaus RK, Grunwald R, Lemasters JJ 1986 Rhodamine 123 as a probe of transmembrane potential in isolated rat-liver mitochondria: spectral and metabolic properties. Biochim Biophys Acta 850:436–448

Glennon MC, Bird GSJ, Takemura H, Thastrup O, Leslie BA, Putney J Jr 1992 *In situ* imaging of agonist-sensitive calcium pools in AR4-2J pancreatoma cells. J Biol Chem 267:25568–25575

Graf P, von Dahl SS, Sies H 1987 Sustained oscillations in extracellular calcium concentrations upon hormonal stimulation of perfused rat liver. Biochem J 241:933–936

Hajnoczky G, Thomas AP 1994 The inositol trisphosphate calcium channel is inactivated by inositol trisphosphate. Nature 370:474–477

Hernandez-Cruz A, Sala F, Adams PR 1990 Subcellular calcium transients visualized by confocal microscopy in a voltage-clamped vertebrate neuron. Science 247:858–862

Hua SY, Nohmi M, Kuba K 1993 Characteristics of $Ca^{2+}$ release induced by $Ca^{2+}$ influx in cultured bullfrog sympathetic neurons. J Physiol 464:245–272

Loew LM, Carrington W, Tuft RA, Fay FS 1994 Physiological cytosolic $Ca^{2+}$ transients evoke concurrent mitochondrial depolarizations. Proc Natl Acad Sci USA 91:12579–12628

Nathanson MH, Burgstahler AD 1992 Subcellular distribution of cytosolic $Ca^{2+}$ in isolated rat hepatocyte couplets: evaluation using confocal microscopy. Cell Calcium 13:89–98

Nathanson MH, Burgstahler AD, Fallon MB 1994a Multistep mechanism of polarized $Ca^{2+}$ wave patterns in hepatocytes. Am J Physiol 267:G338–G349

Nathanson MH, Fallon MB, Padfield PS, Maranto AJ 1994b Localization of the type 3 inositol 1,4,5-trisphosphate receptor to the $Ca^{2+}$ wave trigger zone in pancreatic acinar cells. J Biol Chem 269:4693–4696

Rizzuto R, Brini M, Murgia M, Pozzan T 1993 Microdomains with high $Ca^{2+}$ close to $IP_3$-sensitive channels that are sensed by neighbouring mitochondria. Science 262:744–747

Sanchez-Bueno A, Cobbold PH 1993 Agonist-specificity in the role of $Ca^{2+}$-induced $Ca^{2+}$ release in hepatocyte $Ca^{2+}$ oscillations. Biochem J 291:169–172

Schofl C, Brabant G, Hesch RD, Zurmuhlen AV, Cobbold PH, Cuthbertson KSR 1993 Temporal patterns of $\alpha_1$-receptor stimulation regulate amplitude and frequency of $Ca^{2+}$ transients. Am J Physiol 265:C1030–C1036

Shears SB, Evans WH, Kirk CJ, Michell RH 1988 Preferential localization of rat-liver D-*myo*-inositol 1,4,5-trisphosphate/1,3,4,5-tetrakisphosphate 5-phoshatase in bile canalicular plasma membrane and late endosomal vesicles. Biochem J 256:363–369

Smiley ST, Reers M, Mottola-Hartshorn C et al 1991 Intracellular heterogeneity in mitochondrial membrane potentials revealed by a J-aggregate-forming lipophilic cation JC-1. Proc Natl Acad Sci USA 88:3671–3675

Stucki JW 1991 Non-equilibrium thermodynamic sensitivity of oxidative phosphorylation. Proc R Soc Lond Ser B Biol Sci 244:197–202

Terasaki M, Jaffe LA 1993 Imaging endoplasmic reticulum in living sea urchin eggs. Methods Cell Biol 38:211–220

Thorn P, Petersen OH 1993 Calcium oscillations in pancreatic acinar cells, evoked by the cholecystokinin analogue JMV-180, depend on functional inositol 1,4,5-trisphosphate receptors. J Biol Chem 268:23219–23221

Thorn P, Gerasimento O, Petersen OH 1994 Cyclic ADP-ribose regulation of ryanodine receptors involved in agonist evoked cytosolic $Ca^{2+}$ oscillations in pancreatic acinar cells. EMBO (Eur Mol Biol Organ) J 13:2038–2043

# Calcium puffs in *Xenopus* oocytes

Ian Parker and Yong Yao

*Laboratory of Cellular and Molecular Neurobiology, Department of Psychobiology, University of California, Irvine, CA 92717-4550, USA*

*Abstract.* The second messenger inositol 1,4,5-trisphosphate ($InsP_3$) functions in large part by liberating calcium ions from intracellular stores. This release process is highly non-linear and shows a regenerative characteristic that allows production of all-or-none calcium spikes which propagate as waves. However, at low concentrations of $InsP_3$ an additional mode of calcium liberation is seen in *Xenopus* oocytes, transient 'puffs' of cytosolic calcium that last for a few hundred milliseconds and are restricted to within a few micrometres. Puffs are generally of similar size and the amount of calcium released (about $3 \times 10^{-18}$ mol) suggests that they arise through the concerted opening of several $InsP_3$-gated calcium release channels. Puff sites are present at a density of about one per 30 $\mu m^2$ in the animal hemisphere of the oocyte. Each site functions autonomously, producing puffs at largely random intervals. We conclude that calcium puffs represent 'quantal' units of $InsP_3$-evoked calcium liberation, which may result from local regenerative feedback by cytosolic calcium ions at functionally discrete release sites.

*1995 Calcium waves, gradients and oscillations. Wiley, Chichester (Ciba Foundation Symposium 188) p 50–65*

A primary action of the intracellular messenger molecule inositol 1,4,5-trisphosphate ($InsP_3$) is to liberate calcium ions from intracellular organelles into the cytosol (Berridge 1993). The properties of this release process are highly non-linear, permitting the generation of temporally and spatially complex patterns of cytosolic calcium signals. Specifically, many different cell types show repetitive calcium spikes and propagating calcium waves in response to agonist activation of the phosphoinositide messenger pathway and to direct elevation of intracellular $InsP_3$ concentrations (Berridge 1993, Rooney & Thomas 1993, Jaffe 1993). Oocytes of *Xenopus laevis* are a favourable system in which to study these phenomena, because their enormous size (1 mm diameter) offers a large 'playing field' over which calcium waves may propagate without boundaries, and interpretation of results is simplified by the absence in these cells of a parallel calcium release pathway mediated through ryanodine-sensitive channels (DeLisle & Welsh 1992).

Imaging of oocytes loaded with fluorescent calcium indicator dyes has revealed complex patterns of repetitive planar, circular and spiral waves (Lechleiter et al

1991, Parker & Yao 1991, Lechleiter & Clapham 1992). Because these responses can be evoked by intracellular injections of non-metabolizable $InsP_3$ analogues (Lechleiter & Clapham 1992, Yao & Parker 1994), and because the period of the calcium waves can be short (about 4 s) in comparison with the metabolism of $InsP_3$ (about 18 s; Parker & Ivorra 1992), it is unlikely that calcium spikes and waves arise because of fluctuating $InsP_3$ concentrations. Instead, $InsP_3$ appears to sensitize the oocyte cytoplasm such that it becomes an excitable medium in which calcium ions serve as a diffusible messenger, causing calcium stores to be sequentially triggered to release their contents in a regenerative manner (Parker & Yao 1991, DeLisle & Welsh 1992, Yao & Parker 1992).

An important clue to the functioning and spatial arrangement of $InsP_3$-sensitive stores in the oocyte came from the observation that concentrations of $InsP_3$ that were too low to evoke waves caused liberation of calcium ions in the form of sporadic 'puffs', transient (a few hundred milliseconds) increases in cytosolic calcium concentration that remained localized within a few micrometres (Parker & Yao 1991, Yao & Parker 1994). The existence of puffs provides direct evidence that $InsP_3$-mediated calcium liberation has quantal characteristics and indicates that calcium stores are arranged as discrete and functionally independent units. Here, we describe some of our recent work characterizing calcium puffs in *Xenopus* oocytes, review other results relating to their origin and discuss how the spatial arrangement of calcium stores may influence cellular responses mediated by $InsP_3$.

## Characteristics of calcium puffs

### *Puffs are evoked by photoreleased inositol trisphosphate and by a poorly metabolized inositol trisphosphate analogue*

We first observed calcium puffs in oocytes that were loaded with caged $InsP_3$ and then stimulated by flashes of UV light to photorelease free $InsP_3$ (Parker & Yao 1991). Brief flashes of light caused transient, localized liberation of calcium, whereas slightly stronger stimuli evoked propagating calcium waves of longer duration. However, it was still possible that the localized puffs had arisen artefactually through non-uniform photorelease of $InsP_3$. To circumvent this problem, we loaded oocytes with 3-F-$InsP_3$, a poorly metabolized analogue which is about equipotent to $InsP_3$ in liberating calcium (Kozikowski et al 1990). This caused a sustained activation of $InsP_3$ signalling lasting a few hours, during which time the 3-F-$InsP_3$ ought to have diffused homogeneously throughout the cell (Yao & Parker 1993). The resulting patterns of calcium liberation were then visualized using a video-rate confocal laser-scanning microscope (Odyssey, Noran Instruments Inc.) to image fluorescence signals reported by the calcium indicator calcium green-1 (Yao & Parker 1994). Intracellular concentrations of 3-F-$InsP_3$ greater than about 50 nM gave rise to propagating calcium waves, whereas concentrations between about 5 and 30 nM produced calcium puffs at different sites within the imaging field (Fig. 1).

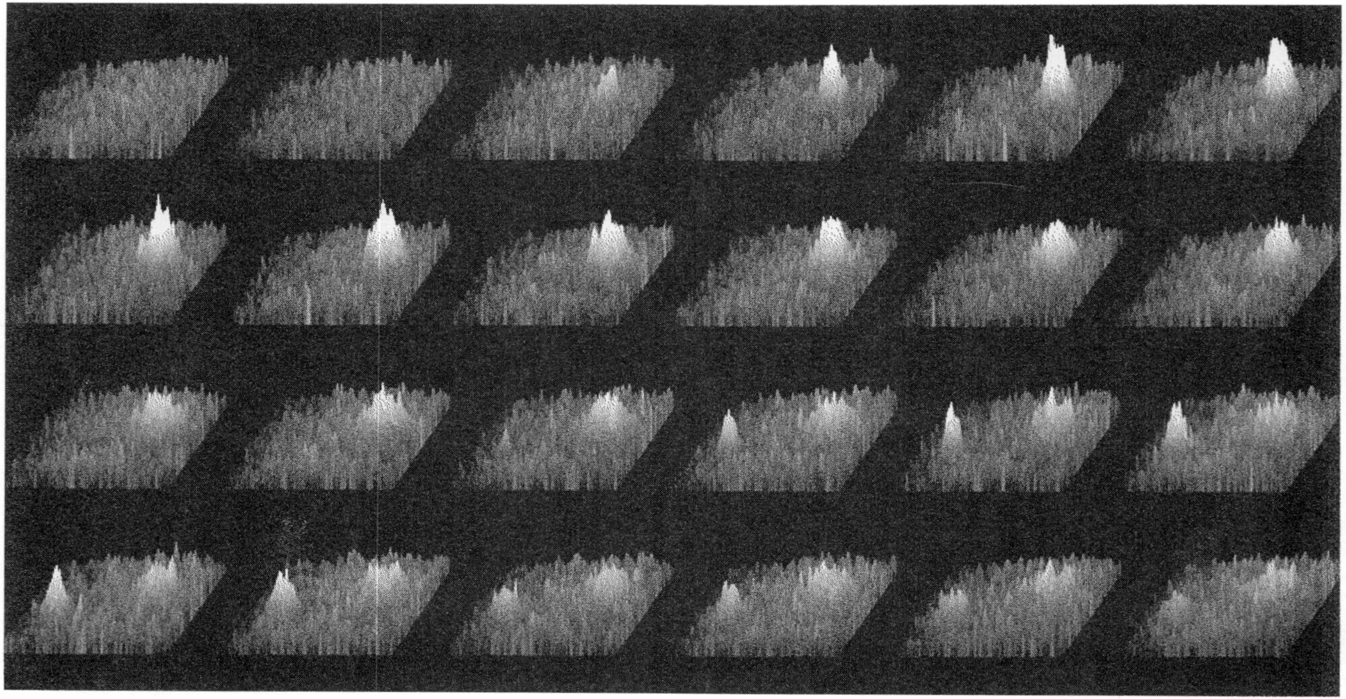

FIG. 1. Spontaneous $Ca^{2+}$ puffs in a *Xenopus* oocyte loaded with 3-F-inositol 1,4,5-trisphosphate (about 30 nM intracellular concentration) visualized using a video-rate confocal laser-scanning microscope. Successive frames (left to right, top to bottom) were captured at 16 ms intervals and show calcium release from two puff sites within a $40 \times 40$ $\mu$m region of the oocyte. Cytosolic free calcium was monitored by fluorescence of calcium green-1; increasing calcium concentrations are depicted by both increasing brightness and increasing height of each pixel.

*Magnitude and kinetics of puffs*

Figure 2 shows the time course of calcium changes during a puff, monitored by recording fluorescence from a minute spot centred on a puff site. Puffs began abruptly from the baseline level of calcium, without any detectable preceding 'foot'. The rising phase was brief (about 50 ms), with the calcium concentration then declining over several hundred milliseconds after the peak at a rate determined primarily by diffusion of calcium ions from their localized site of release, rather than by re-sequestration. Within the limits of our resolution (about 1 $\mu$m), calcium appeared to originate from a point source and diffusional spread was restricted such that little rise in calcium could be detected beyond about 6 $\mu$m from the source.

Figure 3 shows simultaneous reecordings over several minutes of puff activity at two adjacent sites. One feature which is evident is that all the puffs were of similar peak size. This was confirmed by constructing amplitude histograms of puffs recorded over long (several minute) periods at given sites, which showed a roughly Gaussian distribution around a mean size. The amount of calcium released during a puff was estimated to be $4\times10^{-18}$ mol, a value obtained by measuring the volume over which calcium was distributed and estimating how much calcium would be bound to the indicator dye and how much to endogenous buffers in addition to that measured free in solution. Because it takes about 50 ms for the calcium puff to reach its peak, this amount corresponds to a flux of $8\times10^{-17}$ mol $s^{-1}$, or ionic current of about 14 pA. By comparison, single channel currents recorded from reconstituted $InsP_3$ receptors under conditions

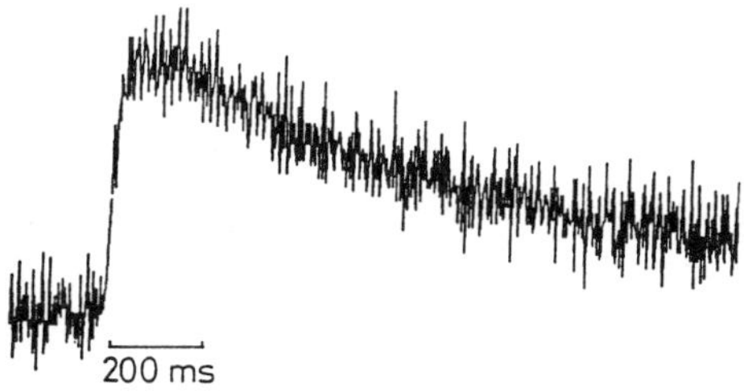


FIG. 2. Kinetics of cytosolic calcium during a puff. The trace shows calcium-dependent fluorescence measured for a virtual point source by confocal microfluorimetry (Parker & Ivorra 1993). The ooctye was loaded with 3-F-inositol 1,4,5-trisphosphate to a final intracellular concentration of about 30 nM, and puff sites were identified by imaging with a confocal microscope. The laser scan was then stopped and the stationary laser spot was positioned over a puff site to obtain records with improved time resolution. The trace shows calcium-dependent changes in fluorescence of calcium green-1, with the peak signal corresponding to an increase in fluorescence of about 60% above the resting baseline.

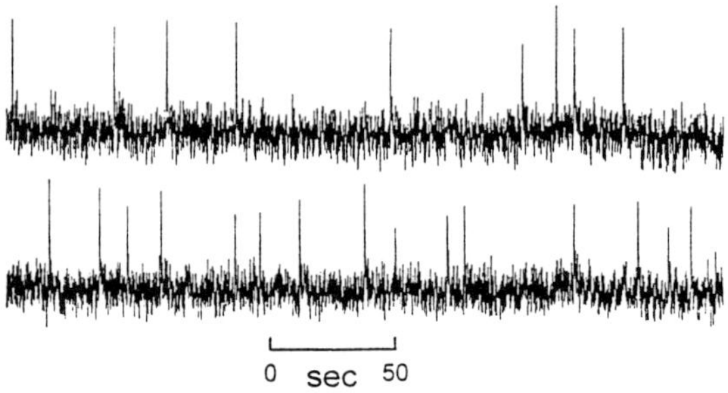


FIG. 3. Simultaneous records of puff activity at two adjacent sites. An oocyte loaded with 3-F-inositol 1,4,5-trisphosphate (about 30 nM) was imaged by video-rate confocal microscopy. The traces show simultaneous records of calcium-dependent fluorescence monitored from small ($3 \times 3$ $\mu$m) regions centred over two puff sites about 10 $\mu$m apart.

approximating the physiological environment are only about 0.25 pA (Bezprozvanny & Ehrlich 1994). Thus, it seems likely that puffs result not from the activity of single channels, but rather from the concerted opening of several channels. Also, puff amplitudes would be expected to be exponentially, not normally, distributed if they arose through the opening of single channels with stochastically determined open durations.

A second feature illustrated by Fig. 3 is that the intervals between successive puffs at each site are highly irregular, in contrast to the regular periodicity seen as calcium waves sweep past a measuring spot (Girard & Clapham 1993, Yao & Parker 1994). Histograms of inter-puff intervals showed that relatively few puffs were seen at intervals shorter than about 8 s, indicating that they show a refractory period similar to that observed when paired light flashes are used to photorelease $InsP_3$ from a caged precursor (Parker & Ivorra 1991a). Once this inhibition had worn off, after about 8 s, the occurrence of puffs at progressively longer intervals declined exponentially, suggesting that the triggering of puffs may be stochastically determined.

### *The spatial distribution of functionally autonomous puff sites*

A final point demonstrated by Fig. 3 is that puffs at one site were not obviously associated with the occurrence of puffs at the other site, even though this was only 10 $\mu$m distant (see also Parker & Yao 1991). Thus, the $InsP_3$-sensitive calcium stores that give rise to puffs are arranged as distinct and functionally independent units. However, it is not yet clear whether this functional autonomy derives from the physical separation of the stores or whether, for example, different puff sites represent clusterings of $InsP_3$ receptors within a continuous reticulum.

As already noted, puffs arose from specific sites. Particular spots within the oocyte gave puffs repeatedly, whereas other regions showed no activity even when monitored for long (tens of minutes) periods. The density of puff sites within the animal hemisphere was about one per 30–60 $\mu m^2$, corresponding to a mean spacing between sites of roughly 6 $\mu$m. Furthermore, confocal recordings of calcium signals at different depths in the cell indicated that release sites were restricted to a narrow band lying 5–7 $\mu$m below the plasma membrane. Thus, a single oocyte may contain about 100 000 puff sites. These sites do not, however, represent the only $InsP_3$-sensitive stores in the cell, because strong photorelease of $InsP_3$ evoked calcium liberation beginning both at puff sites and at regions that failed to show puffs. The puff sites may, therefore, be 'hot spots' of high sensitivity to $InsP_3$, with intervening regions showing a lower sensitivity.

## Mechanisms underlying calcium puffs and waves

### *Inositol trisphosphate-mediated calcium liberation is a regenerative process*

The rapid upstroke and all-or-none characteristics of puffs are consistent with calcium liberation arising through a regenerative process (Parker & Yao 1991); such a mechanism has also been implicated in the generation of calcium spikes and waves (Meyer 1991, Meyer & Stryer 1991, Berridge 1993, Rooney & Thomas 1993). Further support for this idea comes from experiments measuring responses to photoreleased $InsP_3$ in the oocyte (Parker & Ivorra 1990b, 1992, 1993). No calcium liberation was induced by $InsP_3$ concentrations below a certain threshold, but above this threshold, the photoreleased $InsP_3$ abruptly evoked responses which increased little as $InsP_3$ concentrations were further increased (Fig. 4). This threshold phenomenon and steepness of the kinetics of calcium liberation cannot readily be explained by a cooperativity of $InsP_3$ molecules gating channel opening but, instead, probably arise as a result of positive feedback in the release process (Parker & Ivorra 1993).

Several models have been proposed to explain how raised cytosolic calcium levels may feed back to promote further liberation of sequestered calcium (for reviews see: Tsien & Tsien 1990, Berridge 1993, Rooney & Thomas 1993), but current evidence strongly favours the idea that calcium acts as a co-agonist at the $InsP_3$ receptor to promote channel opening. Specifically, modest elevations of calcium increase the opening probability of reconstituted $InsP_3$ receptor–$Ca^{2+}$ channels (Bezprozvanny et al 1991, Callamaras & Parker 1994), and in intact oocytes raised cytosolic calcium concentrations enhance $InsP_3$-mediated calcium liberation (Yao & Parker 1992, 1993), increase the frequency and velocity of $InsP_3$-dependent calcium waves (Girard & Clapham 1993, Yao & Parker 1994) and increase the frequency of calcium puffs (Yao & Parker 1994). Thus, the $InsP_3$ receptor–$Ca^{2+}$ channel can mediate a process of calcium-induced calcium release, *provided* that $InsP_3$ is present.

As mentioned above, the amount of calcium involved in a puff appears too great to be accounted for by the opening of a single $InsP_3$ receptor–$Ca^{2+}$

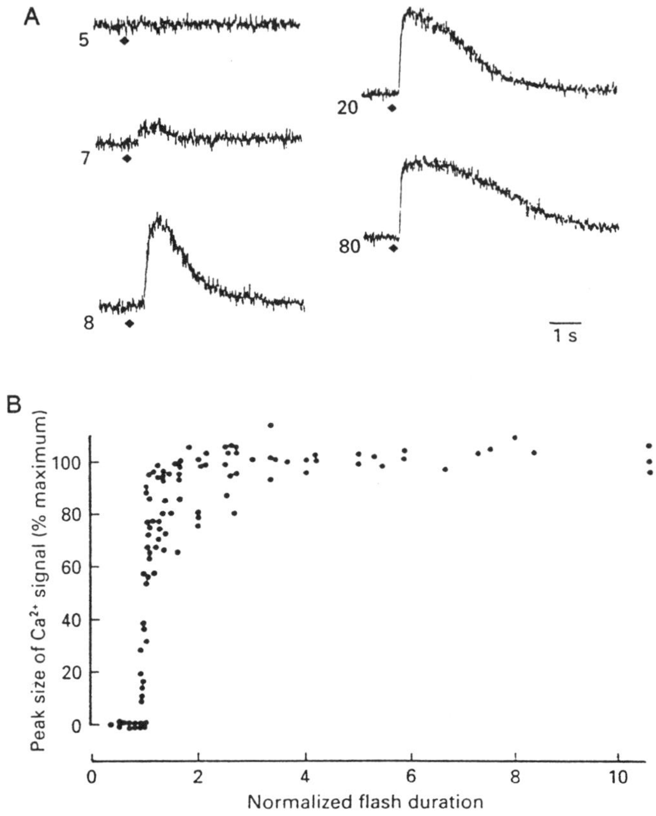


FIG. 4. Calcium liberation in *Xenopus* oocytes shows a nearly all-or-none characteristic at a threshold level of inositol 1,4,5-trisphosphate ($InsP_3$). (A) Traces show point confocal recordings of calcium transients evoked by photorelease of increasing amounts of $InsP_3$ from a caged precursor. The time of the photolysis flash is marked by a diamond, and the duration of each flash (which gives a linear indication of the amount of photoreleased $InsP_3$) is shown in milliseconds. Note that a 5 ms flash gave no detectable response, whereas an 8 ms flash gave a clear response that increased little even when the flash was further lengthened to 80 ms. (B) The graph shows pooled data from several oocytes showing the dependence of peak amplitude of confocal calcium signals on duration of the photolysis flash. Flash durations are normalized to that duration estimated to produce a half-maximal response in each oocyte, and amplitudes are expressed as a percentage of the maximal response in each cell. Adapted from Parker & Ivorra (1993).

channel, suggesting that regenerative activation of several closely clustered channels is involved. In contrast, Cheng et al (1993) report calcium 'sparks' in cardiac muscle, which are analogous to the puffs we describe but were interpreted to arise from the opening of single ryanodine receptor–$Ca^{2+}$

channels. The difference may be because opening of one ryanodine receptor channel allows release of 10- to 100-fold more calcium than opening of one $InsP_3$ receptor channel (Bezprozvanny & Ehrlich 1994).

*Local and propagating calcium signals*

Positive feedback of cytosolic calcium ions on the $InsP_3$ receptor provides an attractive explanation for the generation of all-or-none calcium puffs and waves. For example, calcium puffs may arise when the random opening of an $InsP_3$ receptor–$Ca^{2+}$ channel releases sufficient calcium to trigger the opening of closely adjacent (less than 1 $\mu$m distant) channels, with the resulting calcium liberation in turn feeding back to stimulate yet further release. This process subsequently terminates within a few tens of milliseconds, either because of depletion of store contents, or through the development of an inhibitory process controlling channel gating (Parker & Ivorra 1990a, Payne et al 1990). The question then arises of why calcium liberation remains localized as puffs at low concentrations of $InsP_3$, whereas higher concentrations allow active propagation of calcium waves over large distances.

As summarized diagrammatically in Fig. 5, the answer probably lies in the relatively large distance (about 6 $\mu$m) between puff sites. Measurements of the spatial spread of calcium during a puff are consistent with a diffusion coefficient of about 25 $\mu m^2 s^{-1}$ (J. Choi, unpublished data; see also Allbritton et al 1992), and the space constant for calcium diffusion during a puff (the distance at which the peak calcium concentration attained falls to $1/e$ of that at the source) is about 3 $\mu$m. Thus, the maximal calcium increase produced at one site by a puff occurring 6 $\mu$m away would be only about one-tenth of that measured at the site of the puff. The sensitivity of puff sites at low concentrations of $InsP_3$ may be insufficient to allow them to be triggered by this small calcium signal, so that calcium liberation remains highly localized. However, at higher concentrations of $InsP_3$, adjacent sites may become sufficiently sensitive for them to be triggered to release their contents in a regenerative manner and, in turn, trigger yet further sites to produce a calcium wave that propagates without decrement.

In this model, calcium waves are propagated in a saltatory manner, jumping from hot spot to hot spot. We have been able to resolve such behaviour at concentrations of $InsP_3$ just sufficient to allow wave propagation, but higher concentrations give wave fronts that appear to propagate uniformly. At present it is not clear whether diffusion of calcium simply fills in the gaps between discrete release sites under these conditions, or whether $InsP_3$ receptors present at lower density between puff sites are able to sustain continuous wave propagation.

## The significance of calcium puffs

Our results demonstrate that $InsP_3$-mediated calcium liberation in *Xenopus* oocytes can occur both in a quantal, localized manner as calcium puffs and,

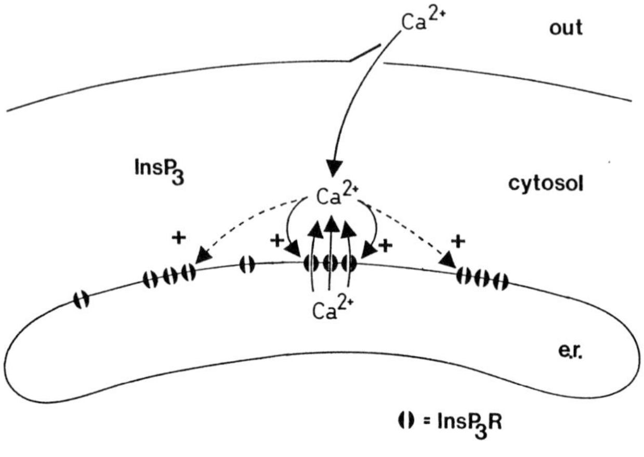


FIG. 5. Schematic diagram illustrating possible mechanisms underlying local calcium puffs and propagating calcium waves. We postulate that inositol 1,4,5-trisphosphate ($InsP_3$) receptor–$Ca^{2+}$ channels ($InsP_3R$) are present in clusters within a continuous reticulum (e.r.). In the presence of $InsP_3$, local increases in cytosolic calcium resulting from stochastic opening of one, or a few, channels within a cluster result in positive feedback on other channels in the cluster, causing regenerative calcium liberation. However, at relatively low concentrations of $InsP_3$ this response remains localized as a calcium puff, because the sensitivity of adjacent clusters is too low for them to be triggered by calcium diffusing from an active site (dashed arrows). Higher concentrations of $InsP_3$ increase the sensitivity of neighbouring sites, allowing them to be activated by diffusing calcium ions and thus support wave propagation. Increased influx of extracellular calcium (for example, through voltage-gated plasma membrane channels) sensitizes these processes, increasing the frequency of puffs and promoting the ability of puffs to trigger waves.

at higher concentrations of $InsP_3$, in a more global manner as calcium waves that propagate across the cell. Two recent reports (Thorn et al 1993, Kasai et al 1993) describe a similar dual mode of local and global calcium signalling evoked by $InsP_3$ in pancreatic acinar cells. In those cells, the production of local calcium spikes in the secretory pole region may avoid undesirable acitivation of calcium-dependent processes elsewhere in the cell and may minimize the energetic cost associated with membrane transport of the large amounts of calcium released during global waves (Thorn et al 1993). Similar arguments about physiological significance are less convincing in the oocyte, where puffs originate at sites scattered apparently at random over the cell, rather than being concentrated at functionally specialized regions. However, one possibility is that puff sites act as sensitive hot spots—as foci to trigger propagating waves—and thus entrain extensive surrounding regions of the cell. For example, when oocytes loaded with low concentrations of 3-F-$InsP_3$ are hyperpolarized to enhance influx of extracellular calcium through an

$InsP_3$-dependent pathway, the frequency of puffs increases and waves are often triggered at puff sites (Yao & Parker 1994).

Whatever their physiological role, the observation of puffs is likely to be significant in elucidating the mechanisms underlying $InsP_3$ action and the spatiotemporal aspects of calcium liberation, rather as the discovery of miniature end-plate potentials provided clues to the nature of synaptic transmission (Fatt & Katz 1952). The existence of puffs demonstrates that $InsP_3$-sensitive stores are functionally (though not necessarily physically) discontinuous and facilitates study of calcium liberation from individual stores. Furthermore, it is already clear that the spatial arrangement of stores must play a large part in determining whether calcium signals remain localized or propagate and is also likely to be important for the characteristics of wave propagation (Yagodin et al 1994).

## *Acknowledgement*

This work was supported by grant GM48071 from the US Public Health Service.

## References

Allbritton NL, Meyer T, Stryer L 1992 Range of messenger action of calcium ion and inositol 1,4,5-trisphosphate. Science 258:1812–1815

Berridge MJ 1993 Inositol trisphosphate and calcium signalling. Nature 361:315–325

Bezprozvanny I, Ehrlich BE 1994 How small is the 'quantum' of inositol 1,4,5-trisphosphate induced $Ca^{2+}$ release. Biophys J 65:A146

Bezprozvanny I, Watras J, Ehrlich BE 1991 Bell-shaped calcium-response curves of Ins(1,4,5)$P_3$- and calcium-gated channels from endoplasmic reticulum of cerebellum. Nature 351:751–754

Callamaras N, Parker I 1994 Inositol 1,4,5-trisphosphate receptors in *Xenopus laevis* oocytes: localization and modulation by $Ca^{2+}$. Cell Calcium 15:60–72

Cheng H, Lederer WJ, Cannell MB 1993 Calcium sparks: elementary events underlying excitation–contraction coupling in heart muscle. Science 262:740–744

DeLisle S, Welsh MJ 1992 Inositol trisphosphate is required for the propagation of calcium waves in *Xenopus* oocytes. J Biol Chem 267:7963–7966

Fatt P, Katz B 1952 Spontaneous subthreshold activity at motor nerve endings. J Physiol 117:109–128

Girard S, Clapham D 1993 Acceleration of intracellular calcium waves in *Xenopus* oocytes by calcium influx. Science 260:229–232

Jaffe LF 1993 Classes and mechanisms of calcium waves. Cell Calcium 14:736–745

Kasai H, Li YX, Miyashita Y 1993 Subcellular distribution of $Ca^{2+}$ release channels underlying $Ca^{2+}$ waves and oscillations in exocrine pancreas. Cell 74:669–677

Kozikowski AP, Fauq AH, Askoy IA et al 1990 Synthesis of the first optically pure fluorinated inositol 1,4,5-trisphosphate of *myo*-inositol. Stereochemistry and its effect on $Ca^{2+}$ release in Swiss 3T3 cells. J Am Chem Soc 112:7403–7404

Lechleiter JD, Clapham DE 1992 Molecular mechanisms of intracellular calcium excitability in *X. laevis* oocytes. Cell 69:283–294

Lechleiter JD, Girard S, Peralta E, Clapham D 1991 Spiral calcium wave propagation and annihilation in *Xenopus laevis* oocytes. Science 252:123–126
Meyer T 1991 Cell signalling by second messenger waves. Cell 64:675–678
Meyer T, Stryer L 1991 Calcium spiking. Annu Rev Biophys Biophys Chem 20:153–174
Parker I, Ivorra I 1990a Inhibition by $Ca^{2+}$ of inositol trisphosphate-mediated $Ca^{2+}$ liberation: a possible mechanism for oscillatory release of $Ca^{2+}$. Proc Natl Acad Sci USA 87:260–264
Parker I, Ivorra I 1990b Localized all-or-none calcium liberation by inositol trisphosphate. Science 250:977–979
Parker I, Ivorra I 1992 Characteristics of membrane currents evoked by photorelease of inositol trisphosphate in *Xenopus* oocytes. Am J Physiol 263:C154–C165
Parker I, Ivorra I 1993 Confocal microfluorimetry of $Ca^{2+}$ signals evoked in *Xenopus* oocytes by photoreleased inositol trisphosphate. J Physiol 461:133–165
Parker I, Yao Y 1991 Regenerative release of calcium from functionally discrete subcellular stores by inositol trisphosphate. Proc R Soc Lond Ser B Biol Sci 246:269–274
Payne R, Flores TM, Fein A 1990 Feedback inhibition by calcium limits the release of calcium by inositol trisphosphate in *Limulus* ventral photoreceptors. Neuron 4:547–555
Rooney TA, Thomas AP 1993 Intracellular calcium waves generated by Ins(1,4,5)P(3)-dependent mechanisms. Cell Calcium 14:674–690
Thorn P, Lawrie AM, Smith PM, Gallacher DV, Petersen OH 1993 Local and global cytosolic $Ca^{2+}$ oscillations in exocrine cells evoked by agonists and inositol trisphosphate. Cell 74:661–668
Tsien RW, Tsien RY 1990 Calcium channels, stores and oscillations. Annu Rev Cell Biol 6:715–760
Yagodin SV, Holtzclaw L, Sheppard CA, Russell JT 1994 Nonlinear propagation of agonist-induced cytoplasmic calcium waves in single astrocytes. J Neurobiol 25:265–280
Yao Y, Parker I 1992 Potentiation of inositol trisphosphate-induced $Ca^{2+}$ mobilization in *Xenopus* oocytes by cytosolic calcium. J Physiol 458:319–338
Yao Y, Parker I 1993 Inositol trisphosphate-mediated $Ca^{2+}$ influx into *Xenopus* oocytes triggers $Ca^{2+}$ liberation from intracellular stores. J Physiol 468:275–296
Yao Y, Parker I 1994 $Ca^{2+}$ influx modulates temporal and spatial patterns of inositol trisphosphate-mediated $Ca^{2+}$ liberation in *Xenopus* oocytes. J Physiol 476:17–28

## DISCUSSION

*Michell:* In your final model (Fig. 5), you drew the three adjacent inositol 1,4,5-trisphosphate ($InsP_3$) foci on the same vesicle compartment. Is that compatible with the fact that adjacent puff sites behave independently? If they could be drawing on the same $Ca^{2+}$ pool, you might expect them to influence one another.

*Parker:* We have thought about that. What you're saying is that a puff at one site might draw down the local pool in the lumen, so that a puff in an adjacent site coming soon after would, on average, be a little smaller than it otherwise would have been. We haven't tested that, and the effect might be hard to detect given the variability in puff responses.

*Michell:* Also, as you slowly bring $[InsP_3]$ up to a level at which you just reach the point at which a puff can turn into the initiation of a wave, the wave doesn't always seem to move in an entirely symmetrical way in all directions.

If you map the preceding puff sites around the origin, is there any evidence that as the wave begins to invade the domain of an adjacent site the wave intensifies locally and thus spreads more rapidly?

*Parker:* Yes. We did an experiment with an oocyte containing no 3-F-$InsP_3$ and used pinhole photorelease of $InsP_3$ with the light turned on for several seconds to generate a gradual increase in [$InsP_3$] at the centre point. Although $InsP_3$ should diffuse out as a circular wavefront, we got what I call a 'broccoli wave' of $Ca^{2+}$ release. We photorelease $InsP_3$ at a point in the centre, get a puff near there, that dies away, and we get another puff some distance away from the source of $InsP_3$. If we keep going we start to get a more sustained release of $Ca^{2+}$, which spreads in a very irregular manner. So, rather than the circular spreading that one might expect, the pattern of $Ca^{2+}$ release shows protrusions and this progressive release spreads to some of the sites that first give puffs, as a bulge rather like a broccoli floret.

*Lechleiter:* With this wave form what percentage of your receptors are bound by $InsP_3$ at 10 or at 100 nM?

*Parker:* The percentage is low. We estimate from other experiments with photoreleased $InsP_3$ that the threshold [$InsP_3$] needed to invoke a local wave is about 60 nM. The [$InsP_3$] giving a half-maximal rate of $Ca^{2+}$ release is about 10-fold greater than that (i.e., 600 nM). We wouldn't expect full receptor occupancy until [$InsP_3$] is in the micromolar range.

*Cannell:* When you raise [$InsP_3$], the frequency of puffs goes up. Do any other properties of the puffs change? Do they get larger, do they spread further? You are suggesting there is a functional unit, which must occupy some space. If you raise the probability of release from each of those individual channels, you might see a change in the geometry of the puff.

*Parker:* A plot of puff frequency, or, more correctly, how many puffs are observed in a given period, against [$InsP_3$] shows that puff frequency rises as quite a steep function of [$InsP_3$], about as a third power function. The amplitude of the puffs seems to have slight, if any, dependence on [$InsP_3$]. The puffs do get bigger (that is to say, spread further) but we don't have quantitative information about this. Over a narrow range of [$InsP_3$] just below the wave threshold we stop getting puffs as such and start seeing abortive $Ca^{2+}$ waves that can spread for variable distances of a few tens of micrometres.

*Berridge:* Would you call that a graded response? In other words, is it the summation of separate responses which gives you a graded response?

*Parker:* We regard the puffs as quantal events; I am using the word quantal in a rather different sense from the way it was first used in the context of $InsP_3$ signalling.

*Meyer:* Are you saying that when you integrate these small events ($4 \times 10^{-18}$ mol $Ca^{2+}$ ions), assuming a spatial separation of 6 $\mu$m, you would get enough $Ca^{2+}$ to explain the [$Ca^{2+}$] increases during wave propagation? Your calibration suggested to me that more $Ca^{2+}$ is produced when the waves are generated than the sum of all the individual releases.

*Parker:* Our estimate is that the free [$Ca^{2+}$] at the peak of the wave is around 2 $\mu$M, appreciably higher than during the puff. In part, that may be simply because the puff is a highly localized event and we can't estimate how high the free [$Ca^{2+}$] is actually at the release site. Also, in our summary model (Fig. 5), you may have noticed that I scattered some $InsP_3$ receptor channels in between clusters of puff sites. We can't yet be sure, but our feeling is that there are $InsP_3$ receptors between the puff sites, but at a much lower density, and that during an active propagating wave these receptors also are recruited. The propagation, therefore, is not truly saltatory; the wave may not have to leap from one site to another.

*Smith:* Are true clusters necessary? Could the clustering you would get with randomly distributed channels suffice? Could you even see this behaviour with a hypothetical regular lattice of channels?

*Parker:* I don't see how with a regular lattice you could account for the fact that particular sites give repetitive puffs, whereas in intervening areas we don't see the puffs.

*Smith:* But the $Ca^{2+}$ triggering does give you some cooperativity?

*Cannell:* That bears on my earlier question (p 61). With a uniformly distributed system you expect the spatial size of the event to increase as you raise the sensitivity of the system to $Ca^{2+}$, which isn't what Dr Parker sees; that has to suggest clustering at dimensions no larger than the puffs themselves.

*Lu:* Dr Parker, how far from the plasma membrane did you release the $InsP_3$?

*Parker:* When we used a pinhole to photorelease the $InsP_3$, the pinhole was focused at the same depth as our confocal section, about 5 $\mu$m into the cell. Of course, the UV light is not in any sense confocal; we have a beam waist focused about 5 $\mu$m below the plasma membrane, but there would be some photorelease above and below that.

*Lu:* Does releasing $InsP_3$ at different depths reveal differences?

*Parker:* We have done the inverse of that experiment, photoreleasing at a given depth using not a pinhole but a wide area stimulus so that the $InsP_3$ should be photoreleased uniformly into the oocyte (except that the oocyte interior is so murky that the light falls off appreciably after 50 $\mu$m or so), then with a confocal microscope recording the responses at different depths. We see the largest signals, and those showing the fastest rise, at a depth of about 5–7 $\mu$m into the cell; above or below that the magnitude of the signal decreases and, more significantly, the rising phase becomes slowed. This is consistent with the idea that the release sites are in a narrow band and the $Ca^{2+}$ we see above or below that may simply reflect $Ca^{2+}$ ions diffusing from that localized area of release.

*Jaffe:* What is the velocity of the fully fledged waves?

*Parker:* The 'normal' velocity is about 20 $\mu$m/s, but varies with basal $[Ca^{2+}]_i$ and [$InsP_3$]. If we enhance $Ca^{2+}$ influx into the oocyte to increase the basal [$Ca^{2+}$], the waves become more frequent and the velocity goes up. Our record velocity is 80 μm/s.

*Thomas:* I was struck by the low amplitude of the $Ca^{2+}$ puffs compared with the fully fledged $Ca^{2+}$ waves, and by the rapid termination of the puffs. Is it possible that the $InsP_3$ receptors continue to release $Ca^{2+}$ for a longer period during the $Ca^{2+}$ waves, causing a larger and broader $Ca^{2+}$ release than in the single puffs?

*Parker:* It's certainly possible that during the wave the release process does not terminate so rapidly. The kinetics of a wave—the rising phase and the tail—are much slower than those of a puff.

*Thomas:* So, either the puff goes on for a longer time during the wave, or there are intermediate amplifying release channels, which may be $InsP_3$ receptors, causing the $Ca^{2+}$ to move at a slower rate between the puff sites—assuming that the puff sites are responsible for maintaining the propagation.

*Cannell:* The slowing of the decay is exactly what you would expect because of the contribution of diffusion to the kinetics of decay. A wave has an almost spatially uniform structure, so diffusion plays little role in the decay of the transient, whereas with a puff there is also diffusion into an essentially infinite absorbing medium. The decay rate of a puff will automatically be faster, even if the uptake systems are the same.

*Thomas:* My point was that they seem to terminate early, lasting only 100 ms, whereas at a particular point a wave could be measured for a second or two.

*Parker:* As we increase [$InsP_3$], we see an all-or-none type of response, with a threshold for $Ca^{2+}$ release. Above that threshold the response will increase further as a graded function of [$InsP_3$], partly, at least, because the release is not terminating as rapidly; there's a continued $Ca^{2+}$ release over several hundred milliseconds.

*Cobbold:* Presumably, the termination of a puff can't be due to depletion of the stores.

*Parker:* That's correct.

*Wier:* Stern (1992) proposed that the release of $Ca^{2+}$ from clusters of release channels in cardiac muscle can die away by a process that he called 'stochastic attrition'. What he meant was that because the channels tend to close, the local [$Ca^{2+}$] gradient dissipates rapidly so that when they open again there's no $Ca^{2+}$. Would a mechanism like that be sufficient to terminate the activity in your cluster?

*Parker:* I would think not. In certain conditions we can see puffs and waves co-existing. If we have a puff occurring just before a wave, when the wave sweeps across, a little black hole is apparent in the wavefront as it hits the puff site. There is a genuine refractoriness; those channels are not going to respond again for several seconds, even if they are hit by increased [$Ca^{2+}$].

*Irvine:* If we projected your oocytes onto a normal-sized cell, such as a pancreatic acinar cell, we would end up with one or two puff centres per cell.

I'm sure we would all be tempted to draw the analogy between these puff spots and the spots of sensitivity to $InsP_3$ that Dr Petersen and Dr Kasai will be talking about later, which we regard as sites of initiation of $Ca^{2+}$ waves in response to a continuous level of $InsP_3$; this is, in a way, what you are looking at. What troubles me, then, is the stochastic nature of the phenomenon in comparison with the beautiful, regular, dose-related pulsing of $Ca^{2+}$ that Peter Cobbold first showed us. Perhaps I'm just looking at this too simplistically, but it seems to me there's a real problem here, that something has to tame these puffs and get them behaving regularly. If you have one puff spot in a normal cell you are not going to get regular $Ca^{2+}$ oscillations, but a random signal.

*Parker:* The distribution of puff intervals has two components; an exponential tail at longer intervals, which presumably reflects stochastic activation, and at shorter intervals, less than 10 s or so, there is a refractory period. We see the puffs only under conditions of low [$InsP_3$]. If you had a cell with a single puff site, with low [$InsP_3$], you might expect to see stochastic behaviour, but if you increased the [$InsP_3$], the mean interval between puffs would shorten. If it shortened sufficiently for it to come into the refractory period, that would dominate and you would tend to get a regular periodicity, because the interval between puffs would be determined by the recovery period.

*Irvine:* And that will be shorter if the $InsP_3$ dose is higher.

*Parker:* If you increase [$InsP_3$], the puff frequency increases and the interval between the puffs gets shorter.

*Goldbeter:* What is the period of the oscillations at high [$InsP_3$]?

*Parker:* The period varies as a function of [$InsP_3$] and basal cytosolic [$Ca^{2+}$]. The range we see in the oocyte is from as short as three to four seconds to as long as 30 s or a minute.

*Petersen:* In pancreatic acinar cells a low dose of $InsP_3$, a threshold dose, produces irregular spiking, but at slightly higher levels of stimulation we get something that looks rather regular. That is equivalent to what you are seeing.

*Stucki:* Because you have these puffs, stochastic events, and then regular waves, could we in a loose way say that the increase of [$InsP_3$] transforms a discrete event into something continuous?

*Parker:* In a sense, yes. Depending on the [$InsP_3$], we have either a discrete localized release or a propagating release.

*Thomas:* If, as you suggested, the stochastic puff site is the initiator of the sequential fixed-frequency oscillations, you have to make the argument that as you increase [$InsP_3$] the refractory period shortens. What would the mechanism be? There's a range with low [$InsP_3$] when you get spontaneous release and no $Ca^{2+}$ wave propagation. Then you come to the point where as you increase the agonist dose, $Ca^{2+}$ waves of increasing frequency are observed. In the liver and, I think, some other cells, the frequency of the spikes increases substantially, and this seems to correlate with increased [$InsP_3$]. If the refractory period is the only thing that is determining frequency, then the refractory period has to get shorter as the frequency gets greater, i.e., as the $InsP_3$ dose goes up. There would have to be a mechanism by which $[InsP_3]$ itself limits the refractory period.

*Parker:* There are observations suggesting that is the case. We've looked at this with paired flash experiments, photoreleasing $InsP_3$, seeing $Ca^{2+}$ release, then monitoring the sensitivity of $Ca^{2+}$ release at different times by giving a second flash, so that we can see the recovery process. If at a given time we increase the intensity of the second flash, to give more $InsP_3$, we can recover the response. In other words, with a stronger test stimulus, we can shorten the refractory period.

*Thomas:* Can you give a double pulse at a low $[InsP_3]$ and measure the refractory period, and then a double pulse at the same site with a higher $[InsP_3]$ and show that the refractory period is shorter? That's the necessary experiment.

*Parker:* Such an experiment is certainly possible, but we have not yet done it.

*Putney:* The mechanism would work if the receptor's refractoriness involves a change in affinity for $InsP_3$.

## Reference

Stern MD 1992 Theory of excitation–contraction coupling in cardiac muscle. Biophys J 63:497–517

# Spiral calcium waves: implications for signalling

Patricia Camacho and James D. Lechleiter

*Department of Neuroscience, University of Virginia Health Sciences Center, Charlottesville, VA 22908, USA*

*Abstract.* Spiral patterns of intracellular $Ca^{2+}$ release demonstrate a direct relationship between increasing wavefront curvature and increasing propagation velocity. An equally important phenomenon is the annihilation of colliding $Ca^{2+}$ waves, which reveals an underlying refractory period during which further $Ca^{2+}$ release is temporarily inhibited. Treatment of intracellular $Ca^{2+}$ release as an excitable medium accounts for both observations. This theoretical framework is analogous to the more familiar concept of electrical excitability in neuronal membranes. In this analogy, the inositol 1,4,5-trisphosphate receptor ion channel plays a role analogous to that of $Na^+$ channels while $Ca^{2+}$-induced $Ca^{2+}$ release provides the mechanism for excitation. Furthermore, $Ca^{2+}$-ATPases play a role similar to that of the $K^+$ channels in neuronal excitation, that is, they return the system to rest. We demonstrated that overexpression of a sarco/endoplasmic reticulum $Ca^{2+}$-ATPase increases the frequency of $Ca^{2+}$ wave activity. More recent experiments reveal a strong dependence of the propagation velocity on wavelength as predicted by the dispersion relation of excitability. This important result accounts for an observed correlation between wave frequency and spatial dominance of $Ca^{2+}$ foci and suggests a new mechanism for the encoding of signal information.

*1995 Calcium waves, gradients and oscillations. Wiley, Chichester (Ciba Foundation Symposium 188) p 66–84*

Inositol 1,4,5-trisphosphate ($InsP_3$)-induced intracellular $Ca^{2+}$ release plays a key role in mediating hormone and neurotransmitter signals (Berridge 1993, Putney & Bird 1993, Jaffe 1991, Tsien & Tsien 1990). Given the multiplicity of receptors which stimulate $InsP_3$ turnover, however, it is not clear how specific signal information is transmitted to different cells, or how single cells distinguish between different receptor inputs. The application of new imaging techniques to the study of intracellular $Ca^{2+}$ signalling has yielded novel insights into this problem and a new understanding of $Ca^{2+}$ signalling is emerging. Here, we focus on the spatial and temporal aspects of intracellular $Ca^{2+}$ release in *Xenopus* oocytes. We describe a framework for $Ca^{2+}$ signalling in terms of an excitable medium (Winfree 1980) and present experimental evidence consistent with this conceptual view. Additionally, we propose a new mechanism for encoding signal information based on an observed correlation between the frequency and spatial dominance of $Ca^{2+}$ foci.

## Methodology

We use molecular techniques to study the contribution of individual components in the $InsP_3$-mediated $Ca^{2+}$ signalling cascade. Briefly, we overexpress proteins which are functionally involved in $InsP_3$-induced $Ca^{2+}$ release by injecting messenger RNAs (mRNAs) encoding them into *Xenopus laevis* oocytes (Camacho & Lechleiter 1993). Oocytes expressing sarco/endoplasmic reticulum $Ca^{2+}$-ATPases (SERCAs) are now routinely tested at six to eight days after mRNA injections when expression of these proteins is highest. Expression levels are further enhanced by inserting the SERCA complementary DNAs between the 5′ and 3′ untranslated regions of *Xenopus* β-globin (Liman et al 1992). When SERCA2b (Gunteski-Hamblin et al 1988) is coexpressed with the bombesin receptor (Spindel et al 1990), SERCA2b mRNA is injected three days before bombesin receptor mRNA injection so that the oocytes are tested when there is peak expression of both proteins, around Day 6 for SERCA2b and Day 3 for the bombesin receptor.

Intracellular $Ca^{2+}$ release is measured optically with the aid of $Ca^{2+}$ indicator dyes. A cell-impermeant analogue of the fluorophore $Ca^{2+}$ green I (Molecular Probes) is injected into individual oocytes and is then excited and imaged with a BioRad 600 confocal laser-scanning microscope (Camacho & Lechleiter 1993). This technique eliminates out-of-focus fluorescence and permits $Ca^{2+}$ release to be recorded within a thin optical slice. Images are collected on an inverted microscope while the oocyte rests on a glass coverslip within a fluid-filled chamber (96 mM NaCl, 2 mM KCl, 2 mM $MgCl_2$, 5 mM Hepes, pH 7.5, 1 mM EGTA, no added $Ca^{2+}$). In the chamber, the oocyte flattens at the polar (bottom) region, providing a larger circular area near the surface for confocal imaging. All confocal images of $Ca^{2+}$ release are within about 30 μm of the oocyte surface so that optical sections do not cut deep into the oocyte. Images of the same optical slice are collected at one-second intervals.

## When are spiral calcium waves observed?

Different types of $Ca^{2+}$ release are observed when $InsP_3$ is injected into an oocyte. We can roughly classify the $Ca^{2+}$ release patterns into four levels of activity according to the concentration of $InsP_3$. When the final $InsP_3$ concentration is very low (typically 10–50 nM), $Ca^{2+}$ release occurs at discrete sites, with no significant propagation beyond the release site. Parker and co-workers refer to this type of release as a $Ca^{2+}$ puff (Parker & Yao 1991). In cardiac cells, where the analogous intracellular release channel is the ryanodine receptor, these discrete events are referred to as $Ca^{2+}$ sparks (Cheng et al 1993). When the $InsP_3$ concentration is increased slightly (to around 50–100 nM), $Ca^{2+}$ is released from larger patches (10–50 $\mu$m diameter) and there is propagation over a short distance. These abortive, short-lived waves are asymmetrical and frequently propagate in only one direction before fading (Lechleiter et al 1991a, Yao & Parker 1993). With very high levels of stimulation, either with $InsP_3$ injections (about 10 $\mu$M) or through hormonal activation of receptors strongly coupled to $InsP_3$ production (Lechleiter et al 1991a), we typically see only one large wave of $Ca^{2+}$ release with no subsequent waves of regenerative release. Only with intermediate concentrations of $InsP_3$ (0.1–1 $\mu$M) are complex spatiotemporal waves of $Ca^{2+}$ release observed (DeLisle & Welsh 1992, Lechleiter & Clapham 1992). Among these waves, spirals are frequently generated (Fig. 1). Additionally, we see planar and circular $Ca^{2+}$ waves and broken arcs. The experiments described below are carried out at intermediate $InsP_3$ concentrations where complex wave patterns predominate.

Three important properties of $Ca^{2+}$ waves are considered here. (1) $Ca^{2+}$ waves can take the form of spirals. The stability of this pattern of $Ca^{2+}$ release suggests that a curvature–velocity relationship exists, that small curvatures propagate more slowly than larger curvatures. If this were not the case, spiral waves would either unravel or wind themselves up over time. (2) $Ca^{2+}$ waves propagate with undiminished amplitude. This observation suggests that the $Ca^{2+}$ signal is continually being regenerated as the wave moves forward. (3) Colliding waves annihilate one another. This indicates that $Ca^{2+}$ release is completely inhibited for a period after a wave has occurred. The same region will support wave propagation at later times, so the ability to release $Ca^{2+}$ must eventually be recovered.

## What is an excitable medium?

An excitable process is one which undergoes a large excursion away from the steady state in response to a suprathreshold stimulus. For regenerative excitability, there must be recovery and return to the initial steady state. Stimuli that are subthreshold are quickly damped out. An excitable medium is a system in which individual excitable processes are coupled together by a diffusible autocatalytic reagent (Winfree 1980). The properties of $Ca^{2+}$ waves can be accounted for when the intracellular $Ca^{2+}$ release system is treated as an excitable medium (Lechleiter et al 1991b). We consider the elementary excitatory event to be $Ca^{2+}$-induced release of $Ca^{2+}$ via the $InsP_3$-bound $InsP_3$ receptor ($InsP_3R$) ion channel (Lechleiter & Clapham 1992). $Ca^{2+}$ itself appears to be the catalyst which diffuses between neighbouring excitable (i.e., $InsP_3$-bound) $InsP_3Rs$. The system recovers and returns to rest because of the $Ca^{2+}$-sensitive activation of $Ca^{2+}$-ATPases (Camacho & Lechleiter 1993).

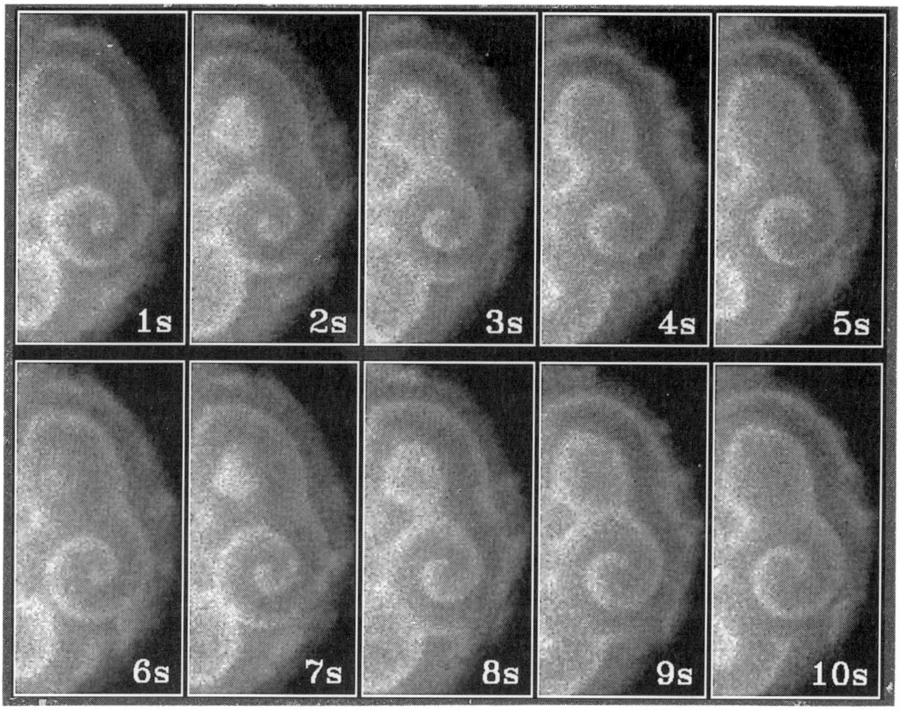


FIG. 1. $Ca^{2+}$ wave activity at intermediate concentrations of inositol 1,4,5-trisphosphate ($InsP_3$). Ten consecutive images collected at one-second intervals are shown. Both the pulsating focus of $Ca^{2+}$ release and the spiral wave go through about two cycles. Note the collisions (3 s and 9 s) and annihilations (4 s and 10 s) of $Ca^{2+}$ waves. $InsP_3$ was injected as a 50 nl bolus of 20 × concentration (6 $\mu$M) to yield a final concentration of around 300 nM, assuming an oocyte volume of 1 $\mu$l. Final image size is 540 × 865 $\mu$m.

Intracellular $Ca^{2+}$ excitability is analogous to electrical excitability in neuronal and cardiac cells. Instead of voltage-gated $Na^+$ channels, there are $Ca^{2+}$-sensitive $InsP_3R$ $Ca^{2+}$ channels. For sodium action potentials, the autocatalytic event is the depolarizing influx of $Na^+$, which increases the probability of the $Na^+$ channel opening, leading to further $Na^+$ influx. In the case of $Ca^{2+}$ excitability, $Ca^{2+}$-induced $Ca^{2+}$ release is the autocalalytic event which generates a rapid, non-linear increase in the intracellular $Ca^{2+}$ concentration. After depolarization, $Na^+$ channels are inactivated and the cell must be repolarized before they can be excited again. Similarly, the $InsP_3R$ channels become inactivated at high $Ca^{2+}$ concentrations and can be excited only after the cytosolic [$Ca^{2+}$] has been lowered (Bezprozvanny et al 1991, Finch et al 1991, Iino 1990). We extend this analogy to include $Ca^{2+}$-ATPases, which lower cytosolic [$Ca^{2+}$] and therefore have a role analogous to that of the repolarizing $K^+$ channels.

It is important to distinguish between excitable and oscillatory media. By definition, an excitable medium is quiescent until it receives sufficient stimulus—sufficient in this case to generate a massive release of $Ca^{2+}$. An oscillator, in contrast, is never at rest, but cycles endlessly—through $Ca^{2+}$ release and re-uptake, here. At intermediate concentrations of $InsP_3$, small regions of the oocyte (foci) release $Ca^{2+}$ in a pulsatile manner and are never fully at rest. These foci drive (excite) neighbouring $Ca^{2+}$ release sites which, in turn, excite their neighbours, thereby generating $Ca^{2+}$ waves. These pulsating foci of $Ca^{2+}$ release are analogous to pacemaker regions in the heart. The sino-atrial node, for example, is an electrical oscillator which drives the surrounding excitable tissue, generating waves of depolarization and, ultimately, the heartbeat.

**What sets the all-or none threshold?**

The concept of a threshold is important in an excitable medium. It is defined simply as that stimulus strength needed to excite the system. To understand $Ca^{2+}$ excitability, it is crucial to determine which processes set the threshold level. We argue that there are two basic processes at work: $Ca^{2+}$ pumping and $Ca^{2+}$-induced $Ca^{2+}$ release. Simply stated, a stimulus is subthreshold when $Ca^{2+}$-ATPases are able to remove more $Ca^{2+}$ from the cytosol than $Ca^{2+}$-induced $Ca^{2+}$ release produces. A stimulus is suprathreshold when $Ca^{2+}$-induced $Ca^{2+}$ release exceeds the rate of $Ca^{2+}$ sequestration by the $Ca^{2+}$-ATPases. Notably, $Ca^{2+}$ pumping and $Ca^{2+}$-induced $Ca^{2+}$ release are dynamic processes.

The efficiency of removal of $Ca^{2+}$ from the cytosol is dependent on the $Ca^{2+}$ sensitivity and the number of $Ca^{2+}$ pumps. Some $Ca^{2+}$ is extruded through the activation of plasma membrane ATPases and exchangers. However, the bulk of the cytosolic $Ca^{2+}$ in the oocyte appears to be resequestered into internal stores by SERCAs. Three genes encoding these $Ca^{2+}$ pumps have been isolated, SERCA1, SERCA2 and SERCA3. SERCA1 is expressed in skeletal muscle and SERCA2a in cardiac cells. SERCA2b and SERCA3 are ubiquitous, appearing both in smooth muscle and in non-muscle tissues (Grover & Khan 1992, Missiaen et al 1991a). The activity of all $Ca^{2+}$-ATPases is increased when [$Ca^{2+}$] increases in the cytosol, but the $Ca^{2+}$ dependency of pump activation varies among different isoforms. In COS cells overexpressing $Ca^{2+}$-ATPase isoforms, SERCA2b appears to be the most sensitive, with an affinity constant of about 270 nM. SERCA1 and SERCA2a are slightly less sensitive, with equal affinities of around 400 nM. SERCA3 is the least sensitive, with a $Ca^{2+}$ affinity of about 1.1 $\mu$M (Lytton et al 1992).

$Ca^{2+}$-induced $Ca^{2+}$ release mediated by the $InsP_3R$ arises through the co-agonist properties of $InsP_3$ and $Ca^{2+}$ (Iino 1990, Bezprozvanny et al 1991, Finch et al 1991). When $InsP_3$ is bound to its receptor, the channel opening probability increases with increasing [$Ca^{2+}$] up to an optimal concentration of 300–500 nM; at higher $Ca^{2+}$ concentrations, the channel opening probability decreases (Bezprozvanny et al 1991, Parys et al 1992). This is presumably due to an increasing likelihood, with increasing [$Ca^{2+}$], that $Ca^{2+}$-binding sites, which govern activation and inactivation of the $InsP_3R$, are occupied. However, Zhang et al (1993) have reported the involvement of a $Ca^{2+}$-dependent kinase and phosphatase. $Ca^{2+}$ release studies in microsomal preparations indicate that $InsP_3Rs$ are incrementally activated with increasing concentrations of $InsP_3$ (Taylor 1992). There are also reports suggesting that the sensitivity of $InsP_3Rs$ to $InsP_3$ is dependent on the luminal concentration of $Ca^{2+}$ stores (Missiaen et al 1991a). At present, the mechanisms responsible for differential sensitivity to $InsP_3$ are not clear. Luminal [$Ca^{2+}$] may indirectly control the $Ca^{2+}$ sensitivity of the $InsP_3R$ (Missiaen et al 1992). However, Iino & Endo (1992) have proposed that increased sensitivity to $InsP_3$ reflects an increased driving force from a $Ca^{2+}$ gradient across the endoplasmic reticulum, with a resultant increase in $Ca^{2+}$-induced $Ca^{2+}$ release.

Propagation of the $Ca^{2+}$ wave is due to the regenerative release of $Ca^{2+}$. For regeneration to be sustained, a critical number of $InsP_3Rs$ must be excited each time to generate (via $Ca^{2+}$-induced $Ca^{2+}$ release) a suprathreshold stimulus in neighbouring sites. When the number of activated $InsP_3Rs$ falls below this limit, wave propagation is aborted or, in the case of stimulation with very low concentrations of $InsP_3$, never initiated. The speed with which cytosolic $Ca^{2+}$ concentration builds will also affect wave propagation. For an expanding circular wave, the neighbouring concentric ring of $Ca^{2+}$ release sites occupies a larger volume. Consequently, a suprathreshold concentration of $Ca^{2+}$ takes longer to build up, resulting in slower wave propagation. In comparison, planar waves propagate faster because there is no volume change between neighbouring sites. Positive curvatures, which can occur at the intersection of two colliding waves, propagate faster because of a reduction in volume between neighbours and a subsequent focusing of the autocatalytic reagent.

## What is the refractory period?

Annihilation of colliding $Ca^{2+}$ waves indicates that $Ca^{2+}$ release is completely inhibited just after a wave has passed. Slow removal of inhibition defines a refractory period during which $Ca^{2+}$ release can be stimulated but only with a larger stimulus. We consider the refractory period to be a collection of states in which $Ca^{2+}$ release is inhibited to different degrees, depending on the time that has elapsed since the preceding wave of excitation. The underlying cause for inhibition of $Ca^{2+}$ release is unknown, but there are two likely possibilities. First, it has been shown that high $Ca^{2+}$ concentrations decrease the opening probability of the $InsP_3$-bound $InsP_3Rs$ (Bezprozvanny et al 1991). Thus, a $Ca^{2+}$ wave should temporarily inactivate $InsP_3R$ channels, until the cytosolic [$Ca^{2+}$] is lowered to resting levels. James Sneyd and co-workers (Atri et al 1993) have modelled most of the observed $Ca^{2+}$ wave activity in *Xenopus* oocytes assuming this mechanism of inhibition. Second, luminal stores could be temporarily depleted of releasable $Ca^{2+}$ by a passing wave. The refractory period would then correspond to the time required to refill the $Ca^{2+}$ stores. This mechanism has also been used to model $Ca^{2+}$ wave activity (Dupont et al 1991, Girard et al 1992).

## Calcium-ATPases reduce the refractory period and increase calcium wave frequency

Activation of $Ca^{2+}$-ATPases results in the removal of $Ca^{2+}$ from the cytoplasm either across the plasma membrane (plasma membrane $Ca^{2+}$-ATPase [PMCA] family) (Carafoli 1991) or into the sarco/endoplasmic reticulum stores (SERCA family) (Grover & Khan 1992, Missiaen et al 1991b). Overexpression of a SERCA would be expected to reduce cytoplasmic [$Ca^{2+}$] more efficiently and to fill $Ca^{2+}$ stores more rapidly during a wave; both these effects should reduce the refractory period of $Ca^{2+}$ inhibition. Thus, at equivalent concentrations of $InsP_3$, one should observe a higher $Ca^{2+}$ wave frequency in oocytes expressing a SERCA than in control oocytes. Additionally, increased $Ca^{2+}$-ATPase activity could lower resting $Ca^{2+}$ levels, depending on the $Ca^{2+}$ sensitivity of the individual $Ca^{2+}$-ATPases (Lytton et al 1992). Current models of $Ca^{2+}$ oscillations predict that this effect would decrease wave frequency, because a larger $Ca^{2+}$ stimulus would then be necessary to exceed the threshold for $Ca^{2+}$ excitation (Atri et al 1993, Girard et al 1992, Goldbeter et al 1990). Finally, increased $Ca^{2+}$-ATPase activity could overload luminal $Ca^{2+}$ stores, increasing the sensitivity of the $Ca^{2+}$-releasing receptor to $InsP_3$ (Missiaen et al 1991a). This effect would lower the threshold and again increase the wave frequency.

To test these hypotheses, we increased the total $Ca^{2+}$-ATPase activity in *Xenopus* oocytes by overexpressing an avian form of the fast twitch skeletal muscle SERCA1 (Camacho & Lechleiter 1993). Two to six days after mRNA injections $InsP_3$-induced $Ca^{2+}$ wave activity was examined. $InsP_3$ (1 $\mu$M) produces responses in SERCA1-expressing oocytes markedly different from those in control (non-SERCA1-expressing) oocytes (Fig. 2). First, the initial injection of $InsP_3$, which usually stimulates a large wave of $Ca^{2+}$ enveloping the entire oocyte in controls, is absent in the expressing oocytes. Instead, pulsatile $Ca^{2+}$ release immediately follows the $InsP_3$ injection. Second, the frequency of $Ca^{2+}$ waves is significantly increased, with

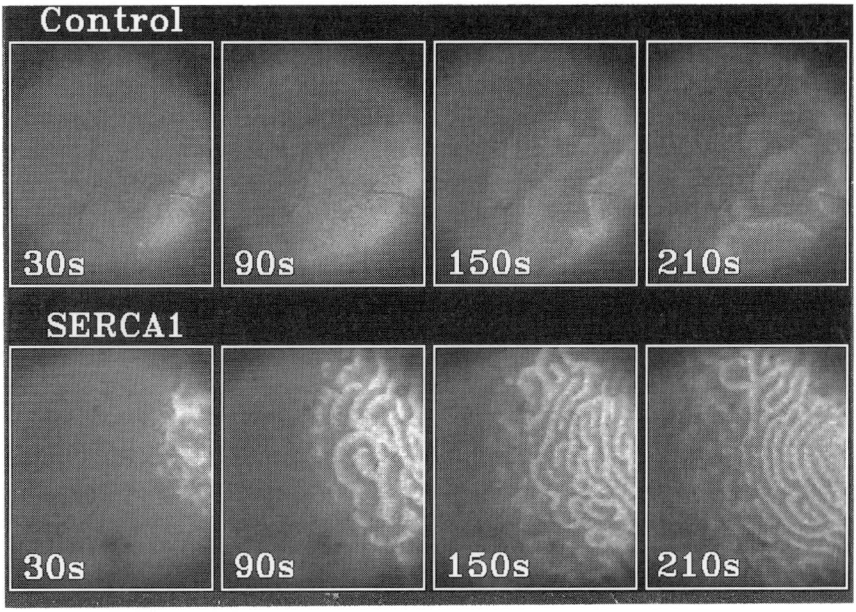


FIG. 2. Increased frequency of $Ca^{2+}$ waves in oocytes expressing messenger RNA for the sarco/endoplasmic reticulum $Ca^{2+}$-ATPase SERCA1 in comparison with control oocytes. The final $InsP_3$ concentration is about 1 $\mu$M. Final image size is $620 \times 750$ $\mu$m.

no significant effect on their velocity. Third, individual $Ca^{2+}$ waves are narrower in SERCA1-expressing oocytes. These findings are consistent with the hypothesis that increased $Ca^{2+}$-ATPase activity decreases the refractory period of inhibition, with minimal effects of resting $Ca^{2+}$ concentration.

Petersen and co-workers have reported that low concentrations of thapsigargin (a $Ca^{2+}$-ATPase inhibitor) increase the frequency of $Ca^{2+}$ oscillations (Petersen et al 1993). Partial inhibition of $Ca^{2+}$-ATPases presumably increases the net flux of $Ca^{2+}$ from the endoplasmic reticulum, which may saturate some of the cytosolic $Ca^{2+}$-binding sites. Consequently, free $Ca^{2+}$ could build more quickly to threshold concentration, resulting in an increase in $Ca^{2+}$ wave frequency. In agreement with this observation, increasing influx of extracellular $Ca^{2+}$ into the cytoplasm increases the frequency of $Ca^{2+}$ oscillations in AR42J (pancreatic) cells (Zhang et al 1992) and increases the frequency and velocity of $Ca^{2+}$ waves in *Xenopus* oocytes (Girard & Clapham 1993).

## Calcium waves exhibit dispersion

Dispersion is a characteristic property of a classical excitable medium (Dockery et al 1988, Miller & Rinzel 1981). Simply stated, this means that the maximum frequency of wave activity of an excitable medium is limited by the time needed to reset the excitatory mechanisms. When the stimulus frequency is too high, waves begin to impinge on the refractory period of the preceding wave. This results in a reduction in wave velocity because of the additional time needed to generate a suprathreshold stimulus. In neuronal excitability, this relationship was demonstrated by plotting increased frequencies of depolarizing stimuli at the nerve cell body and against the action potential velocity. In our initial experiments, we did not detect a dependence of $Ca^{2+}$ wave velocity on wave frequency (Camacho & Lechleiter 1993). However, after significantly increasing the expression of SERCA1 (see Methodology, p 67), we successfully demonstrated a dispersion effect in oocytes (Fig. 3). Under these experimental conditions, $InsP_3$-induced $Ca^{2+}$ waves are of such high frequency that each wave impinges on the refractory period left by the preceding wave. Consequently, the waves stack up and propagate with decreasing velocity and amplitude. The dependency of the velocity ($v$) on distance between successive waves (wavelength, $\lambda$) is not pronounced for $\lambda > 50\ \mu m$ and $v = 21 \pm 4\ \mu m/s$ ($n = 21$). However, by the time $\lambda$ values are between 20 and 30 $\mu m$, $v$ is reduced to $9 \pm 2\ \mu m/s$ ($n = 12$). At $\lambda < 20\ \mu m$, individual waves break up and fail to propagate (i.e., wave dispersion).

### The implications of dispersion for the encoding of signal information

Recent experiments demonstrate a potentially important principle for the encoding of signal information which we attribute to the refractory period of $Ca^{2+}$ release. SERCA2b was co-expressed with the bombesin receptor, which is coupled to the $InsP_3$ signalling pathway. $InsP_3$ production was stimulated over the entire oocyte surface by bath application of bombesin (0.3 nM). Owing

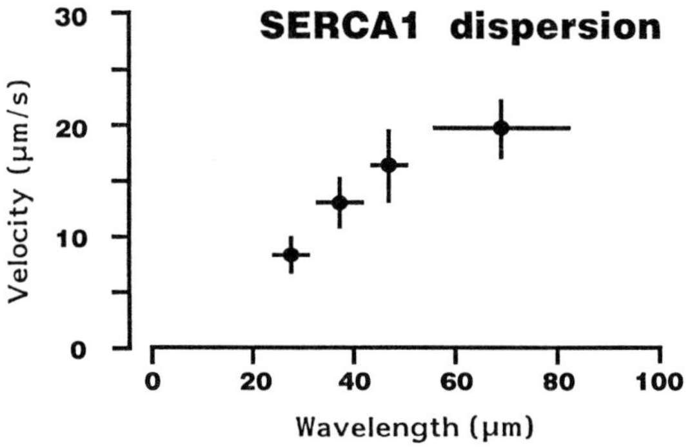


FIG. 3. Dependence of velocity on the distance between successive $Ca^{2+}$ waves. Data were collected from an oocyte expressing sarco/endoplasmic reticulum $Ca^{2+}$-ATPase SERCA1 messenger RNA injected with $InsP_3$ (final concentration 300 nM). While the velocity and wavelength decreased (means $\pm$ SD), the frequency of $Ca^{2+}$ waves was unchanged during this recording period.

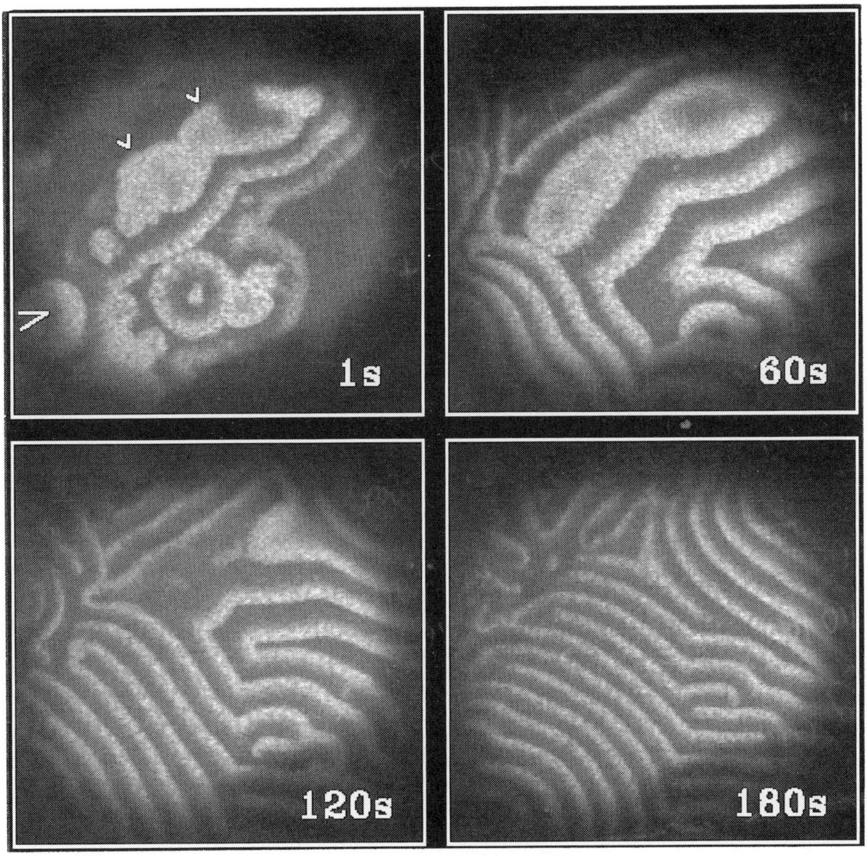


FIG. 4. Transduction of a temporal $Ca^{2+}$ wave signal into a spatial $Ca^{2+}$ signal. The images show bombesin (0.3 nM)-induced $Ca^{2+}$ wave activity in an oocyte expressing bombesin receptors and the sarco/endoplasmic reticulum $Ca^{2+}$-ATPase SERCA2b. Two foci of $Ca^{2+}$ release are indicated: that in the lower left of the oocyte (large arrowhead) is pulsating at ~0.35 cycle/s, whereas the foci at the upper centre of the oocyte (small arrowheads) are pulsating at ~0.33 cycle/s. Final image size is $162 \times 216\ \mu m$.

to the high expression of SERCA2b, $InsP_3$ production brings the cell into immediate regenerative $Ca^{2+}$ activity (Fig. 4). There are two main foci of $Ca^{2+}$ release pulsating at slightly different frequencies. Over time, the waves emanating from the faster focal releasing site gain spatial control over half of the cell. We conclude from this observation that a *temporal* $Ca^{2+}$ signal (frequency) can be transduced into a *spatial* $Ca^{2+}$ signal. Therefore, mechanisms which alter the refractory period of $Ca^{2+}$ release through physiological regulation of $Ca^{2+}$-ATPase expression levels (Magnier et al 1992, Zarain-Herzberg et al 1994) should greatly affect $InsP_3$-mediated $Ca^{2+}$ signalling.

*Acknowledgements*

We thank Dr Emily Liman for the *Xenopus* expression vector and Drs Kunio Takeyasu, Gary Shull and Tom Segerson for the complementary DNAs encoding SERCA1, SERCA2b and the bombesin receptor, respectively. This work was supported by NIH grant GM48451.

## References

Atri A, Amundson J, Clapham D, Sneyd J 1993 A single-pool model for intracellular calcium oscillations and waves in the *Xenopus laevis* oocyte. Biophys J 65:1727–1793

Berridge MJ 1993 Inositol trisphosphate and calcium signalling. Nature 361:315–325

Bezprozvanny I, Watras J, Ehrlich BE 1991 Bell-shaped calcium-response curves of Ins(1,4,5)$P_3$- and calcium-gated channels from endoplasmic reticulum of cerebellum. Nature 351:751–754

Camacho P, Lechleiter JD 1993 Increased frequency of calcium waves in *Xenopus laevis* oocytes that express a calcium-ATPase. Science 260:226–229

Carafoli E 1991 The calcium pumping ATPase of the plasma membrane. Annu Rev Physiol 53:531–547

Cheng H, Lederer WJ, Cannell MB 1993 Calcium sparks: elementary events underlying excitation–contraction coupling in heart muscle. Science 262:740–744

DeLisle S, Welsh MJ 1992 Inositol trisphosphate is required for the propagation of calcium waves in *Xenopus* oocytes. J Biol Chem 267:7963–7966

Dockery JD, Keener JP, Tyson JJ 1988 Dispersion of traveling waves in the Belousov–Zhabotinskii reaction. Physica D 30:177–191

Dupont G, Berridge MJ, Goldbeter A 1991 Signal-induced $Ca^{2+}$ oscillations: properties of a model based on $Ca^{2+}$-induced $Ca^{2+}$ release. Cell Calcium 12:73–85

Finch EA, Turner TJ, Goldin SM 1991 Calcium as a coagonist of inositol 1,4,5-trisphosphate-induced calcium release. Science 252:443–446

Girard S, Clapham D 1993 Acceleration of intracelular calcium waves in *Xenopus* oocytes by calcium influx. Science 260:229–232

Girard S, Luckhoff A, Lechleiter JD, Sneyd J, Clapham D 1992 Two-dimensional model of calcium waves reproduces the patterns observed in *Xenopus* oocytes. Biophys J 61:509–517

Goldbeter A, Dupont G, Berridge MJ 1990 Minimal model for signal-induced $Ca^{2+}$ oscillations and for their frequency encoding through protein phosphorylation. Proc Natl Acad Sci USA 87:1461–1465

Grover AK, Khan I 1992 Calcium pump isoforms: diversity, selectivity and plasticity. Cell Calcium 13:9–17

Gunteski-Hamblin AM, Greeb J, Shull GE 1988 A novel $Ca^{2+}$ pump expressed in brain, kidney, and stomach is encoded by an alternative transcript of the slow-twitch muscle sarcoplasmic reticulum Ca-ATPase gene. J Biol Chem 263:15032–15040

Iino M 1990 Biphasic $Ca^{2+}$ dependence of inositol 1,4,5-trisphosphate-induced Ca release in smooth muscle cells of the guinea pig *Taenia caeci.* J Gen Physiol 95:1103–1122

Iino M, Endo M 1992 Calcium-dependent immediate feedback control of inositol 1,4,5-trisphosphate-induced $Ca^{2+}$ release. Nature 360:76–78

Jaffe LF 1991 The path of calcium in cytosolic calcium oscillations: a unifying hypothesis. Proc Natl Acad Sci USA 88:9883–9887

Lechleiter JD, Clapham DE 1992 Molecular mechanisms of intracellular calcium excitability in *X. laevis* oocytes. Cell 69:283–294

Lechleiter JD, Girard S, Clapham D, Peralta E 1991a Subcellular patterns of calcium release determined by G protein-specific residues of muscarinic receptors. Nature 350:505–508
Lechleiter JD, Girard S, Peralta E, Clapham D 1991b Spiral calcium wave propagation and annihilation in *Xenopus laevis* oocytes. Science 252:123–126
Liman ER, Tygat J, Hess P 1992 Subunit stoichiometry of mammalian $K^+$ channel determined by construction of multimeric cDNAs. Neuron 9:861–871
Lytton J, Westlin M, Burk SE, Shull GE, MacLennan DH 1992 Functional comparisons between isoforms of the sarcoplasmic or endoplasmic reticulum family of calcium pumps. J Biol Chem 267:14483–14489
Magnier C, Papp B, Corvazier E et al 1992 Regulation of sarco-endoplasmic reticulum $Ca^{2+}$-ATPases during platelet-derived growth factor-induced smooth muscle cell proliferation. J Biol Chem 267:15808–15815
Miller RN, Rinzel J 1981 The dependence of impulse propagation speed on firing frequency, dispersion, for the Hodgkin–Huxley model. Biophys J 34:227–259
Missiaen L, Wuytack F, Raeymaekers L et al 1991a $Ca^{2+}$ extrusion across plasma membrane and $Ca^{2+}$ uptake by intracellular stores. Pharmacol & Ther 50:191–232
Missiaen L, Taylor CW, Berridge MJ 1991b Spontaneous calcium release from inositol trisphosphate-sensitive calcium stores. Nature 352:241–244
Missiaen L, De Smedt H, Droogmans G, Casteels R 1992 Luminal $Ca^{2+}$ controls the activation of the Ins(1,4,5)$P_3$ receptor by cytosolic calcium. J Biol Chem 267: 22961–22966
Parker I, Yao Y 1991 Regenerative release of calcium from functionally discrete subcellular stores by inositol trisphosphate. Proc R Soc Lond Ser B Biol Sci 246:269–274
Parys B, De Smedt SW, DeLisle S et al 1992 Isolation, characterization, and localization of the inositol 1,4,5-trisphosphate receptor protein in *Xenopus laevis* oocytes. J Biol Chem 267:18776–18782
Petersen CCH, Petersen OH, Berridge MJ 1993 The role of endoplasmic reticulum calcium pumps during cytosolic calcium spiking in pancreatic acinar cells. J Biol Chem 268:22262–22264
Putney JW Jr, Bird GS 1993 The inositol phosphate–calcium signaling system in nonexcitable cells. Endocr Rev 14:610–631
Spindel ER, Giladi E, Brehm P, Goodman RH, Segerson TP 1990 Cloning and functional characterization of a complementary DNA encoding the murine fibroblast bombesin/gastrin-releasing peptide receptor. Mol Endocrinol 4:1956–1963
Taylor C 1992 Kinetics of inositol 1,4,5-trisphosphate-stimulated $Ca^{2+}$ mobilization. In: Putney JW (ed) Inositol polyphosphates and calcium signalling. Raven Press, New York (Adv Second Messenger Phosphoprotein Res 26) p 109–142
Tsien RW, Tsien RY 1990 Calcium channels, stores and oscillations. Annu Rev Cell Biol 6:715–760
Winfree AT 1980 Excitable kinetics and excitable media. In: Winfree AT (ed) Geometry of biological time. Springer-Verlag, New York, p 239–250
Yao Y, Parker I 1993 Inositol trisphosphate-mediated $Ca^{2+}$ influx into *Xenopus* oocytes triggers $Ca^{2+}$ liberation from intracellular stores. J Physiol 468:275–296
Zarain-Herzberg A, Marques J, Sukovich D 1994 Thyroid hormone receptor modulates the expression of the rabbit cardiac (endo)plasmic reticulum calcium-ATPase gene. J Biol Chem 269:1460–1467
Zhang BX, Zhao H, Loessberg P, Muallem S 1992 Regulation of agonist evoked $[Ca^{2+}]_i$ oscillation by intracellular $Ca^{2+}$ and $Ba^{2+}$ in AR42J cells. Am J Physiol 262: C1125–C1133
Zhang BX, Zhao H, Muallem S 1993 $Ca^{2+}$ dependent kinase and phosphatase control inositol 1,4,5-trisphosphate-mediated $Ca^{2+}$ release. J Biol Chem 268:10997–11001

**DISCUSSION**

*Meyer:* Did you see any differences in the $Ca^{2+}$ load in the stores of transfected oocytes? You could use $^{45}Ca^{2+}$ to determine whether there is more in the stores.

*Lechleiter:* We have not done this experiment yet.

*Meyer:* What about the free $Ca^{2+}$ level? A lot of people are interested in the effect of store loading on $Ca^{2+}$ influx into cells. One would predict that a change in influx would alter the free $[Ca^{2+}]$.

*Lechleiter:* The resting $[Ca^{2+}]$ is difficult to measure because we use visible wavelength dyes. We are trying to circumvent this problem by normalizing with another dye that is insensitive to $Ca^{2+}$.

*Whitaker:* In the immature oocytes, you have a system which, as I understand it, is not fully functional as a $Ca^{2+}$ signalling system. You can stimulate it with $InsP_3$ and $InsP(S)_3$ or you can put acetylcholine on the outside, but those aren't physiological triggers. The $Ca^{2+}$ release mechanism is maturing as the oocyte matures, so that it can produce a fully functional fertilization wave. It's clear that there are changes in the properties of the $Ca^{2+}$ release mechanism during oocyte maturation. What do you think happens to produce the robust wave in the fertilized egg, and how does this relate to the $Ca^{2+}$ puffs seen by Ian Parker and the spacing between them?

*Lechleiter:* I consider the mature oocyte to be somewhat abnormal compared to immature oocytes. The mature egg is a specialized cell in which a component of the endoplasmic reticulum (ER) wraps around cortical granules resulting in an entirely different ER distribution.

*Jaffe:* But it doesn't give a fertilization wave. It doesn't give the phenomenon that you're studying.

*Lechleiter:* We are not studying fertilization: we're studying $Ca^{2+}$ release. I think the processes in our situation are more comparable to those in other cells. Fertilization is a different question. I wouldn't even try to extrapolate from our work to fertilization. Larabell & Nuccitelli (1992) have shown that there's a massive release of $Ca^{2+}$ during fertilization, which is outside the range in which we observe regeneration activity. Moreover, their estimate of the resting cytosolic $[Ca^{2+}]$ is 200 nM, probably too high for the regenerative activity, at least in our system. The mature egg is highly specialized and $Ca^{2+}$ release is not necessarily comparable to what you see in normal cells.

*Berridge:* Dr Jaffe, you should be excited about this work because it is consistent with your earlier studies demonstrating that there are $Ca^{2+}$ currents surrounding the immature egg. It is conceivable that this spontaneous activity is responsible for setting up the spatial organization recorded by your vibrating electrode. One mustn't confuse it with fertilization; these events occur pre-fertilization in the immature oocyte and may well be helping to organize the egg to get it ready for fertilization. Perhaps the spontaneous waves function to establish polarity in the egg.

*Miyazaki:* Mikoshiba's group (Kume et al 1993) showed the uneven distribution of $InsP_3$ receptors in the immature *Xenopus* oocyte. If I remember correctly, receptor density is higher in the animal pole. Do spiral waves occur only in the animal pole? I have the same question to Dr Parker regarding the 'puff' phenomenon.

*Parker:* We see more intense $Ca^{2+}$ signalling in the animal hemisphere, which matches the distribution of the $InsP_3$ receptor. Also, the staining for the receptor was much more intense in a narrow band just under the plasma membrane, which would fit with the localization of these sites axially in the oocyte.

*Lechleiter:* It's not always clear in our oocytes where the animal hemisphere is, since we work with albino oocytes. When we see a translucent body, presumably the germinal vesicle, it's fairly clear that there is a little more activity compared with the vegetal pole. We have looked for $InsP_3$ receptor clustering but haven't seen anything significant yet. However, we may not have the right sensitivity. It could be that there's simply a stochastic arrangement of $InsP_3$ receptors in which small clustered receptors give rise to hot spots. I recall a poster presented at a meeting (Max-Planck, Dortmund, 1992: Spatial–Temporal Organization in Non-Equilibrium Systems) where the authors attempted to study the effects of random distribution of $InsP_3$ receptors on $Ca^{2+}$ signalling. They found small hot spots, but I have never seen the work published.

*Goldbeter:* This work was done by Overboom & Hogeweg in Utrecht. You mentioned that the sensitivity of the $InsP_3$ receptor changes when you overexpress the pump. What exactly do you mean by that? Is there a change in the affinity?

*Lechleiter:* It means that we can get the same frequency of regenerative activity across the entire oocyte, at a much lower [$InsP_3$]. We think that the $Ca^{2+}$ stores become overloaded. This is in contrast to some work that Girard & Clapham (1993) reported. If you increase the amount of $Ca^{2+}$ in the cytoplasm, you get faster wave velocities, but not increased sensitivity, only an increase in the $Ca^{2+}$ diffusion constant. In contrast, we see a frequency increase without a change in velocity.

*Parker:* Do you have a feeling whether it's the more rapid fall in cytosolic [$Ca^{2+}$] or the more rapid refilling of the stores that's responsible for the refractory period? Have you tried expressing surface membrane pumps to dissociate these two effects?

*Leichleiter:* We have expressed the plasma membrane $Ca^{2+}$-ATPase to try to answer that exact question. The problem is that when you overexpress it so extensively, the protein is no longer located solely at the plasma membrane; you get some ER distribution too. Currently, we are trying to overexpress the $Na^+/Ca^{2+}$ exchanger, so we can control movement across the plasma membrane only.

*Jaffe:* Do you know of any natural cases of spiral $Ca^{2+}$ waves?

*Lechleiter:* We consider this to be natural. If you're asking about specific function, we have no data. However, it would be a mistake to get lost in the pattern formation, though there may be something there. What the patterns tells us is that the system is acting like an excitable system—that's the important take-home fact. The first initial wave of $InsP_3$ primes the system, creates an excitable medium. We are not aware of a functional consequence.

*Dictyostelium discoideum* can also be described as an excitable medium (Devreotes et al 1983). There's a suggestion that the centre filament of the cyclic AMP vortex, which is a spiral in cross-section and represents a constant concentration, provides a direction for the stalk formation. I couldn't judge how accurate the cyclic AMP measurements were, but the constant concentrations at the core of a spiral could have some physiological consequence in that multicellular system.

*Goldbeter:* In *Dictyostelium* you can get concentric waves of cyclic AMP as well, as you can with $Ca^{2+}$.

*Berridge:* Dr Lechleiter wasn't talking about the aggregation phase but about the stalk. In the slug itself there are spiral waves in addition to those recorded during the process of aggregation.

Another example of a spiral, although not a $Ca^{2+}$ one, is in heart activation. In the sino-atrial node there are spirals of activation spinning out from localized initiation sites.

*Galione:* You can see beautiful spiral waves of cell swelling during spreading depression across whole chick retina (Gorelova & Bures 1983). That's presumably a multicellular situation, and it would be fascinating if a $Ca^{2+}$ wave underlay that phenomenon; it could be due to $Ca^{2+}$ activation of $K^+$ and $Cl^-$ channels and consequent movement of water.

*Lechleiter:* That has also been modelled as an excitable medium quite successfully (Gorelova & Bures 1983). However, it is a multicellular system and may be somewhat different.

*Galione:* Do you think that you could get spiral waves across the whole aggregate if you had a two-dimensional matrix of cells in culture?

*Lechleiter:* It is possible, but I have seen no data in support of this.

*Sanderson:* We have never seen intercellular spiral waves.

*Lechleiter:* The difference between your data and ours is that in oocytes, the initial stimulis ($InsP_3$) is just setting the stage, priming the system. For the oocyte to be excitable, you must have a bound $InsP_3$ receptor that's ready for $Ca^{2+}$-induced $Ca^{2+}$ release. Once you have a system that's excitable, any subsequent stimulus may set up spiral activity. There may be a second stimulus that you're not seeing.

*Sanderson:* In glial cells, it appears that an intercellular $Ca^{2+}$ wave mimicks the behaviour of oocyte cytoplasm. One initially observes a surge or a wave of $Ca^{2+}$ followed by a period of excitability when the spiral waves occur, i.e.,

$InsP_3$ diffusion sets the stage for subsequent bouts of $Ca^{2+}$-induced $Ca^{2+}$ release. In individual cells of small size, full spiral waves are not observed but their beginnings are in the form of independent oscillations. The large, long wave fronts of the spiral waves are not observed because the gap junctions restrict $Ca^{2+}$ movement and thus the propagation of $Ca^{2+}$-induced $Ca^{2+}$ release.

*Whitaker:* I was interested in the scale of these spiral waves in comparison with the average distance between the puff sites. It seemed to me that the spiral waves actually have quite a lot of fine structure.

*Lechleiter:* There is fine structure, but when you go up to a higher magnification the fine structure turns out not to be ER but a lot of yolk platelets, disappointingly. These yolk platelets actually cluster and give some inhomogeneities. We've been looking for a discontinuity in the leading wavefront. It seems to be homogeneous. At some point, you would expect to see individual release sites. We do see a phase of activity in which there are ratty, irregular fronts, but that's the abortive wave state, in which it appears that the wave is trying to jump to the next $Ca^{2+}$ recharging station.

*Whitaker:* Obviously, there comes a point towards the origin of the spiral wave where there has to be a smaller theoretical radius of curvature than could be accounted for by the distance between the release sites. There would be a radius of curvature beyond which you could not go because of the microscopic inhomogeneities in $Ca^{2+}$ release or, put the opposite way, if the waves are continuous and have a radius of curvature of less than 6 $\mu$m, then the wave can't be due to release from discrete sites 6 $\mu$m apart .

*Lechleiter:* Theoretically, yes, but it's hard to predict because it depends a lot on the underlying kinetics. The first thing we did was try to plot the curvature–velocity relationship, but the linear extrapolation is oversimplified, at best. It does, however, give an interesting value of about 9 $\mu$m for a radius of curvature, which is in the right range. Dr Parker, do you have a critical area where you go from a puff to an abortive wave, where $Ca^{2+}$ waves begin to propagate?

*Parker:* We wouldn't say there is such a thing as a critical radius. The abortive waves can be very variable. A 10 $\mu$m abortive wave might terminate or it might grow to 100 $\mu$m.

*Lechleiter:* There are clearly transitions going from a puff to an abortive wave. Do you have an estimate for this? It would seem to us to be of the order of tens of micrometres.

*Parker:* We've tried essentially repeating your experiment, plotting the radius versus velocity, but with the better time resolution at our system we can go much further back to small radii. In fact, we didn't get a straight line extrapolating down towards 9 $\mu$m. We saw no evidence of the velocity decreasing as the radius became smaller.

*Lechleiter:* You have to look at very small curvatures and look at symmetrical foci.

*Parker:* We could get measurements down to a radius of about 15 $\mu$m.

*Kasai:* Have you ever studied the fine structural changes occurring after the expression of the $Ca^{2+}$-ATPase?

*Lechleiter:* No. Yong Yao in Ian Parker's lab has done some work on ER but not with overexpressed $Ca^{2+}$-ATPases. They looked at the propagation of a wave along an ER-like structure, stained with a molecule of the carbocyanine dye DiI.

*Smith:* Do you and Dr Parker feel that you're looking at the same thing? I see many similarities, but there are also a lot of things in your work that I didn't see in Dr Parker's, yet you're discussing fine points about the radius of curvature and velocity.

*Lechleiter:* I would say we're looking at exactly the same thing. Dr Parker is just using a lower [$InsP_3$].

*Smith:* You two have a better feel than we do for the concentrations, the objective magnifications, etc, so are better placed to judge whether you're studying the same thing.

*Parker:* We're seeing the same thing but under different conditions. We work at different concentrations, and our work tends to be at higher magnification, looking at finer structure.

*Putney:* I'm sitting here and trying to understand spikes and oscillations in salivary glands and liver cells, and learning a lot about the kinetics in oocytes, and I'm wondering, is it correct for me to think about a 40 $\mu$m ball of oocyte cytoplasm and organelles behaving in the same way as a smaller mammalian cell?

*Lechleiter:* We certainly think so.

*Putney:* The question came into my mind when you started talking about the size of an abortive wave. An abortive wave is bigger than a hepatocyte, so you may be dealing with a level of organization that might be irrelevant to small cells.

*Lechleiter:* The $InsP_3$ receptor is not any larger in the oocyte.

*Putney:* But is the fine structure, the organization of the ER and the space between stacks and other structures the same?

*Berridge:* The diameter of one of Dr Parker's little puffs was about 6 $\mu$m. Isn't that roughly the same size as the initiation site in hepatocytes?

*Thomas:* At our resolution, it would be difficult to know if it was that size or smaller.

*Berridge:* So a single puff in a *Xenopus* oocyte might be equivalent to the initiation site we see in a normal cell.

*Thomas:* But the location of this puff site must be functionally organized, following the polarization of the cell.

*Putney:* We have never seen a puff in a hepatocyte. We see nothing or we see a wave.

*Berridge:* Other people have seen the equivalent to puffs and have referred to them as abortive spikes, i.e., they represent iniation events that don't actually fire a wave.

*Thomas:* One other difference between the $Ca^{2+}$ puff sites in oocytes and in hepatocytes, with which I'm most familiar, is that we often see an increase in cytosolic [$Ca^{2+}$] prior to the main $Ca^{2+}$ wave in the initiation region—the so-called pacemaker—whereas you don't see anything like that in the oocyte. We occasionally see a small elevation of $Ca^{2+}$, lasting for three to four seconds, then the cytosolic [$Ca^{2+}$] rises rapidly to a peak in this region and we get a wave. We have considered that this pacemaker relates to the initiation phase—a partial cytosolic response. Perhaps this is a puff and then something else picks that up.

*Lechleiter:* Dr Wier, didn't you distinguish between a diffusional mechanism and wave propagation in cardiac myocytes? Were you able to distinguish between the two possibilities? Were the speeds comparable?

*Wier:* We concluded that it was difficult to distinguish those two. The experiment we did was to wound the cell, which we assumed would provide a constant influx of $Ca^{2+}$, which we then compared with the theoretical distribution of $Ca^{2+}$ around that site if it were spreading only by passive diffusion. We concluded that the waves in the cultured smooth muscle cells we studied were propagated by an active mechanism rather than by diffusion.

*Whitaker:* One difference between an oocyte and a smaller cell is that the oocyte has an infinite volume into which the puff can diffuse whereas the cell has a plasma membrane which tends to constrain that diffusion. A phenomenon in which the [$Ca^{2+}$] begins to rise and hovers around a little before causing the spike is more likely to occur in a small cell than in an oocyte.

*Lechleiter:* In a small cell, the plasma membrane pumps will also play a larger role. All our experiments are done without external $Ca^{2+}$. The oocyte can function for a very long period without extracellular $Ca^{2+}$.

*Berridge:* But there is a paradox in that oocytes will respond quickly to variations in $Ca^{2+}$ influx, yet they will function for a long time in the absence of external $Ca^{2+}$.

*Sanderson:* Dr Parker described that $Ca^{2+}$ puffs seem to be generated from a little cluster of receptors, so that $Ca^{2+}$-induced $Ca^{2+}$ release is working on a small scale within the puff. However, it seems reasonable, as has been suggested by Steve Smith, that there is a more uniform distribution and perhaps a higher density of $InsP_3$ receptors in smaller, differentiated cells. A result of this high receptor density is that the $Ca^{2+}$ puff of the oocyte will appear as a full cellular $Ca^{2+}$ wave in a smaller cell. Thus, scale is important in the spatial description of $Ca^{2+}$ changes.

*Jaffe:* Dr Lechleiter, you have spoken repeatedly but a little vaguely about waves being related to a curvature–velocity relationship. I didn't really understand what you meant.

*Lechleiter:* For a wave to propagate in an excitable medium, $Ca^{2+}$-induced $Ca^{2+}$ release must be stimulated in the next tier of $Ca^{2+}$ release sites to continually regenerate the wave. A certain [$Ca^{2+}$] has to be built up to get above the threshold. At small curvatures, the next tier of sites has a slightly larger volume, so it takes a little longer to build up to the threshold concentration. A planar wave has a one-to-one relationship with the next tier of release sites. With a positive curvature, for example at the intersection of two $Ca^{2+}$ waves, you focus the autocatalytic agent, $Ca^{2+}$. This produces a very rapid wave propagation.

*Michell:* Dr Sanderson tried to compare the annulus at the surface of an oocyte with a tissue cell of another sort, and asked whether the tissue cell is more complicated, with more $InsP_3$ receptors that are more widely distributed. Does anybody know whether or not the densities per unit volume of $InsP_3$ channels in the annulus around the egg and in a typical mammalian tissue cell are similar?

*Berridge:* I don't think we have the tools to answer that question yet.

## References

Devreotes PN, Potel MJ, Mackay SA 1983 Quantitative analysis of cyclic AMP waves mediating aggression in *Dictyostelium discoideum*. Dev Biol 96:405–415

Girard S, Clapham D 1993 Acceleration of intracellular calcium waves in *Xenopus* oocytes by calcium influx. Science 260:230–232

Gorelova NA, Bures J 1983 Spiral waves of spreading depression in the isolated chicken retina. J Neurobiol 14:353–363

Kume S, Muto A, Aruga J et al 1993 The *Xenopus* $IP_3$ receptor: structure, function, and localization in oocytes and eggs. Cell 73:555–570

Larabell C, Nuccitelli R 1992 Inositol lipid hydrolysis contributes to the $Ca^{2+}$ wave in the activating egg of *Xenopus laevis*. Dev Biol 153:347–355

# Local calcium spiking in pancreatic acinar cells

O. H. Petersen

*MRC Secretory Control Research Group, The Physiological Laboratory, University of Liverpool, Crown Street, PO Box 147, Liverpool L69 3BX, UK*

*Abstract.* A few years ago, my laboratory obtained evidence for local agonist-evoked repetitive $Ca^{2+}$ spikes in single pancreatic acinar cells. We have now confirmed this and shown that regular cytosolic $Ca^{2+}$ spikes evoked by low concentrations of acetylcholine or inositol 1,4,5-trisphosphate ($InsP_3$) are confined to the secretory granule area at the luminal pole of the cells. The local subcellular signals probably arise because the first internal messenger ($InsP_3$), generated from the basolateral membrane, can diffuse rapidly, whereas the $Ca^{2+}$ released from the more responsive secretory granule region has a very restricted mobility. Local $Ca^{2+}$ spikes are useful from an energetic point of view and also help to avoid undesirable activation of $Ca^{2+}$-dependent processes. Another messenger, cyclic ADP-ribose, may also regulate intracellular $Ca^{2+}$ release. In pancreatic acinar cells cyclic ADP-ribose induces repetitive $Ca^{2+}$ spikes localized in the secretory granule area; these spikes are blocked by ryanodine, but also by the $InsP_3$ receptor antagonist heparin. Ryanodine abolishes or markedly inhibits agonist-evoked $Ca^{2+}$ spiking, but enhances the frequency of spikes evoked by internal $InsP_3$ application. These results indicate that both ryanodine and $InsP_3$ receptors are involved in $Ca^{2+}$ spike generation in pancreatic acinar cells, and that both $InsP_3$ and cyclic ADP-ribose may act as internal messengers.

*1995 Calcium waves, gradients and oscillations. Wiley, Chichester (Ciba Foundation Symposium 188) p 85–103*

It is well established that hormones and neurotransmitters can evoke release of $Ca^{2+}$ from intracellular stores followed by entry of $Ca^{2+}$ from the extracellular fluid (Nielsen & Petersen 1972). In recent years, the application of methods for high-resolution intracellular calcium measurements (Tsien & Poenie 1986) has given us insights into the complex spatiotemporal patterns of hormone-evoked changes in the cytosolic $Ca^{2+}$ concentration, $[Ca^{2+}]_i$. Subtle spatial and temporal regulations of $[Ca^{2+}]_i$ can induce ordered sequences of events (dynamic decoding) (Petersen et al 1994) and agonist-specific

$[Ca^{2+}]_i$ signatures (Woods et al 1987, Osipchuk et al 1990, Petersen et al 1991a, Toescu et al 1993).

## Local and global calcium spikes

Cytosolic $Ca^{2+}$ signals induced via $Ca^{2+}$ release channels usually take the form of repetitive spikes; $[Ca^{2+}]_i$ rises sharply from a threshold level and recovers to the resting level during sustained stimulation (Petersen et al 1994). In pancreatic acinar cells, these $Ca^{2+}$ spikes can occur locally at a specific subcellular domain and form $Ca^{2+}$ micro-gradients (Fig. 1). Under certain conditions the local $Ca^{2+}$ spike can subsequently spread as a wave towards other cellular areas. Such global rises in $[Ca^{2+}]_i$ can also be induced transiently and can be described as global $Ca^{2+}$ spikes (Fig. 1). Pancreatic acinar cells are thus able to control the spatial extent of the $Ca^{2+}$ signal, which is functionally and energetically important (Petersen et al 1994).

## Local calcium spikes occur in the secretory granule area

The first evidence for local agonist-evoked cytosolic $Ca^{2+}$ spikes in pancreatic acinar cells was obtained in 1990 when Osipchuk et al (1990) showed that a low concentration of acetylcholine could evoke repetitive spikes of $Ca^{2+}$-sensitive $Cl^-$ current without concomitant global rises in $[Ca^{2+}]_i$ (as measured by single-cell microfluorimetry using fura-2—photon counting). It was later shown with simultaneous patch-clamp whole-cell current recording and digital imaging of $Ca^{2+}$-sensitive fura-2 fluorescence that the short-lasting spikes of $Ca^{2+}$-dependent ionic currents evoked by agonists or intracellular infusion of inositol 1,4,5-trisphosphate ($InsP_3$) are due to rises in $[Ca^{2+}]_i$ specifically localized in the secretory pole (secretory granule area) of the acinar cell (Thorn et al 1993). These results could be explained most easily by a localization of $InsP_3$ receptors with a particularly high affinity in the secretory pole region (Thorn et al 1993). Kasai et al (1993) reached exactly the same conclusion on the basis of digital imaging studies of cells injected selectively with $InsP_3$ in the secretory granule area or the basal area. The conclusion reached by Kasai et al (1993) and Thorn et al (1993) has recently been supported by the demonstration that the type 3 $InsP_3$ receptor in pancreatic acinar cells is specifically localized in the trigger zone in the secretory granule area (Nathanson et al 1994). It seems reasonable to propose that the agonist-evoked repetitive $Ca^{2+}$ spikes in the secretory pole arise because the first internal messenger generated from the basolateral membrane—$InsP_3$ (the hormone receptors are specifically localized on the surface of the basolateral membrane [Rosenzweig et al 1983])—can diffuse rapidly throughout the cytosol, whereas

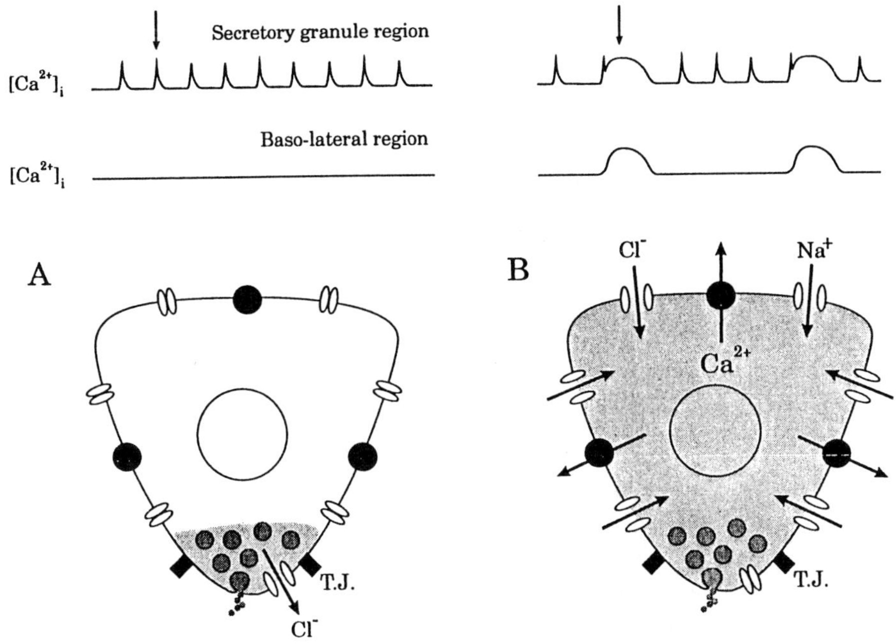


FIG. 1. Local and global cytosolic $Ca^{2+}$ spikes illustrated diagrammatically on the basis of data obtained from pancreatic acinar cells. A and B show single acinar cells in which tight junctions (T.J.) separate a small luminal area from a much larger basolateral plasma membrane area. Secretory (zymogen) granules are shown in the area close to the luminal membrane (secretory granule region). $Ca^{2+}$-activated $Cl^-$ and non-selective cation channels are shown together with $Ca^{2+}$ pumps. The nucleus is also indicated. The shaded areas represent regions in which cytosolic [$Ca^{2+}$] is elevated at the times corresponding to the arrows above the cytosolic [$Ca^{2+}$] traces shown in the upper part of the figure. In A the $Ca^{2+}$ signal (generated, for example, by a low concentration of acetylcholine or intracellular infusion of inositol 1,4,5-trisphosphate) consists of repetitive $Ca^{2+}$ spikes exclusively in the secretory granule region, whereas in B a more complex signal pattern is shown, composed of short-lasting spikes in the secretory granule region followed, in some cases, by larger global transients (such as would be generated, for example, by stimulation with the hormone cholecystokinin). Reproduced with permission from Petersen et al (1994). Copyright Annual Reviews Inc.

the $Ca^{2+}$ released from the more responsive secretory granule area has a very restricted mobility (Kasai & Petersen 1994).

## How much stored calcium is released per spike

Agonists can evoke $Ca^{2+}$ spikes in the absence of external $Ca^{2+}$ (Wakui et al 1989) so the primary event must be release of stored $Ca^{2+}$. The number of $Ca^{2+}$ spikes that can be evoked by a constant hormone concentration, in the absence of external $Ca^{2+}$, varies between cell types and depends on the type and strength of stimulation. In pancreatic acinar cells, many short-lasting local $Ca^{2+}$ spikes in the secretory granule region can be evoked by $InsP_3$ or a low concentration of acetylcholine when the external $Ca^{2+}$ concentration is very low (Petersen et al 1994). It has not yet been possible to estimate the amount of intracellular $Ca^{2+}$ released during these short-lasting spikes. In the same cells, cholecystokinin additionally evokes broader global $Ca^{2+}$ transients (Fig. 1); in experiments with a low external $Ca^{2+}$ concentration only a few (two to eight) of such events occur (Tepikin et al 1992a). The amount of $Ca^{2+}$ extruded during a cholecystokinin-evoked spike, measured with the droplet method (Tepikin et al 1992b), corresponds to about 40% of the total mobilizable intracellular $Ca^{2+}$ pool (Tepikin et al 1992a). The amount of stored $Ca^{2+}$ released during a spike must be at least equal to the amount extruded and, therefore, at least 40% of the $Ca^{2+}$ stores are emptied. It seems likely that a large proportion of the stored $Ca^{2+}$, perhaps all, is released during each spike, and that in pancreatic acinar cells, at least, about 40% of the $Ca^{2+}$ released may be extruded while the remaining 60% is taken back into the endoplasmic reticulum (ER). The large extrusion of $Ca^{2+}$ explains the need for compensatory $Ca^{2+}$ entry and why only a few global spikes can be generated in the absence of external $Ca^{2+}$.

## Repetitive calcium spikes evoked by a constant intracellular $InsP_3$ concentration

Infusion of non-metabolizable $InsP_3$ analogues by patch pipettes into pancreatic acinar cells (Wakui et al 1989, Thorn et al 1993) evokes repetitive cytosolic $Ca^{2+}$ signals, demonstrating that fluctuating concentrations of $InsP_3$ are not required for the production of $Ca^{2+}$ oscillations.

The repetitive $Ca^{2+}$ spikes in many cells (including pancreatic acinar cells) do not depend in the immediate term on extracellular $Ca^{2+}$ (Wakui et al 1989). The primary mechanism generating calcium spikes must thus be one that depends on only intracellular processes.

Each $Ca^{2+}$ spike is most probably generated by a mechanism of $Ca^{2+}$-induced $Ca^{2+}$ release (Berridge 1988, Wakui et al 1990). Positive feedback operates on the $Ca^{2+}$ release channels in such a way that an increase in $[Ca^{2+}]_i$ leads to a greater release of $Ca^{2+}$ from the intracellular stores (Bezprozvanny et al 1991). $Ca^{2+}$ release is inhibited by the depleted state of the stores and by direct inhibition of the release channels at high $[Ca^{2+}]_i$. The spike is finally terminated by refilling of the stores by the $Ca^{2+}$-ATPases of the ER.

Infusion of a low-affinity $Ca^{2+}$ buffer into pancreatic acinar cells greatly reduces the spiking frequency (Petersen et al 1991a,b), indicating that cytosolic $Ca^{2+}$ buffers have to be saturated before there can be an explosive release of

$Ca^{2+}$. Partial inhibition of the ER $Ca^{2+}$-ATPase in the presence of a non-metabolizable $InsP_3$ analogue increases the spiking frequency (or initiates spiking in cells with subthreshold concentrations of $InsP_3$) (Petersen et al 1993). The simplest explanation is that considerable amounts of stored intracellular $Ca^{2+}$ are released in the inter-spike period but are rapidly buffered and resequestered into the $Ca^{2+}$ stores. Partial inhibition of the $Ca^{2+}$ pump makes the cytosolic buffers saturated more rapidly, leading to an earlier explosive release of $Ca^{2+}$ and, thus, a higher spiking frequency. The events leading to repetitive cytosolic $Ca^{2+}$ spiking can be explained most simply by the model shown in Fig. 2.

The ER pumps have a much wider role than merely resequestering $Ca^{2+}$after a spike. In fact, the ER $Ca^{2+}$ATP-pase is intimately involved in generating long inter-spike periods and is thus able to control cytosolic $Ca^{2+}$ spiking.

**Are agonist-evoked calcium oscillations generated by repetitive intracellular calcium release always due to an increase in the intracellular concentration of inositol trisphosphate?**

Cytosolic $Ca^{2+}$ oscillations can be evoked in several different ways, but most agonists are considered to act via the $Ca^{2+}$-releasing messenger $InsP_3$ (Petersen et al 1994). Nevertheless, it has been shown that with exceptionally low intracellular $Ca^{2+}$ buffering, vasoactive intestinal polypeptide and its messenger cyclic AMP (cAMP) can stimulate cytosolic $Ca^{2+}$ spiking in pancreatic acinar cells (Kase et al 1991). The thiol reagent thimerosal can also induce $Ca^{2+}$ oscillations in pancreatic acinar cells (Thorn et al 1992) and in HeLa cells (Bootman et al 1992). These oscillations are not a result of $InsP_3$ generation, but can be blocked by the $InsP_3$ receptor antagonist heparin. The available evidence suggests that both sulphydryl reagents and cAMP-dependent kinases increase the sensitivity of the $InsP_3$ receptor for $InsP_3$ (Bootman et al 1992, Thorn et al 1992, Bird et al 1993, Hilly et al 1993), possibly to such an extent that the resting $[InsP_3]$ is sufficient to evoke $Ca^{2+}$ oscillations. These findings raise the question of whether all $Ca^{2+}$ oscillations evoked by neurotransmitters or hormones are necessarily a result of $InsP_3$ generation. The synthetic heptapeptide cholecystokinin analogue JMV-180 releases $[Ca^{2+}]_i$ in pancreatic acinar cells, and Yule & Williams (1992) found that the phospholipase C inhibitor U73122 cannot inhibit cytosolic $Ca^{2+}$ oscillations evoked by JMV-180. These findings suggest that JMV-180 evokes $Ca^{2+}$ signals via mechanisms that do not include generation of $InsP_3$.

More recently, it has been shown that the $InsP_3$ receptor antagonist heparin blocks both cholecystokinin-induced and JMV-180-induced $Ca^{2+}$ oscillations, whereas de-*N*-sulphated heparin, which does not affect binding of $InsP_3$ to its receptor, fails to inhibit the responses to the two agonists (Thorn & Petersen 1993). It seems that the cytosolic $Ca^{2+}$ oscillations produced by both

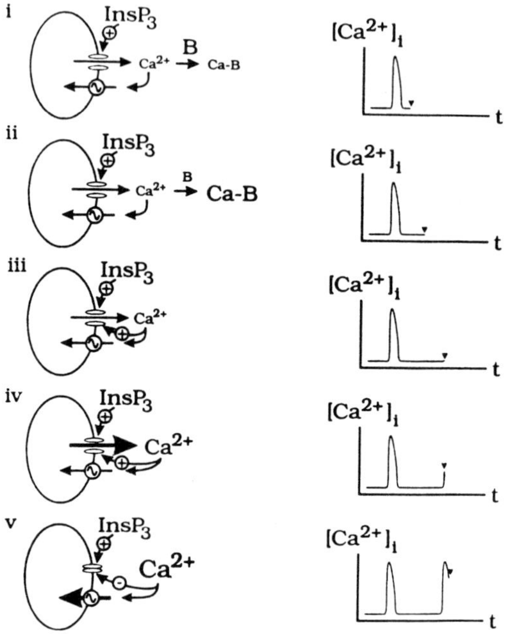


FIG. 2. A simple one-pool model explaining repetitive cytosolic $Ca^{2+}$ spiking at a constant inositol 1,4,5-trisphosphate ($InsP_3$) concentration. The left side illustrates an intracellular $Ca^{2+}$ store, with an $InsP_3$ receptor and a $Ca^{2+}$ pump, at five stages in the spiking cycle. The right side depicts changes in cytosolic $Ca^{2+}$ concentration ($[Ca^{2+}]_i$) with time, and shows an idealized spiking pattern. This model shows the importance of active and passive $Ca^{2+}$ buffering (B) in the generation of $Ca^{2+}$ spikes. (i) $Ca^{2+}$ released at a slow rate from the $InsP_3$-sensitive stores is rapidly buffered so that no increase in $[Ca^{2+}]_i$ is observed. The buffering consists of both cytosolic calcium-binding proteins and $Ca^{2+}$-ATPases on other $Ca^{2+}$ stores. (ii) As the $Ca^{2+}$ release continues during the inter-spike period, the buffers begin to become saturated. (iii) The decrease in available $Ca^{2+}$ buffering capacity (owing to local saturation of cytosolic $Ca^{2+}$-binding proteins and the filling of other $Ca^{2+}$ stores) slows the rate at which $Ca^{2+}$ is removed from the cytosol, and at a given time the release rate overtakes the maximum rate of buffering. At this point $[Ca^{2+}]_i$ begins to rise slowly. (iv) Because of positive feedback ($Ca^{2+}$-induced $Ca^{2+}$ release) on the $Ca^{2+}$ release channels, an explosive rise in $[Ca^{2+}]_i$ is initiated. (v) As $[Ca^{2+}]_i$ reaches its maximum, further release is inhibited, allowing the $Ca^{2+}$ to be resequestered by the $Ca^{2+}$-ATPases in the endoplasmic reticulum. This process can then be repeated, generating repetitive $Ca^{2+}$ spikes. Reproduced with permission from Petersen et al (1993). Copyright The American Society for Biochemistry and Molecular Biology.

cholecystokinin (5–20 pM) and JMV-180 (10–40 nM) are dependent on functional $InsP_3$ receptors, but because it has not been possible to demonstrate $InsP_3$ formation in response to concentrations of cholecystokinin below 100 pM or to concentrations of JMV-180 below 1 $\mu$M (Matozaki et al 1989, 1990), it is possible that the cytosolic $Ca^{2+}$ oscillations are initiated by a $Ca^{2+}$ release process not dependent on $InsP_3$ generation. An action mediated entirely by sensitization of $InsP_3$ receptors to the resting $InsP_3$ concentration, similar to that described above for thimerosal, is one possiblilty.

## Cyclic ADP-ribose regulation of ryanodine receptors may be involved in agonist-evoked calcium oscillations

In pancreatic acinar cells there is strong evidence for the involvement of $InsP_3$ receptors in agonist-evoked $Ca^{2+}$ oscillations (Wakui et al 1989, Thorn et al 1993), but there are also indications that an $InsP_3$-insensitive $Ca^{2+}$ release channel that is sensitive to caffeine and to $Ca^{2+}$ is operational (Wakui et al 1990, Kasai et al 1993). The secretory pole of the acinar cell contains both channels and plays, as already discussed, a particularly important role in $Ca^{2+}$ signalling (Thorn et al 1993, Kasai et al 1993). The functional interrelationship between the two types of $Ca^{2+}$ release channel is unclear and there has, until recently, been little information about the nature and control of the $InsP_3$-insensitive $Ca^{2+}$-release channel.

Cyclic adenosine 5′-diphosphoribose (cADP-ribose) is a putative second messenger regulating the ryanodine receptor (Galione 1992). It has recently been shown that this agent can by itself induce repetitive $Ca^{2+}$ spikes localized in the secretory pole of pancreatic acinar cells and that these spikes are abolished by ryanodine (Thorn et al 1994). The cADP-ribose-induced $Ca^{2+}$ spiking is also blocked by the $InsP_3$ receptor antagonist heparin, as are the responses to agonists, $InsP_3$ and caffeine. Ryanodine rapidly abolishes or severely inhibits agonist-evoked $Ca^{2+}$ spiking, but enhances the frequency of the spikes evoked by intracellular $InsP_3$ infusion (Thorn et al 1994).

In contrast to the results recently obtained in sea urchin eggs, where there is clear evidence of redundancy because selective inhibition of either pathway had no effect on $Ca^{2+}$ signalling (Galione et al 1993), the results obtained in pancreatic acinar cells (Thorn et al 1994) show that both $InsP_3$- and cADP-ribose-sensitive $Ca^{2+}$ release channels are required for agonist-evoked $Ca^{2+}$ spiking (Fig. 3). Because the cADP-ribose evoked, short-lasting $Ca^{2+}$ spikes are localized in the secretory pole of the acinar cell, like those evoked by agonists and $InsP_3$ (Thorn et al 1994), both the ryanodine and $InsP_3$ receptors are likely to be present in this part of the cell. This is in agreement with the results of Kasai et al (1993) showing that the secretory pole is the site of a heparin-insensitive $Ca^{2+}$-induced $Ca^{2+}$ release.

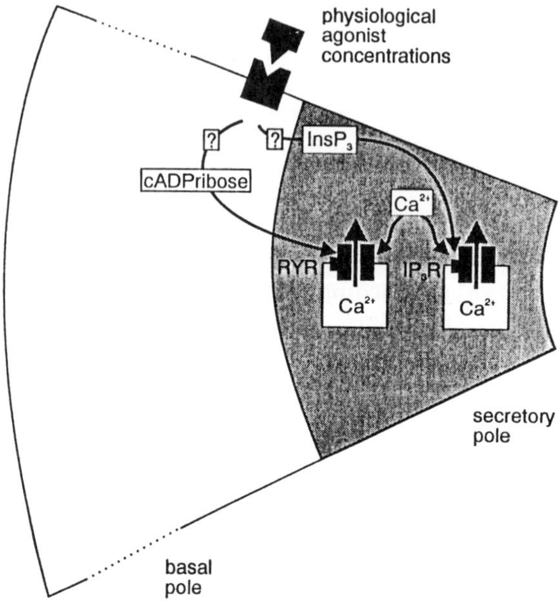


FIG. 3. Working model for agonist-evoked $Ca^{2+}$ spiking in the secretory pole of a pancreatic acinar cell. The question marks indicate that there is at present no evidence showing regulation of $InsP_3$ and cyclic ADP-ribose (cADPribose) concentrations in these cells by low (physiological) concentrations of the main agonists acetylcholine and cholecystokinin. RYR, ryanodine receptor; $Ip_3R$, $InsP_3$ receptor. Reproduced from Thorn et al (1994) by permission of Oxford University Press.

The finding that ryanodine blocks agonist- and cADP-ribose-evoked $Ca^{2+}$ spiking but enhances $InsP_3$-induced $Ca^{2+}$ spiking (Thorn et al 1994) shows that although intracellular $InsP_3$ application can mimic the effect of agonists (Petersen et al 1991b), it is unlikely that agonists act by generating the relatively high $InsP_3$ concentrations ($>3\,\mu M$) required to induce spiking in intracellular infusion experiments.

A messenger in addition to $InsP_3$ that can activate the ryanodine receptor may, therefore, be needed to explain the generation of agonist-evoked $Ca^{2+}$ spikes. cADP-ribose could be this messenger, but it is not clear how generation of cADP-ribose could be controlled in pancreatic acinar cells.

*Acknowledgement*

This work was supported by a Medical Research Council programme grant.

## References

Berridge MJ 1988 Inositol trisphosphate-induced membrane potential oscillations in *Xenopus* oocytes. J Physiol 403:589–599

Bezprozvanny I, Watras J, Ehrlich BE 1991 Bell-shaped calcium-response curves of Ins(1,4,5)$P_3$- and calcium-gated channels from endoplasmic reticulum of cerebellum. Nature 351:751–754

Bird GSJ, Burgess GM, Putney JW 1993 Sulfhydryl reagents and cAMP-dependent kinase increase the sensitivity of the inositol 1,4,5-trisphosphate receptor in hepatocytes. J Biol Chem 268:17917–17923

Bootman M, Taylor CW, Berridge MJ 1992 The thiol reagent, thimerosal, evokes $Ca^{2+}$ spikes in HeLa cells by sensitizing the inositol 1,4,5-trisphosphate receptor. J Biol Chem 267:25113–25119

Galione A 1992 $Ca^{2+}$-induced $Ca^{2+}$ release and its modulation by cyclic ADP-ribose. Trends Pharmacol Sci 13:304–306

Galione A, McDougall A, Busa WB, Willmott N, Gillot I, Whitaker MJ 1993 Redundant mechanisms of calcium-induced calcium release underlying calcium waves during fertilization of sea urchin eggs. Science 261:348–352

Hilly M, Pietri-Rouxel F, Coquil J-F, Guy M, Mauger J-P 1993 Thiol reagents increase the affinity of the inositol 1,4,5-trisphosphate receptor. J Biol Chem 268:16488–16494

Kasai H, Petersen OH 1994 Spatial dynamics of second messengers: $IP_3$ and cAMP as long-range and associative messengers. Trends Neurosci 17:95–101

Kasai H, Li YX, Miyashita Y 1993 Subcellular distribution of $Ca^{2+}$ release channels underlying $Ca^{2+}$ waves and oscillations in exocrine pancreas. Cell 74:669–677

Kase H, Wakui M, Petersen OH 1991 Stimulatory and inhibitory actions of VIP and cyclic AMP on cytoplasmic $Ca^{2+}$ signal generation in pancreatic acinar cells. Pfluegers Arch Eur J Physiol 419:668–670

Matozaki T, Martinez J, Williams JA 1989 A new CCK analogue differentiates two functionally distinct CCK receptors in rat and mouse pancreatic acini. Am J Physiol 257:G594–G600

Matozaki T, Goke B, Tsunoda Y, Rodriguez M, Martinez J, Williams JA 1990 Two functionally distinct cholecystokinin receptors show different modes of actions on $Ca^{2+}$ mobilization and phospholipid hydrolysis in isolated rat pancreatic acini. J Biol Chem 265:6247–6254

Nathanson MH, Fallon MB, Padfield PJ, Maranto AR 1994 Localization of the type 3 inositol 1,4,5-trisphosphate receptor in the $Ca^{2+}$ wave trigger zone of pancreatic acinar cells. J Biol Chem 269:4693–4696

Nielsen SP, Petersen OH 1972 Transport of calcium in the perfused submandibular gland of the cat. J Physiol 223:685–697

Osipchuk YV, Wakui M, Yule DI, Gallacher DV, Petersen OH 1990 Cytoplasmic $Ca^{2+}$ oscillations evoked by receptor stimulation, G-protein activation, internal application of inositol trisphosphate or $Ca^{2+}$: simultaneous microfluorimetry and $Ca^{2+}$-dependent $Cl^-$ current recording in single pancreatic acinar cells. EMBO (Eur Mol Biol Organ) J 9:697–704

Petersen CCH, Toescu EC, Petersen OH 1991a Different patterns of receptor-activated cytoplasmic $Ca^{2+}$ oscillations in single pancreatic acinar cells: dependence on receptor type, agonist concentration and intracellular $Ca^{2+}$ buffering. EMBO (Eur Mol Biol Organ) J 10:527–533

Petersen CCH, Toescu EC, Potter BVL, Petersen OH 1991b Inositol trisphosphate produces different patterns of cytoplasmic $Ca^{2+}$ spiking depending on its concentration. FEBS (Fed Eur Biochem Soc) Lett 293:179–182

Petersen CCH, Petersen OH, Berridge MJ 1993 The role of endoplasmic reticulum calcium pumps during cytosolic calcium spiking in pancreatic acinar cells. J Biol Chem 268:22262–22264

Petersen OH, Petersen CCH, Kasai H 1994 Calcium and hormone action. Annu Rev Physiol 56:297–319

Rosenzweig SA, Miller LJ, Jamieson JD 1983 Identification and localization of cholecystokinin-binding sites on rat pancreatic plasma membranes and acinar cells: a biochemical and autoradiographic study. J Cell Biol 96:1288–1297

Tepikin AV, Voronina SG, Gallacher DV, Petersen OH 1992a Pulsatile $Ca^{2+}$ extrusion from single pancreatic acinar cells during receptor-activated cytosolic $Ca^{2+}$ spiking. J Biol Chem 267:14073–14076

Tepikin AV, Voronina SG, Gallacher DV, Petersen OH 1992b Acetylcholine-evoked increase in the cytoplasmic $Ca^{2+}$ concentration and $Ca^{2+}$ extrusion measured simultaneously in single mouse pancreatic acinar cells. J Biol Chem 267:3569–3572

Thorn P, Petersen OH 1993 Calcium oscillations in pancreatic acinar cells, evoked by the cholecystokinin analogue JMV-180, depend on functional inositol 1,4,5-trisphosphate receptors. J Biol Chem 268:23219–23221

Thorn P, Brady P, Llopis J, Gallacher DV, Petersen OH 1992 Cytosolic $Ca^{2+}$ spikes evoked by the thiol reagent thimerosal in both intact and internally perfused single pancreatic acinar cells. Pfluegers Arch Eur J Physiol 422:173–178

Thorn P, Lawrie AM, Smith PM, Gallacher DV, Petersen OH 1993 Local and global cytosolic $Ca^{2+}$ oscillations in exocrine cells evoked by agonists and inositol trisphosphate. Cell 74:661–668

Thorn P, Gerasimenko O, Petersen OH 1994 Cyclic ADP-ribose regulation of ryanodine receptors involved in agonist-evoked cytosolic $Ca^{2+}$ oscillations in pancreatic acinar cells. EMBO (Eur Mol Biol Organ) J 13:2038–2043

Toescu EC, Lawrie AM, Gallacher DV, Petersen OH 1993 The pattern of agonist-evoked cytosolic $Ca^{2+}$ oscillations depends on the resting intracellular $Ca^{2+}$ concentration. J Biol Chem 268:18654–18658

Tsien RY, Poenie M 1968 Fluorescence ratio imaging: a new window into intracellular ionic signalling. Trends Biochem Sci 11:450–455

Wakui M, Potter BVL, Petersen OH 1989 Pulsatile intracellular calcium release does not depend on fluctuations in inositol trisphosphate concentration. Nature 339:317–320

Wakui M, Osipchuk YV, Petersen OH 1990 Receptor-activated cytoplasmic $Ca^{2+}$ spiking mediated by inositol trisphosphate is due to $Ca^{2+}$-induced release. Cell 63:1025–1032

Woods NM, Cuthbertson KSR, Cobbold PH 1987 Agonist-induced oscillations in cytoplasmic free calcium concentration in single rat hepatocytes. Cell Calcium 8:79–100

Yule DI, Williams JA 1992 U73122 inhibits $Ca^{2+}$ oscillations in response to cholecystokinin and carbachol but not to JMV-180 in rat pancreatic acinar cells. J Biol Chem 267:13830–13835

## DISCUSSION

*Cannell:* I was puzzled by the rapid reversibility of your ryanodine effect (Thorn et al 1994). That isn't what is usually seen with the ryanodine receptor.

*Petersen:* It was a surprise to us too. I suspect that the ryanodine receptor in pancreatic acinar cells is slightly different from those characterized so far.

To my knowledge, there has been no molecular characterization of ryanodine receptors in these cells.

*Nathanson:* You are suggesting that agonist-induced $InsP_3$ formation is not sufficient to initiate $Ca^{2+}$ signals in these cells.

*Petersen:* We need to infuse $InsP_3$ to a minimum concentration of 3–5 $\mu$M to see $Ca^{2+}$ spiking in our cells. That $Ca^{2+}$ spiking looks very similar to that evoked with low concentrations of acetylcholine. However, ryanodine is unable to block the $InsP_3$-induced effect, although it can block the effect of agonists. This concerns us because it seems to indicate that the effects of the agonist are not mediated by such high $InsP_3$ concentrations but probably by much lower concentrations of $InsP_3$ in addition to something else. That is why I put the question mark in Fig. 3.

*Galione:* Is the effect of caffeine blocked by heparin?

*Petersen:* The effect of caffeine is blocked by heparin, as well as by ryanodine. All the effects we have found are blocked both by heparin and by ryanodine, except the effects of $InsP_3$, which are blocked only by heparin.

*Putney:* You have shown that the effect of high [$Ca^{2+}$] is not blocked by heparin (Wakui et al 1990). Is that a problem? What is high [$Ca^{2+}$] doing, where is it acting, differently from cyclic ADP-ribose (cADP-ribose)?

*Petersen:* This is a problem. I now suspect that the experiments we have done with $Ca^{2+}$ infusion (Wakui et al 1990) may be slightly unphysiological. It was clear in those experiments that you can get repetitive $Ca^{2+}$-induced $Ca^{2+}$ spiking only under conditions of exceptionally low intracellular buffering. It may be that in that situation the $Ca^{2+}$ stores are overloaded.

*Berridge:* Barbara Ehrlich showed that heparin can stimulate the ryanodine receptor. Couldn't that explain your results?

*Petersen:* In fact, if you look at the data in Thorn et al (1993), although we didn't comment on it, you will see that heparin dramatically reduces the frequency of spiking in response to cholecystokinin, but the individual spikes are actually larger. This fits in well with Ehrlich's observation.

*Thomas:* So, potentially, the inhibition of $Ca^{2+}$ spiking by ryanodine could be a result of the stores being drained.

*Petersen:* Yes, that is one interpretation.

*Jaffe:* How dependent are these phenomena on extracellular $Ca^{2+}$?

*Petersen:* Short-lasting spikes can continue for at least 20 min in the complete absence of external $Ca^{2+}$. If you continue recording for hours, you eventually come to a stage when you can finally show that they depend on external $Ca^{2+}$. The relative independence could tie in with Dr Thomas's concept of a continuous store. The local spikes are definitely not dependent on $Ca^{2+}$ influx.

*Nathanson:* The $Ca^{2+}$ trigger zone Dr Kasai has identified is only a small part of the granular region. Is your spatial resolution adequate to resolve whether cADP-ribose causes $Ca^{2+}$ release specifically from the trigger zone or from a nearby region within the secretory granule area?

*Petersen:* Our resolution is not yet as good as Dr Kasai's. As far as cADP-ribose is concerned, all we can say is that the spikes are in the secretory granule area.

*Meyer:* In your measurements, you could compare the local $Ca^{2+}$ concentration close to the plasma membrane with the membrane depolarization caused by opening of $Ca^{2+}$-dependent $Cl^-$ channels. Does the $Ca^{2+}$ dependence of the $Cl^-$ channels match up with the measured calcium concentrations?

*Petersen:* The available data do not indicate a dramatic difference between the effects of internal $Ca^{2+}$ infusion and intracellular $Ca^{2+}$ release. Evans & Marty (1986) infused well-buffered solutions into lacrimal cells and looked at the $Ca^{2+}$ dependency of the $Cl^-$ channels; they are activated mainly at 0.1–1 $\mu$M. So, there does not seem to be a huge difference, but these experiments were not in pancreatic acinar cells.

*Smith:* Dr Kasai's paper with George Augustine a few years ago (Kasai & Augustine 1990) suggested a definite functional role for oscillations and gradients *per se*, namely the driving of $Cl^-$ and fluid across the cell. Do you think that hypothesis is still viable?

*Petersen:* The hypothesis is certainly viable, but Dr Kasai and I have some problems concerning the general applicability of this to all exocrine cells, because there are a number of puzzling points with regard to the different types of $Ca^{2+}$-sensitive channels in different exocrine cells (Kasai & Augustine 1990, Petersen 1992).

*Thomas:* I gather that cyclic GMP (cGMP) does not seem to play a role in pancreatic acinar cells.

*Petersen:* We have some old data about this, though there's no definite evidence against a role for cGMP. We used dibutyryl cGMP extensively some years ago because it so happens that it is a very good competitive antagonist at the cholecystokinin receptor. We incubated many slices with dibutyryl cGMP, which completely antagonizes cholecystokinin's actions but does not affect responses to acetylcholine and bombesin, and does not by itself induce any conductance changes in these cells (Philpott & Petersen 1979). From this, I feel that cGMP is unlikely to be a messenger generating $Ca^{2+}$ signals. Having said that, these results were obtained in a completely different context, not in isolated cells. There's certainly no positive evidence that cGMP is the messenger, but the cGMP concentration does go up in these cells when you stimulate them with acetylcholine (Christophe et al 1976).

*Galione:* Increasing [cGMP], either by using dibutyryl cGMP or by injecting cGMP, produces nice $Ca^{2+}$ waves in sea urchin eggs that mimic the fertilization $Ca^{2+}$ transient better than the waves produced by any other putative intracellular messenger.

*Whitaker:* That's true. The kinetics of the response curve is much like that of fertilization, whereas the response to $InsP_3$, for example, is much shorter.

*Galione:* Yes; there is latency and the transient is prolonged. You can mimic some features of this in a homogenate of sea urchin eggs. The latency may be due to the conversion of NAD to cADP-ribose, and cGMP promotes that conversion, probably through a cGMP-dependent protein kinase because kinase inhibitors block the effect (Galione et al 1993). 8-Amino-cADP-ribose completely blocks the release of $Ca^{2+}$ induced by cGMP. Nitric oxide can also induce $Ca^{2+}$ waves, and we've found that nitric oxide increases [cGMP] in the egg (A. Galione, unpublished work). I don't know whether this has physiological significance, but there is some indication that [cGMP] goes up at fertilization (B. Ciapa, personal communication).

*Whitaker:* That's true, but I wouldn't put my money on cGMP.

*Petersen:* There is evidence from several groups that cGMP does not affect $Ca^{2+}$ signalling in pancreatic acinar cells (Petersen 1982).

*Galione:* In the old literature, cGMP was thought to be an excitatory agent whereas cyclic AMP (cAMP) was often inhibitory. The fact that cGMP relaxes smooth muscle put people off that idea, but perhaps the idea should be revived.

*Lechleiter:* Dr Petersen, thapsigargin treatment usually uncovers a leak current from the endoplasmic reticulum, whereas you think the leak is the $InsP_3$ receptor itself (Petersen et al 1993)?

*Petersen:* Low doses of thapsigargin have no effect in cells not infused with $InsP_3$ (Petersen et al 1993). The leak was uncovered only in the presence of $InsP_3$, suggesting that the leak is through the $InsP_3$ channel. There is a recirculation of $Ca^{2+}$ through the $InsP_3$ receptor and the $Ca^{2+}$ pump, and you can uncover the release through the $Ca^{2+}$ release channels by putting a slight break on the $Ca^{2+}$ pump.

*Putney:* There is an alternative explanation. Normally with thapsigargin there is a little leak, which is enough to raise [$Ca^{2+}$], but in the presence of $InsP_3$ that little leak causes $Ca^{2+}$-induced $Ca^{2+}$ release. The $InsP_3$ is not, in the alternative model, releasing much $Ca^{2+}$. The latent period may be due to resensitization of the $InsP_3$ receptor. If you raise either [$Ca^{2+}$] or [$InsP_3$], you reach a triggering threshold. Thapsigargin could do that and cause a premature discharge of $Ca^{2+}$.

*Petersen:* There are several interpretations, but the data have actually shown that there is recirculation through the $InsP_3$ channel and the pump.

*Putney:* The system could be completely quiescent, and if you raised [$Ca^{2+}$] just a little with thapsigargin in the presence of $InsP_3$, $Ca^{2+}$-induced $Ca^{2+}$ release would begin.

*Petersen:* Because the effect is seen only in the presence of $InsP_3$, it is simplest to assume that the $Ca^{2+}$ leak is through that channel.

*Jaffe:* Is there any evidence for a standing $Ca^{2+}$ gradient in these cells?

*Petersen:* That depends on how long you want the gradient to be standing. The micro-gradients that I talked about are not really standing; the short-lasting spikes usually last for only a few seconds.

*Cobbold:* As you need activation of both receptors ($InsP_3$ and ryanodine) to get a good $Ca^{2+}$ response, have you tried patching in cADP-ribose and $InsP_3$ together?

*Petersen:* Peter Thorn (unpublished experiments) made heroic attempts to do something along those lines, but disappointingly didn't get a clear answer. We had hoped to be able to reproduce the different types of $Ca^{2+}$ spike patterns evoked by acetylcholine and cholecystokinin by mixing different doses of cADP-ribose and $InsP_3$. We never succeeded. We saw a slight effect: at a certain dose of $InsP_3$ where we usually see a small level of spiking, when we added a bit of cADP-ribose the frequency of spiking was generally higher. One of the problems with these experiments was that they were done as separate, individual experiments, one cell with $InsP_3$ and another with both $InsP_3$ and cADP-ribose. If one were to pursue this more systematically it would obviously be preferable to change the infusion during an experiment, but such experiments are technically difficult.

*Berridge:* The $Ca^{2+}$-induced $Ca^{2+}$ release pool seems to be juxtaposed between the $InsP_3$-sensitive receptors and those that are less sensitive to $InsP_3$ and are responsible for the global spread. If cADP-ribose is acting on this intermediate region, one would have expected more evidence that cADP-ribose could trigger these global responses.

*Petersen:* We had hoped that in the experiments that Peter Cobbold just asked about we would have seen evidence of this. The secretory granule could be regarded as a sort of barrier between the trigger zone and the basal area. With both cADP-ribose and $InsP_3$ present you might expect to see big waves more frequently at lower doses of $InsP_3$. The complicating factor may be that in the basal pole the sensitivity to $InsP_3$ is rather low. Rather high $InsP_3$ concentrations may be necessary for $Ca^{2+}$ waves to penetrate into that zone.

*Nathanson:* Does co-infusion of $Ca^{2+}$ and cADP-ribose potentiate cADP-ribose-induced $Ca^{2+}$ release?

*Petersen:* We haven't done $Ca^{2+}$ infusion experiments in combination with cADP-ribose.

*Berridge:* Are the spikes that you see related to the puffs that Ian Parker was talking about (p 50–65)? The release of $Ca^{2+}$ might be highly localized within the apical region of the pancreatic cell.

*Petersen:* It is an intuitively attractive idea that Dr Parker's puffs are related to our spikes. Dr Parker mentioned that at low concentrations of $InsP_3$ the puffs are irregular and that certainly is also the case for our spikes. If we went through all our data, I think we would find that at a low, threshold, stimulatory concentration the spikes are irregular. Although the amplitudes of the spikes do vary, they are roughly the same. The ryanodine receptor adds a little complication, but different cells will have different sets of $Ca^{2+}$ release channels.

*Galione:* Infusion of $InsP_3$ can induce oscillations for hours. Is the same true for cADP-ribose?

*Petersen:* I don't think we have addressed that question. We have obviously done more experiments over the years with $InsP_3$. I'm not sure whether Peter Thorn has infused cADP-ribose for several hours, but it has certainly been followed for more than 10 min (Thorn et al 1994). There is a slight tendency for the frequency of spiking to go down over time.

*Galione:* That is the case in dorsal root ganglion cells. cADP-ribose doesn't seem to work on purified sea urchin egg microsomes but it does if supernatant is present (Lee et al 1994). I wonder whether you are losing responsiveness because factors are being lost into the patch pipette.

*Petersen:* That is a possibility. It is likely, although we cannot measure this, that in experiments in which there is no $InsP_3$ in the pipette, the intracellular concentration of $InsP_3$ gradually decreases; since there's some kind of cooperativity between $InsP_3$ and cADP-ribose, that could be important, in addition to the possible loss of other factors. We had difficulty seeing an effect of cADP-ribose, if the cells were not freshly isolated. The most successful experiments were done within one and a half hours of isolation of the cells.

*Thomas:* We find a similar thing with ryanodine. If we add ryanodine to freshly isolated liver cells, within 30 mins of plating, a large proportion of the cells give waves of $Ca^{2+}$. If we wait even for a few hours, ryanodine alone will not induce a $Ca^{2+}$ response but it will still block responses to agonists. I don't know what we're losing, but something is declining with time after isolation.

*Michell:* Studies on hepatocytes in the early 1980s showed that sensitivity to different adrenergic ligands, acting through different types of catecholamine receptors, changes dramatically when you take hepatocytes out of the liver and put them into primary culture under good conditions. The cells will change their relative sensitivity to $\alpha$ and $\beta$ stimulation over only a few hours, so there's no reason why expression of internal channels shouldn't change similarly over time.

*Berridge:* It does seem that the expression of the ryanodine receptor is very labile and seems to disappear quickly when cells begin to de-differentiate.

*Michell:* Does putting the cells onto an extracellular matrix when you culture them preserve the phenotype, as it does with mammary epithelial cells?

*Thomas:* I don't know. We've used only polylysine, which is not very physiological! I gather that the polarization of receptors and various proteins and retention of some of the growth factor receptors in a functional form is enhanced by culturing the cells in a collagen gel matrix, but we haven't tried that because it would make imaging difficult. The hormone receptors we look at are down-regulated over 24 h, and by 48 h after plating very few cells respond.

*Stucki:* We have tried various extracellular matrices, such as collagen, laminin and fibronectin, but we couldn't see big differences. We prepare hepatocytes by EDTA dissociation and then freeze them in liquid nitrogen. After thawing, most of the cells are dead and only the best survive and attach to collagen-coated glass, for example. We are doing a kind of affinity chromatography, selecting the best cells. This may be why our cells are more sensitive to hormones than Peter Cobbold's.

*Cobbold:* Some of the differences between our results might be due to your cells having a higher basal cAMP concentration than ours. We all think we're working with the same cells, isolated rat hepatocytes, but we seem to be working with subtly different cells and subtly different treatments, and the $Ca^{2+}$ responses are different, even qualitatively.

*Williams:* We should discuss buffering. Buffering is important because it determines whether the released $Ca^{2+}$ can innovate specific action locally. In total, it represents the calcium that is not free. All we are discussing at the moment is free $Ca^{2+}$ going in and out of stores, which is rather boring as far as $Ca^{2+}$ is concerned—$Ca^{2+}$ is supposed to do something. Buffering could be a subtle timing device in spatiotemporal $Ca^{2+}$ signalling, where buffering controls free $Ca^{2+}$ in space and time and/or the buffer itself becomes the signal.

*Berridge:* That's true. All cells have a certain degree of buffering, and this is increased by putting in probes such as fura-2. But the point to emphasize is that you still see the waves going through, despite the increase in buffering. It is our view that saturation of the buffers sets the stage for wave propagation.

*Williams:* But we know perfectly well that putting a buffering protein into muscle cells can alter temporal signalling, e.g., it can increase relaxation rates in the case of parvalbumin in muscle or prevent excessive $Ca^{2+}$ increase in the case of neuronal cells. But once $Ca^{2+}$ is bound to the buffer, the 'buffer protein' becomes the possible active entity, not the free calcium.

*Berridge:* What I'm saying is that in a normal cell you may have to saturate the buffers, including the intracellular stores, before a spike will take place. A different situation exists in skeletal or cardiac muscle, where the event is not spontaneous but has to be driven by a stimulus. In these cases the stores pour out an enormous amount of $Ca^{2+}$, which then acts on the calcium-binding proteins such as troponin C in order to stimulate contraction.

*Williams:* A buffer works on a time basis, not just on a concentration basis. In one case the $Ca^{2+}$ can be released first to a certain receiver, which has fast uptake rates. There can then be a slow, more strongly binding buffer which takes the free $Ca^{2+}$ as it is released and prevents any further activity. Here, if your spike lasts for a certain length of time, you have to fit in some activity before the buffer catches the $Ca^{2+}$. In these circumstances you can discriminate between spatiotemporal activities just on the basis of buffers, not just on the basis of any of the release or pumping actions that you have described so far. The question is, do these buffer proteins—and we know what they are—exist in cells described by you? What is the trace left behind in the cell when the free $[Ca^{2+}]$ falls? Following only free $[Ca^{2+}]$ doesn't tell us much about $Ca^{2+}$ activity.

*Berridge:* Most people here would accept that what we actually see when we record a spike is the tip of the iceberg. In fact, the cell has released 95% more $Ca^{2+}$ than we actually see, for the very reason that most of it ends up on the buffers.

*Williams:* The concentration of calbindin in a cell can be as high as 3–5 mM, and it picks up $Ca^{2+}$ more slowly than calmodulin does. It is possible that calmodulin can be triggered quickly, but if you have annexin in the cell as well, because annexin binds slowly, annexin might not be triggered provided the buffer calbindin removes $Ca^{2+}$ at an intermediate rate. This means that a cell can control what it is doing with the help of buffers. The buffers are extremely important, not as thermodynamic buffers but as timing constants for the activity of the $Ca^{2+}$.

*Meyer:* Can you give us an example of such a time constant?

*Williams:* The time on for calmodulin *in vitro* is very fast, virtually diffusion controlled, and time off is in the millisecond range. Troponin C is a 'calmodulin' and controls the whole process of the muscle. Calbindin can keep hold of $Ca^{2+}$ for perhaps a minute but it may take the $Ca^{2+}$ microseconds at least to bind. The same is true of annexin. Thus, calbindin could eliminate annexin action but not that of calmodulin.

*Fay:* You have talked about buffers being a temporal filter. They can also potentially act as a spatial filter if they are not evenly distributed. To what extent might the localization of the $Ca^{2+}$ targets give rise to spatial aspects of the $Ca^{2+}$ response system?

*Williams:* The annexins are associated with membranes and tubulin systems, whereas the calbindins are more generally distributed. So the spatial distribution of calcium entry and exit and protein distribution are critical. It's the class of cell that worries me more. The different cells have different buffering capacities. There would be no point in having a fast, strong buffer in a cell which is supposed to react to annexins, because the buffer would take all the $Ca^{2+}$ straight away. You would see $Ca^{2+}$ release and might see a spike because your dye is fast, but which proteins, if any, are activated?

*Berridge:* Most $Ca^{2+}$-sensitive control systems in cells use calmodulin as a sensor. $Ca^{2+}$ goes on and off calmodulin very quickly. The effect of the other buffers will depend on how they bind $Ca^{2+}$. For example, if $Ca^{2+}$ is sitting on calbindin and it takes a minute to dissociate, whereas there are spikes every two seconds, the effect of calbindin would seem to be irrelevant because it will remain saturated.

*Cobbold:* But if calbindin is not saturated with each spike, could you build up the amount of $Ca^{2+}$-liganded calbindin over several spikes?

*Williams:* Then you would build up another type of signal in the cell, a high concentration of $Ca^{2+}$-bound calbindin, and that calbindin could report to other systems a long time later, after the spikes have disappeared. Here the buffer protein takes on the role of a slow signal.

*Petersen:* The $Ca^{2+}$ signals I have talked about do actually do something. The spikes cause exocytosis and secretion (Maruyama & Petersen 1994). They are physiologically relevant. We do not have a complete understanding of the buffering in pancreatic acinar cells. A. V. Tepikin in my laboratory (unpublished work) has tried to measure the buffering capacity of these cells. With the microdroplet technique, he can measure the absolute amount of $Ca^{2+}$ that is removed from the cell. If you stimulate with a supramaximal agonist dose, you are presumably removing the effect of the endoplasmic reticulum $Ca^{2+}$ pumps because the channels will be completely open, and you get a single transient in the absence of external $Ca^{2+}$. By looking at the amount of $Ca^{2+}$ that has been removed in order to bring the $[Ca^{2+}]$ back to what it was before stimulation, you can correlate the absolute amount of $Ca^{2+}$ taken out of the system with the changes in the free $[Ca^{2+}]$ and calculate the buffer capacity. The $Ca^{2+}$-binding capacity is high in acinar cells, three- to fourfold higher than in muscle cells. The distribution of this buffering capacity could be one factor that helps to confine the signals to the secretory granule area.

*Williams:* The $Ca^{2+}$ is required as a signal all the way through to things like phosphatases, which in the end signal to the DNA. Phosphorylation and dephosphorylation reactions are actually much slower than $Ca^{2+}$ off rates from free calmodulin, but the bound $Ca^{2+}$ must be present if it is to function: it must not be pumped straight out again.

*Berridge:* That's the whole point of spiking. If you want to activate transcription, you have to keep on spiking.

*Williams:* If you keep on spiking, it's effectively like having low $[Ca^{2+}]$ all the time, less than $10^{-6}$ M. Thus, either binding constants must be higher than we measure or the $Ca^{2+}$ is retained for a long period. What is the spiking for? Do we really know?

*Petersen:* In pancreatic acinar cells it is clear what $Ca^{2+}$ spiking is for. It is a mechanism for initiating secretion.

*Williams:* But secretion is slow.

*Petersen:* Secretion occurs every time there's a $Ca^{2+}$ spike (Maruyama & Petersen 1994): that's the point.

*Williams:* How long does the $Ca^{2+}$ spike last?

*Petersen:* A few seconds.

*Berridge:* By making capacitance measurements you can actually show that during each spike a certain number of vesicles fuse, and when there's another spike another few vesicles fuse. You can correlate quite precisely, in a single cell, the relationship between the $Ca^{2+}$ transient and a physiological end response. In smooth muscle cells, $Ca^{2+}$ spiking correlates with contractions. Therefore, some progress is being made concerning the function of spiking. However, one must not underestimate the difficulty of relating a spike to a clear physiological response in a single cell.

*Whitaker:* I'm very conscious that one has to look at what the $Ca^{2+}$ is doing. D. L. Taylor (Hahn et al 1990) has a fluorescent calmodulin and K. Török (Török & Whitaker 1994) has developed another one that has a rather better fluorescence enhancement on calcium binding. You can put these inside living cells and watch calmodulin being activated. We've done that in the sea urchin egg at fertilization and during the cell cycle (unpublished work). That's an important step forward. If you are actually looking at calmodulin activation, rather than $[Ca^{2+}]_i$, it tells you what the calmodulin targets in the cell are seeing, which is, after all, what is functionally important. Then you can try to explain why the calmodulin activation kinetics look like they do.

## References

Christophe JP, Frandsen EK, Conlon TP, Krishna G, Gardner JO 1976 Action of cholecystokinin, cholinergic agents and A-23187 on accumulation of guanosine 3′,5′-monophosphate in dispersed guinea-pig pancreatic acinar cells. J Biol Chem 251:4640–4645

Evans MG, Marty A 1986 Calcium-dependent chloride curents in isolated cells from rat lacrimal glands. J Physiol 378:437–460

Galione A, White A, Willmott N, Turner M, Potter BVL, Watson SP 1993 cGMP mobilizes intracellular $Ca^{2+}$ in sea urchin eggs by stimulating cylic ADP-ribose synthesis. Nature 365:456–459

Hahn KM, Waggoner A, Taylor DL 1990 A calcium-sensitive fluorescent analog of calmodulin based on a novel calmodulin-binding fluorophore. J Biol Chem 265: 20335–20345

Kasai H, Augustine GJ 1990 Cytosolic $Ca^{2+}$ gradients triggering unidirectional fluid secretion from exocrine pancreas. Nature 348:735–738

Lee HC, Aarhus R, Graeff R, Gurnack ME, Walseth TF 1994 Cyclic ADP ribose activation of the ryanodine receptor is mediated by calmodulin. Nature 370:307–309

Maruyama Y, Petersen OH 1994 Delay in granular fusion evoked by repetitive cytosolic $Ca^{2+}$ spikes in mouse pancreatic acinar cells. Cell Calcium 16:419–430

Petersen CCH, Petersen OH, Berridge MJ 1993 The role of endoplasmic reticulum calcium pumps during cytosolic calcium spiking in pancreatic acinar cells. J Biol Chem 268:22262–22264

Petersen OH 1982 Stimulus–excitation coupling in plasma membranes of pancreatic acinar cells. Biochim Biophys Acta 694:163–184

Petersen OH 1992 Stimulus–secretion coupling: cytoplasmic calcium signals and the control of ion channels in exocrine acinar cells. J Physiol 448:1–51

Philpott HG, Petersen OH 1979 Separate activation sites for cholecystokinin and bombesin on pancreatic acini. Pfluegers Arch Eur J Physiol 382:263–267

Thorn P, Lawrie AM, Smith PM, Gallacher DV, Petersen OH 1993 Local and global cytosolic $Ca^{2+}$ oscillations in exocrine cells evoked by agonists and inositol trisphosphate. Cell 74:661–668

Thorn P, Gerasimenko O, Petersen OH 1994 Cyclic ADP-ribose regulation of ryanodine receptors involved in agonist evoked cytosolic $Ca^{2+}$ oscillations in pancreatic acinar cells. EMBO (Eur Mol Biol Organ) J 13:2038–2043

Török K, Whitaker MJ 1994 Taking a long hard look at calmodulin's warm embrace. Bioessays 16:221–224

Wakui M, Osipchuk YV, Petersen OH 1990 Receptor-activated cytoplasmic $Ca^{2+}$ spiking mediated by inositol trisphosphate is due to $Ca^{2+}$-induced $Ca^{2+}$ release. Cell 63:1025–1032

# Pancreatic calcium waves and secretion

Haruo Kasai

*Department of Physiology, Faculty of Medicine, University of Tokyo, Hongo, Bunkyo-ku, Tokyo 113, Japan*

*Abstract.* Pancreatic acinar cells display stereotypic $Ca^{2+}$ waves resulting from $Ca^{2+}$ release from internal stores during stimulation. The $Ca^{2+}$ waves are initiated at the luminal pole, and, at high agonist concentrations, spread towards the basal pole. Two key mechanisms behind the generation of $Ca^{2+}$ waves have been identified. First, the $Ca^{2+}$ waves are composite, mediated by three distinct $Ca^{2+}$ release mechanisms with a polarized distribution: high-sensitivity inositol 1,4,5-trisphosphate ($InsP_3$) receptors at a small trigger zone (T zone) in the secretory granule area, $Ca^{2+}$-induced $Ca^{2+}$ release channels in the granular area and low-sensitivity $InsP_3$ receptors in the basal area. Second, $InsP_3$ can readily diffuse in the cytosol, whereas rises in cytosolic $Ca^{2+}$ concentration ($[Ca^{2+}]_i$) can be confined through strong buffering and sequestration of $Ca^{2+}$. $InsP_3$ is thus used as a long-range messenger to transmit agonist signals to the T zone, and $[Ca^{2+}]_i$ rises at the T zone are used as a local switch. These mechanisms enable preferential activation of the T zone, irrespective of localization of stimuli and agonist receptors. The secretion of enzymes and fluid is a direct consequence of $[Ca^{2+}]_i$ rises at the T zone. The $Ca^{2+}$ waves and oscillations probably boost the T zone functions.

*1995 Calcium waves, gradients and oscillations. Wiley, Chichester (Ciba Foundation Symposium 188) p 104–120*

The key feature of the $Ca^{2+}$ waves occurring in pancreatic acinar cells is their stereotypic appearance along polarized cellular structures (Kasai & Augustine 1990, Toescu et al 1992). Acinar cell plasma membranes are composed of two domains, the basolateral and luminal membranes, separated by a tight junction (Petersen 1992). Agonist receptors are localized in the basolateral membrane, whereas enzymes and ions are secreted from the luminal plasma membrane. The intracellular mediator for the secretion is cytosolic $Ca^{2+}$, which is released from internal stores by another second messenger, inositol 1,4,5-trisphosphate ($InsP_3$). $InsP_3$ is generated by the action of agonist receptors in the basolateral plasma membrane (Petersen 1992). We were therefore surprised when we found that the agonist-induced rise in cytosolic $Ca^{2+}$ concentration ($[Ca^{2+}]_i$) was initiated at the luminal pole and that the $Ca^{2+}$ waves spread towards the basal pole (Kasai & Augustine 1990). The direction of the $Ca^{2+}$ waves is opposite to the expected flow of second messengers generated by agonist receptors. This puzzling, but seemingly effective, $Ca^{2+}$ wave pattern has been reported not only in the pancreas, but also in lacrimal and parotid glands (Tan et al 1992, Toescu et al 1992, Gromada et al 1993), and could be a general phenomenon in exocrine glands and possibly in other secretory epithelia. Here, I introduce two fundamental mechanisms underlying the $Ca^{2+}$ waves and consider the functional implications of $Ca^{2+}$ waves and oscillations in exocrine gland acinar cells.

## Methods

Acinar cells from rat pancreas (or mouse lacrimal gland for Fig. 3) were dissociated enzymically and plated on a thin (0.12 mm) glass cover slip (Kasai et al 1993). The cells were loaded with fura red acetoxymethyl ester (Molecular Probes, Eugene, OR), and fluorescent images were obtained using a confocal laser-scanning microscope (BioRad MRC600). Until now, the T zone (a small zone in the granular area from which $Ca^{2+}$ waves are initiated; see below) has been clearly visualized only in our laboratory. I ascribe this to our choice of confocal optics and oil immersion objective (Olympus, 63 ×, numerical aperture 1.4). The spatial resolution of our confocal imaging is depicted in Fig. 1, where intensity of fluorescence at the edge of an acinar cell is shown quantitatively. A half-decay of the fluorescence was achieved at 0.5 $\mu$m in the plane of focus (Fig. 1B,D) and 2 $\mu$m along the $z$-axis (Fig. 1C,E). With this resolution, we could sometimes identify dark spots corresponding to individual secretory granules. The secretory granules of acinar cells never take up the dye, unlike those of mast cells. The granular area thus appeared dark and could be identified reliably in the fluorescent image. We normally focused on the plane in the middle of the granular area. $Ca^{2+}$ images were obtained by taking the ratio of the fluorescence image during stimulation to that before stimulation (Kasai et al 1993). We studied the $Ca^{2+}$ waves occurring within a relatively short period. The patterns of the $Ca^{2+}$ waves did not depend on the external $[Ca^{2+}]$, indicating that $[Ca^{2+}]_i$ rises are mostly due to $Ca^{2+}$ release from the intracellular stores in our experimental conditions.

## Calcium waves in the three distinct areas of the acinar cell

The pancreatic $Ca^{2+}$ waves appeared as a homogeneous spread of increases in $[Ca^{2+}]_i$ from the luminal to the basal pole when acinar cells were stimulated with a high concentration (10 $\mu$M) of acetylcholine (Fig. 2C; Kasai & Augustine 1990, Toescu et al 1992, Nathanson et al 1992). At lower agonist concentrations (0.2–1 $\mu$M), however, the presence of subcellular compartments became apparent (Kasai et al 1993, Thorn et al 1993). With low acetylcholine concentrations, the $Ca^{2+}$ waves did not actively spread from the granular area (luminal pole)

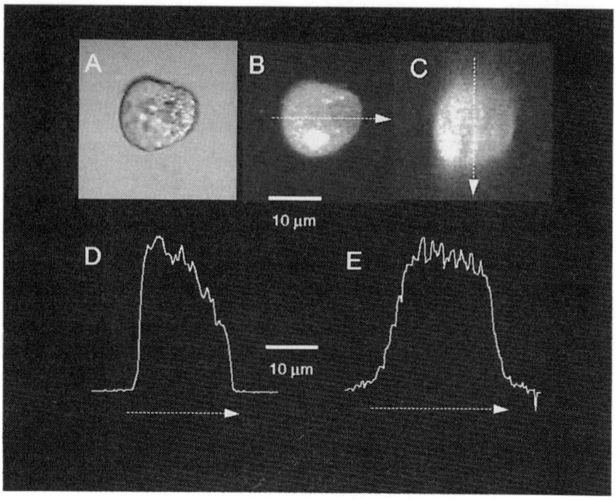


FIG. 1. Spatial resolution of the confocal laser-scanning microscope. (A) Non-confocal transmission image of a pancreatic acinar cell. The basal end is to the left (including the granular area), the luminal end to the right. (B) Confocal fluorescent image of the same cell stained with fura red acetoxymethyl ester. (C) Confocal image of the same cell obtained by scanning in the vertical direction (i.e., the *z*-axis) through the arrow shown in (B). (D) and (E) show intensity of fluorescence along the dotted arrows in (B) and (C), respectively. Fluorescence is represented by an arbitrary scale.

into the basal area (Figs 2A and 3C). Moreover, there seemed to be a small zone within the granular area from which the $Ca^{2+}$ waves were always initiated and at which the $[Ca^{2+}]_i$ rises were highest throughout the single $Ca^{2+}$ spike (see Figs 1 and 2 of Kasai et al 1993). Figure 3C shows that the trigger zone, or T zone, could also be detected in lacrimal acinar cells. Because the $Ca^{2+}$ waves in the basal area were too rapid to be analysed with two-dimensional imaging in our system, we used the line-scanning mode in which one line crossing a cell can be scanned every 4 ms. Figure 4B shows such a quantitative analysis of the spread of $Ca^{2+}$ within the granular area of a pancreatic acinar cell. This indicates that the $Ca^{2+}$ waves were triggered simultaneously within the small zone at the luminal edge of the granular area. The position of the T zone visualized by the line-scanning method correlated well with that in the two-dimensional image of the same cell (see Fig. 3A of Kasai et al 1993). The global $Ca^{2+}$ waves induced by a high concentration (10 $\mu$M) of acetylcholine were shown to occur in three steps (Fig. 4E): first, $Ca^{2+}$ waves were induced at the T zone, then they spread to other parts of the granular area, and finally reached the basal area. Propagation of the $Ca^{2+}$ waves was always faster in the granular area than in the basal area. The maximum rates of $[Ca^{2+}]_i$ increase were highest at the T zone, intermediate in the rest of the granular area and lowest in the

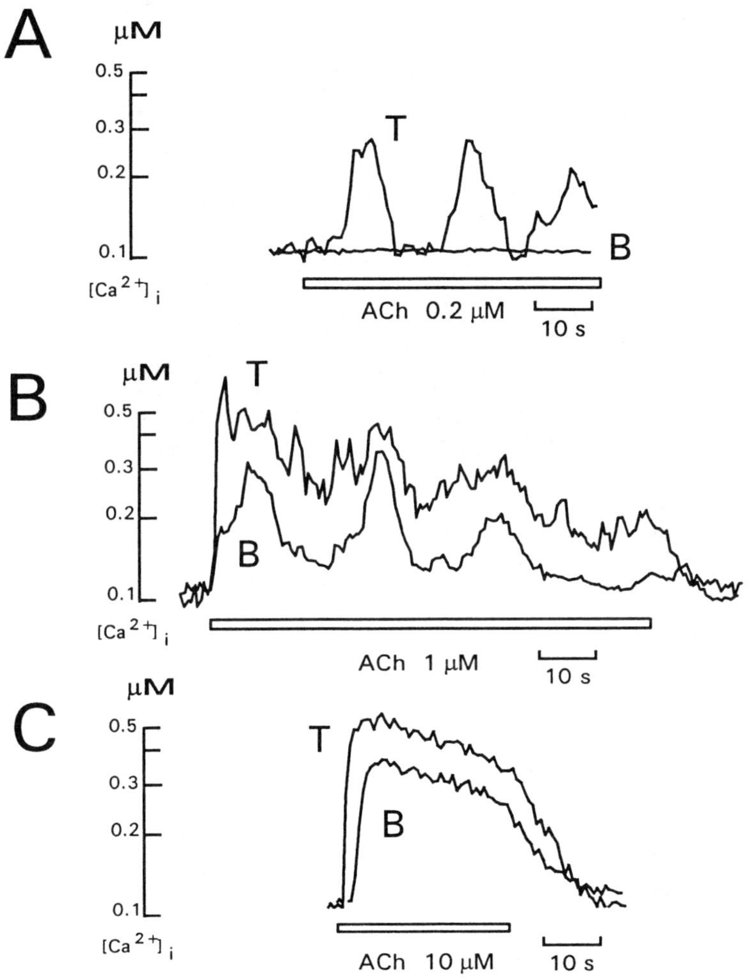


FIG. 2. Agonist concentration dependence of spatial patterns of $[Ca^{2+}]_i$ rises. Increases in $[Ca^{2+}]_i$ evoked with three different concentrations of acetylcholine (ACh) in rat pancreatic acinar cells are shown. Traces marked T show $[Ca^{2+}]_i$ at the T zone, and those marked B $[Ca^{2+}]_i$ in the basal area. (A) and (C) were recorded from the same cell. Adapted from Figs 1 and 2 of Kasai et al (1993), with permission (copyright Cell Press).

basal area (Fig. 4C,F). Table 1 summarizes the properties of $Ca^{2+}$ waves in the three areas of the pancreatic acinar cell.

To investigate the molecular basis of the composite nature of the $Ca^{2+}$ waves, we injected $InsP_3$ directly from a patch pipette (see Fig. 6 of Kasai et al 1993). The $InsP_3$ injection (10 μM) produced $[Ca^{2+}]_i$ changes similar to

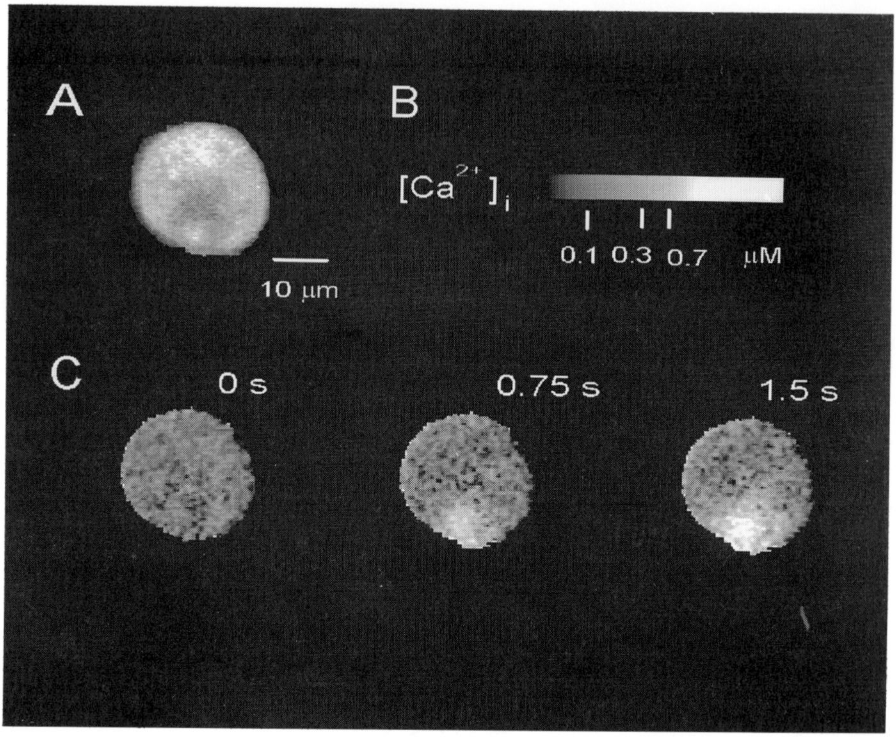


FIG. 3. $[Ca^{2+}]_i$ rises at the trigger (T) zone. (A) A fluorescent image of a mouse lacrimal acinar cell stained with fura red. The dark area at the bottom correlates precisely with the secretory granule area, as in the pancreatic acinar cell. (B) The $[Ca^{2+}]_i$ scale. (C) $Ca^{2+}$ images of the same cell during application of 1 μM acetylcholine.

those produced by a low concentration of acetylcholine, except that there was an additional artefactual $[Ca^{2+}]_i$ rise at the injection site. A sharp $[Ca^{2+}]_i$ rise was detected in a small area at the luminal edge of the acinar cell, even when $InsP_3$ was injected at the basal pole. The $[Ca^{2+}]_i$ rises induced by $InsP_3$ were confined to the granular area, as they were with stimulation by a low agonist concentration. These findings were the key to understanding the 'backwards' spread of $Ca^{2+}$ waves in pancreatic cells. They indicate, first, that the T zone is more sensitive to $InsP_3$ than the rest of the cell and, second, that $InsP_3$ can readily diffuse across the cell, whereas the diffusion of $Ca^{2+}$ can be limited. Values of the diffusion constants ($D$) for buffered $Ca^{2+}$ and $InsP_3$ have been directly estimated in the cytosol of *Xenopus* oocytes to be 13 and 283 $\mu m^2/s$, respectively (Allbritton et al 1992). Assuming the lifetime ($\tau$) of both buffered $Ca^{2+}$ and $InsP_3$ to be one second, we could calculate diffusion lengths ($\sqrt{D\tau}$) for the two messengers as 3.6 and 17 μm, respectively. The diffusion length

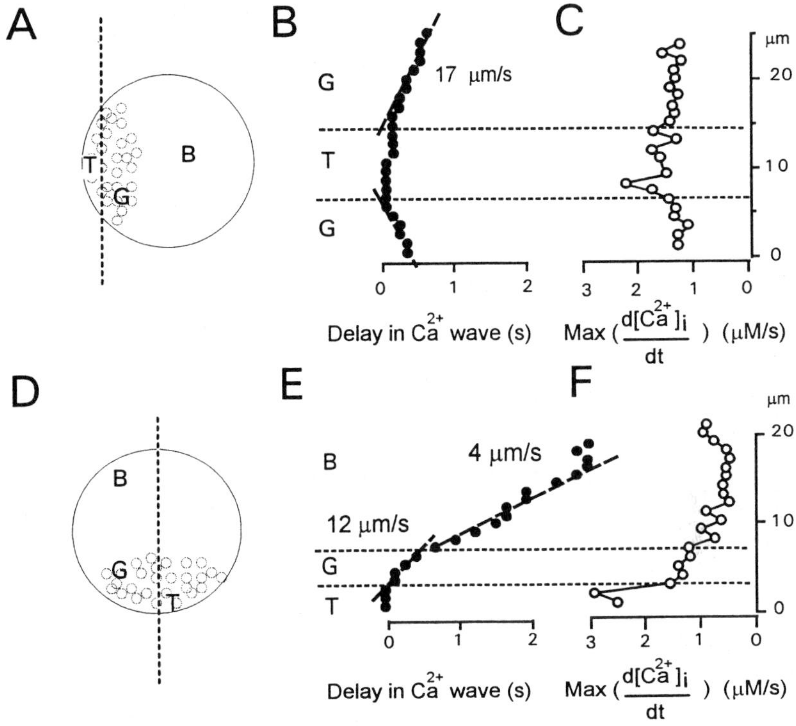


FIG. 4. Speeds of $Ca^{2+}$ waves and maximum rates of $[Ca^{2+}]_i$ increases induced by application of 10 μM acetylcholine. (B) and (E) show the delay in the onset of $Ca^{2+}$ waves plotted against position (dashed line) in the pancreatic acinar cells shown schematically in (A) and (D). The onset was defined as the time at which the half-maximal $[Ca^{2+}]_i$ value was reached at each position. T, G and B denote the trigger (T) zone, the granular area outside the T zone and the basal area, respectively. Dashed oblique lines show least squares fit of points within each area; the slopes of these lines represent the speeds of the $Ca^{2+}$ waves. (C) and (F) show the maximum rate of the $[Ca^{2+}]_i$ rise at each position in the two acinar cells. B, C, E and F reproduced from Kasai et al (1993), with permission (copyright Cell Press).

indicates the distance over which messengers exponentially decay in steady-state conditions (Kasai & Petersen 1994). The theoretical estimates of the range of actions of the two second messengers agree well with the observed differences in the ease with which the two messengers diffuse across the cell. This further supports the idea that $InsP_3$ acts as a long-range messenger to transmit the agonist signal to the opposite side of the cell, whereas $Ca^{2+}$ is used as a local switch at the T zone.

Because $Ca^{2+}$ may act as a second messenger triggering $Ca^{2+}$ release from the intracellular stores, we tested the effects of injecting $Ca^{2+}$ through a patch pipette. When $Ca^{2+}$ was injected into the basal pole, slow loading of $Ca^{2+}$ resulted throughout the cells (see Fig. 7A of Kasai et al 1993), reflecting passive loading of $Ca^{2+}$ from the injection patch pipette. In contrast, when $Ca^{2+}$ was injected into the granular area, regenerative $Ca^{2+}$ spikes were triggered throughout the granule area (see Fig. 7C of Kasai et al 1993). The $Ca^{2+}$ wave did not seem to spread actively into the basal area, consistent with our finding that injection of $Ca^{2+}$ in the basal area did not trigger $Ca^{2+}$ release. Our results therefore suggest that $Ca^{2+}$-induced $Ca^{2+}$ release mechanisms are distributed homogeneously in the granular area. Despite the apparent presence of $Ca^{2+}$-induced $Ca^{2+}$ release mechanisms, the acinar cells showed no or only weak responses to a classical agonist of $Ca^{2+}$-induced $Ca^{2+}$ release, caffeine.

## Molecular correlates in the three distinct areas

The simplest hypothesis to account for all these observations is that three distinct types of $Ca^{2+}$ channel are distributed differently in three subcellular compartments of the acinar cell, as shown in Table 1, with the high-sensitivity $InsP_3$ receptor ($InsP_3R$) being localized in the T zone, the $Ca^{2+}$-induced $Ca^{2+}$ release channel in the granular area and the low-sensitivity $InsP_3R$ in the basal area or throughout the cell. This hypothesis would explain regional differences in sensitivities to acetylcholine and spatiotemporal patterns of acetycholine-induced $Ca^{2+}$ waves, irrespective of the location of agonist receptors.

The high sensitivity to $InsP_3$ at the T zone might result from a higher density of $InsP_3Rs$ in the T zone. This is relevant because the $InsP_3R$ has its own $Ca^{2+}$-induced $Ca^{2+}$ release mechanisms, and positive feedback effects of $Ca^{2+}$ might be greater, making the $InsP_3Rs$ at the T zone more sensitive to $InsP_3$, if

**TABLE 1 Characteristics of the three areas of the pancreatic acinar cell**

| | *T zone* | *Granular area outside T zone* | *Basal area* |
|---|---|---|---|
| Onset of $Ca^{2+}$ wave | 1st | 2nd | 3rd |
| Speed of $Ca^{2+}$ wave | Synchronous | 21 $\mu$m/s | 8 $\mu$m/s |
| Sensitivity to acetylcholine | High | High | Low |
| Sensitivity to $InsP_3$ | High | Low | Low |
| Sensitivity to $Ca^{2+}$ | High | High | Low |
| High-sensitivity $InsP_3$ receptors | + | – | – |
| $Ca^{2+}$-induced $Ca^{2+}$ release channel | + | + | – |
| Low-sensitivity $InsP_3$ receptors | (+) | + | + |

T zone, trigger zone; $InsP_3$, inositol 1,4,5-trisphosphate.

the density of $InsP_3Rs$ recently in the T zone is higher than in other areas. It has been reported recently that immunoreactivity to the type 3 $InsP_3R$ is selectively localized at the luminal pole of the granular area (Nathanson et al 1994). It is very likely, though not proven, that the type 3 $InsP_3R$ is the high-sensitivity $InsP_3R$, and the zone identified by the antibody correlates with the T zone. With this antibody, we may soon be able to pinpoint the organelles carrying the $Ca^{2+}$ stores at the T zone. To investigate receptor density, however, we must also identify other types of $InsP_3R$ in other parts of the acinar cell.

The $Ca^{2+}$-induced $Ca^{2+}$ release found in the granular area may be mediated by $InsP_3Rs$, because they have a $Ca^{2+}$-induced $Ca^{2+}$ release mechanism. However, the following lines of evidence are inconsistent with this possibility. First, the patterns of $[Ca^{2+}]_i$ changes induced by $Ca^{2+}$ injection were different from those induced by $InsP_3$ or acetylcholine. Acetylcholine- and $InsP_3$-induced $[Ca^{2+}]_i$ rises were always highest at the T zone (Fig. 2), whereas $Ca^{2+}$-induced $[Ca^{2+}]_i$ rises occurred more homogeneously throughout the granular area (Fig. 7 of Kasai et al 1993). Second, the prior injection of heparin, a blocker of the $InsP_3R$, did not prevent the $Ca^{2+}$-induced $Ca^{2+}$ release (Kasai et al 1993, Wakui et al 1990). However, this does not exclude the involvement of $InsP_3Rs$ in the granule area that are weakly sensitive to $InsP_3$ and to heparin but more sensitive to $Ca^{2+}$. Another important question that we cannot answer at this moment is whether these hypothetical $Ca^{2+}$ release channels are involved in any way with the acetylcholine-induced $Ca^{2+}$ waves in the granular area. To solve these problems, we need selective antagonists for this type of $Ca^{2+}$-induced $Ca^{2+}$ release process.

Several other mechanisms may also affect the $Ca^{2+}$ waves. The $Ca^{2+}$-buffering capacity of the cell may differ between the three areas, with, for example, the buffering at the T zone being weakest and that in the basal area highest. If this is true, then $Ca^{2+}$ gradients similar to those evoked by acetylcholine should be formed when $[Ca^{2+}]_i$ is raised by any other means, such as by injection of $Ca^{2+}$ into the basal area or application of A23187 or thapsigargin. With all three treatments, $[Ca^{2+}]_i$ rises slowly and homogeneously throughout acinar cells (see Fig. 7 of Kasai et al 1993; Toescu et al 1992, unpublished observations). This excludes the involvement of pumps in the formation of $Ca^{2+}$ gradients. However, it is still possible that there are gradients of $Ca^{2+}$-binding proteins which may contribute to rapid $Ca^{2+}$ spikes and waves. Also, the granular area may have less cytosolic space because it is packed with secretory granules, as demonstrated by fura red staining (Fig. 1A). Hence, the same amount of $Ca^{2+}$ release would induce larger rises in $[Ca^{2+}]_i$ in the granular area than in the basal area. This space factor might speed up the propagation of $Ca^{2+}$ waves in the granular area.

We are now faced with the question of which of the organelles in the three areas are the $Ca^{2+}$ stores. Most simply, our results might imply that rough endoplasmic reticulum is the $Ca^{2+}$ store carrying the low-affinity $InsP_3R$ (Fig. 5).

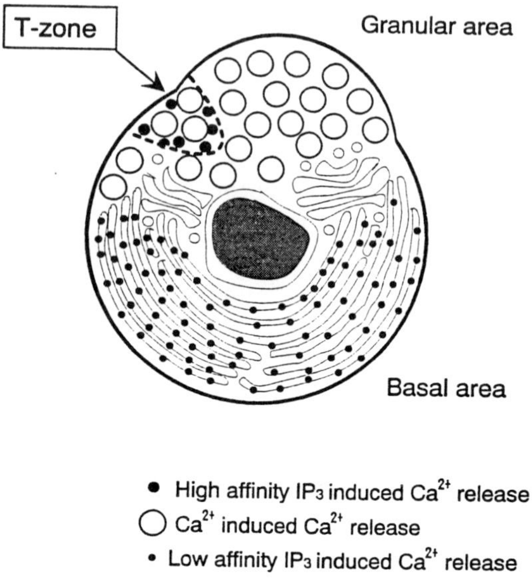


FIG. 5. Hypothetical distribution of $Ca^{2+}$ release channels in an exocrine acinar cell.

Rough endoplasmic reticulum is known to have $Ca^{2+}$ pumps and $InsP_3$-binding capacity (Sharp et al 1992). The lower sensitivity of $InsP_3Rs$ in the basal area might then be due to their immaturity. Secretory granules might contain $Ca^{2+}$-induced $Ca^{2+}$ release channels and/or high-affinity $InsP_3Rs$ (Fig. 5). This hypothesis is particularly attractive, because $Ca^{2+}$ sensors for exocytosis have been proposed to have a low affinity and to require $Ca^{2+}$ domains close to open $Ca^{2+}$ channels, in neurons and in endocrine cells. The involvement of $Ca^{2+}$ domains has also been suggested in exocrine acinar cells (Maruyama et al 1993). The secretory granules would be most effectively placed in the $Ca^{2+}$ domains, if they have $Ca^{2+}$ release channels in their membranes.

**Multiple mechanisms for calcium spikes and calcium oscillations**

Each of the three areas in the acinar cell appears to have its own mechanism for producing $Ca^{2+}$ spikes and oscillations; the mechanisms could be differentially induced depending on agonist concentrations. At a low agonist concentration, the triggering $Ca^{2+}$ spike at the T zone is followed by a $Ca^{2+}$ spike in the granular area, and these composite spikes are repeated during an oscillation (Fig. 2A). The $Ca^{2+}$-induced $Ca^{2+}$ release mechanism in the granular area may partly be involved in $Ca^{2+}$ spiking and the repetitive appearance of composite $Ca^{2+}$ spikes. If so, the $Ca^{2+}$ oscillations may fit into the classical two-pool model (Goldbeter et al 1990), which assumes there are two $Ca^{2+}$ stores, one sensitive to $InsP_3$ and the other to $Ca^{2+}$. In addition to the mechanisms proposed by the two-pool model, the T zone seems to have its own $Ca^{2+}$ spiking mechanism, because at an intermediate agonist concentration the $Ca^{2+}$ spikes confined to the T zone are induced repetitively, whereas the $Ca^{2+}$ release mechanism in other parts of the granular area is refractory (Fig. 2B). This suggests that the $InsP_3R$ itself has a mechanism for initiating $Ca^{2+}$ spiking and oscillation. This also applies to the low-sensitivity $InsP_3R$ in the basal area, because $Ca^{2+}$ waves can actively spread into the basal area at a high agonist concentration.

It is not known how $InsP_3Rs$ can generate $Ca^{2+}$ spikes and oscillations by themselves. The prevailing hypothesis assumes fast positive and slower negative feedback effects of $Ca^{2+}$ (Berridge 1993, Kasai et al 1993). Such effects have been reported with $InsP_3Rs$ in some preparations (Iino 1990, Parker & Ivorra 1990, Bezprozvanny et al 1991, Finch et al 1991). Modelling studies also show that $Ca^{2+}$ spikes, waves and oscillations can be reconstituted on the basis of this hypothesis and other assumptions (Atri et al 1993). The key to this hypothesis is that the negative feedback effect must be slower than the positive feedback effect, otherwise the negative feedback effect cancels out that of the positive feedback. This point, however, has been challenged by an experiment on skinned smooth muscle with caged compounds in which $Ca^{2+}$ had immediate negative feedback effects on the $InsP_3R$ (Iino & Endo 1992). Moreover, we have too little information on the homeostasis of $Ca^{2+}$ and $InsP_3$ in the resting state of the cell to validate the models.

## The functional significance of calcium waves and oscillations

The $[Ca^{2+}]_i$ rises at the T zone can directly trigger the major functions of the acinar cell. They are sufficient to activate $Cl^-$ channels, presumably located in the luminal membrane (Kasai & Augustine 1990, Maruyama et al 1993, Thorn et al 1993). The $Cl^-$ channels push $Cl^-$ into the lumen, leading to fluid secretion. Furthermore, capacitance measurements indicate that the triggering $Ca^{2+}$ rises at the T zone are enough to induce exocytosis (Maruyama et al 1993). Therefore, the high-sensitivity $InsP_3R$ is strategically localized at the T zone to regulate the acinar cell's functions most effectively.

We wonder, then, what the role of $Ca^{2+}$ waves and $Ca^{2+}$ oscillations is. One hypothesis is that $Ca^{2+}$ triggers two distinct events in the luminal and basolateral membranes and that the sequential activation of these processes is necessary for proper cellular functioning. This may be the case for fluid secretion. At the luminal membrane, $Ca^{2+}$ selectively activates $Cl^-$ channels, pushing $Cl^-$ out into the lumen because of the resting potential. In contrast, at the basolateral membrane, $Ca^{2+}$ activates both $Cl^-$ channels and cation channels and $Cl^-$ is therefore pulled from the blood into the acinar cell because of depolarization induced by the cation channels. These two phases occur repeatedly during the $Ca^{2+}$ oscillation and a sustained flow of $Cl^-$ is induced (Kasai & Augustine 1990, Petersen et al 1994). For direct proof of this push–pull model, determination of the precise localization of $Ca^{2+}$-activated channels, their sensitivity to $Ca^{2+}$ and their localization relative to the three $Ca^{2+}$ stores is necessary. The push–pull model can easily account for the otherwise peculiar biphasic dose dependence of fluid secretion (Habara 1980): fluid secretion is maximum at 0.3 $\mu$M $Ca^{2+}$, at which concentration the rate of oscillation is highest, and it is suppressed at higher concentrations, where $[Ca^{2+}]_i$ rises tend to be sustained rather than oscillatory. One can speculate further that the pull phase is not necessary for the most energy-efficient fluid secretion, because $Cl^-$ ions can also be taken up from the blood via the Na–K–Cl symporter in the basal membrane. At low agonist concentration, local $Ca^{2+}$ spikes are triggered at the T zone, and $Cl^-$ may be transported into the lumen by a push–push mechanism. For the selective push phase, the $Ca^{2+}$ rises should be intermittent, to prevent spread of the $[Ca^{2+}]_i$ rises through saturation of the $Ca^{2+}$ buffers and uptake in the secretory granule area.

Another hypothesis proposed to explain the role of $Ca^{2+}$ oscillations is that the $[Ca^{2+}]_i$ rise needs to be intermittent for the appropriate functioning of the cell. This may be so for exocytosis. Capacitance measurements show that exocytosis stops shortly after $[Ca^{2+}]_i$ rises (Maruyama et al 1993), possibly because of depletion of the releasable pool of secretory granules. If the priming of the movement of the secretory granule to the release site requires some intermission between $[Ca^{2+}]_i$ rises, a $Ca^{2+}$ oscillation would be the most effective way of regulating exocytosis. This point should be clarified by combined $[Ca^{2+}]_i$ and capacitance meaurements in the near future.

## Perspectives

The acinar cell has been and will continue to be an attractive preparation for the study of signal transduction mechanisms for many reasons. First, distinct organelles are distributed in a highly polarized manner, making the acinar cell an ideal preparation for studying the functioning of organelles (Palade 1975). We believe that exocrine acinar cells will provide us with a good opportunity to identify functional $Ca^{2+}$ stores, which have never been specified in any type of cell other than muscle cells. The elucidation of the $Ca^{2+}$ stores will be an important step for research into $Ca^{2+}$ signalling, as well as research into sorting of proteins and vesicular trafficking. This is because the amount of $Ca^{2+}$ within the organelles may regulate their functioning (Kuznetsov et al 1992, Gordon et al 1993). Imaging $[Ca^{2+}]$ in organelles of the acinar cell will be the key technique to assess these possibilities directly. Second, the patterns of $Ca^{2+}$ signals have direct links to the acute functions of the acinar cells, namely exocytosis and $Cl^-$ secretion. The molecular mechanisms of exocytosis and fluid secretion have been a major topic of study in a variety of cells. The acinar cell is a good model with which to investigate precisely the dynamics with which $Ca^{2+}$ regulates these cellular functions. Finally, there are reasons to assume the presence of novel signalling mechanisms other than those involving $Ca^{2+}$ and $InsP_3$. For example, one of the types of agonist on which the patterns of $[Ca^{2+}]_i$ rises depend does not even require $InsP_3$ (Saluja et al 1992). The combination of modern biophysical techniques and the new tools provided by molecular biology will certainly contribute to the disclosure of new cellular mechanisms and the further understanding of the functions of acinar cells, as well as other cells.

*Acknowledgements*

This work was supported by Grants-in-Aid of the Ministry of Education, Culture and Science of Japan and a research grant from the Human Frontier Science Program to H. K.

## References

Allbritton NL, Meyer T, Stryer L 1992 Range of messenger action of calcium ion and inositol 1,4,5-trisphosphate. Science 258:1812–1815

Atri A, Amundson J, Clapham D, Sbeyd J 1993 A single-pool model for intracellular calcium oscillations and waves in the *Xenopus laevis* oocyte. Biophys J 65:1727–1739

Berridge MJ 1993 Inositol trisphosphate and calcium signalling. Nature 361:315–325

Bezprozvanny I, Watras J, Ehrlich BE 1991 Bell-shaped calcium-response curves of $Ins(1,4,5)P_3$- and calcium-gated channels from endoplasmic reticulum of cerebellum. Nature 351:751–754

Finch EA, Turners TJ, Goldin SM 1991 Calcium as a coagonist of inositol 1,4,5-trisphosphate-induced calcium release. Science 252:443–446

Goldbeter A, Dupont G, Berridge MJ 1990 Minimal model for signal-induced $Ca^{2+}$ oscillations and for their frequency encoding through protein phosphorylation. Proc Natl Acad Sci USA 87:1461–1465

Gordon PB, Holen I, Fosse M, Rotnes JS, Seglen PO 1993 Dependence of hepatocytic autophagy on intracellularly sequestered calcium. J Biol Chem 268:26107–26112

Gromada J, Jorgensen TD, Tritsaris K, Nauntofte B, Dissing S 1993 $Ca^{2+}$ signalling in exocrine acinar cells: the diffusional properties of cellular inositol 1,4,5-trisphosphate and its role in the release of $Ca^{2+}$. Cell Calcium 14:711–723

Habara Y 1980 Influences of ionic environments on ACh-induced secretory responses in isolated perfused pancreas of rats. Jpn J Physiol 30:561–574

Iino M 1990 Biphasic $Ca^{2+}$ dependence of inositol 1,4,5-trisphosphate-induced Ca release in smooth muscle cells of the guinea pig *Taenia caeci*. J Gen Physiol 95:1103–1122

Iino M, Endo M 1992 Calcium-dependent immediate feedback control of inositol 1,4,5-trisphosphate induced $Ca^{2+}$ release. Nature 360:76–78

Kasai H, Augustine GJ 1990 Cytosolic $Ca^{2+}$ gradients triggering unidirectional fluid secretion from exocrine pancreas. Nature 348:735–738

Kasai H, Petersen OH 1994 Spatial dynamics of second messengers: $InsP_3$ and cAMP as long range and associative messengers. Trends Neurosci 17:95–101

Kasai H, Li YX, Miyashita Y 1993 Subcellular distribution of $Ca^{2+}$ release channels underlying $Ca^{2+}$ waves and oscillations in exocrine pancreas. Cell 74:669–677
Kuznetsov G, Brostrom MA, Brostrom CO 1992 Demonstration of a calcium requirement for secretory protein processing and export. J Biol Chem 267:3932–3939
Maruyama Y, Inooka G, Li Y, Miyashita Y, Kasai H 1993 Agonist-induced localized $Ca^{2+}$ spikes directly triggering exocytotic secretion in exocrine pancreas. EMBO (Eur Mol Biol Organ) J 12:3017–3022
Nathanson MH, Padfield PJ, O'Sullivan AJ, Burgstahler AD, Jamieson JD 1992 Mechanism of $Ca^{2+}$ wave propagation in pancreatic acinar cells. J Biol Chem 267:18118–18121
Nathanson MH, Fallon MB, Padfield PJ, Maranto AR 1994 Localization of the type 3 inositol 1,4,5-trisphosphate receptors in the $Ca^{2+}$ wave trigger zone of pancreatic acinar cells. J Biol Chem 269:4693–4696
Palade G 1975 Intracellular aspects of the process of protein synthesis. Science 189:347–358
Parker I, Ivorra I 1990 Inhibition by $Ca^{2+}$ of inositol trisphosphate-mediated $Ca^{2+}$ liberation: a possible mechanism for oscillatory release of $Ca^{2+}$. Proc Natl Acad Sci USA 87:260–264
Petersen OH 1992 Stimulus–secretion coupling: cytoplasmic calcium signals and the control of ion channels in exocrine acinar cells. J Physiol 448:1–51
Petersen OH, Petersen CCH, Kasai H 1994 Calcium and hormone action. Annu Rev Physiol 56:297–319
Saluja AK, Dawra RK, Lerch MM, Steer ML 1992 CCK-JMV-180, an analog of cholecystokinin, releases intracellular calcium from an inositol trisphosphate-independent pool in rat pancreatic acini. J Biol Chem 267:11202–11207
Sharp AH, Snyder SH, Nigram SK 1992 Inositol 1,4,5-trisphosphate receptors. J Biol Chem 267:7444–7449
Tan YP, Marty A, Trautmann A 1992 High density of $Ca^{2+}$-dependent K and Cl channels on the luminal membrane of lacrimal acinar cells. Proc Natl Acad Sci USA 89:11229–11233
Thorn P, Lawrie AM, Smith PM, Gallacher DV, Petersen OH 1993 Local and global cytosolic $Ca^{2+}$ oscillations in exocrine cells evoked by agonists and inositol trisphosphate. Cell 74:661–668
Toescu EC, Lawrie AM, Petersen OH, Gallacher DV 1992 Spatial and temporal distribution of agonist-evoked cytoplasmic $Ca^{2+}$ signals in exocrine acinar cells analysed by digital image microscopy. EMBO (Eur Mol Biol Organ) J 11:1623–1629
Wakui M, Osipchuk YV, Petersen OH 1990 Receptor-activated cytoplasmic $Ca^{2+}$ spiking mediated by inositol trisphosphate is due to $Ca^{2+}$-induced $Ca^{2+}$ release. Cell 63:1025–1032

## DISCUSSION

*Jaffe:* Figure 5 suggested that there is endoplasmic reticulum (ER) only in the basal area, with none in the secretory granule area. Did you mean that?

*Kasai:* No.

*Jaffe:* So does that mean that the $Ca^{2+}$ that's released even in the apical area comes from the ER?

*Kasai:* That is still possible.

*Smith:* But didn't you also propose that $Ca^{2+}$ is released from the granules?

*Kasai:* Yes, but by a different mechanism; $Ca^{2+}$-induced $Ca^{2+}$ release.

*Putney:* I admit that we do our experiments on cells cultured overnight, which for some reason may have lost the $Ca^{2+}$-induced $Ca^{2+}$ release component, but all the thapsigarin-sensitive stores in all the epithelial cells we've looked at are also $InsP_3$ sensitive. I doubt that a $Ca^{2+}$-induced $Ca^{2+}$ release mechanism, in cells that have one, exists in an anatomically distinct structure. It's more likely that there are components of the ER that are responsive to either $InsP_3$ or a $Ca^{2+}$-induced $Ca^{2+}$ release mechanism.

*Kasai:* I agree, that possibility cannot be excluded.

*Berridge:* Perhaps we should try to use some term other than $Ca^{2+}$-induced $Ca^{2+}$ release, because it implies release from ryanodine receptors. Perhaps we should say ryanodine receptor-sensitive $Ca^{2+}$-induced $Ca^{2+}$ release.

*Putney:* Or $InsP_3$-independent $Ca^{2+}$-induced $Ca^{2+}$ release. My point is that there is probably only one type of $Ca^{2+}$ store but multiple release mechanisms.

*Jaffe:* Dr Kasai, what is the basis for your idea that $Ca^{2+}$ is released from the secretory granules?

*Kasai:* $Ca^{2+}$-induced $Ca^{2+}$ release occurs in the secretory granule area, exactly where the secretory granules are. It is also likely that other organelles associated with the secretory granules are the $Ca^{2+}$ stores for the $Ca^{2+}$-induced $Ca^{2+}$ release channels.

*Meyer:* Your model then makes a simple prediction: if you add an agonist that releases only the ryanodine-sensitive $Ca^{2+}$ stores, this should correlate more with the capacitance change than with $Cl^-$ secretion. I don't know whether or not it's possible for you to look at the same time at the $Cl^-$ conductance, at capacitance and at $Ca^{2+}$ and its localization, but doing so could prove whether or not the $Ca^{2+}$-induced $Ca^{2+}$ release mechanism is responsible for secretion from the granules.

*Lechleiter:* There is a report of a $Ca^{2+}$-ATPase located in the granules that's different from the sarco/endoplasmic reticulum $Ca^{2+}$-ATPase in that it's not sensitive to thapsigargin and is not located in the plasma membrane (Fasolato et al 1991).

*Jaffe:* It is certainly true that secretory granules typically have an extremely high [$Ca^{2+}$], which is surely established by a pump. What is new is the suggestion that $Ca^{2+}$ can release $Ca^{2+}$ from the granules.

*Berridge:* That suggestion has been around for a long time.

*Galione:* Yoo & Albanesi (1990) showed that $InsP_3$ released $Ca^{2+}$ from secretory chromaffin granules in a preparation claimed to be clean and to contain no ER markers. The $Ca^{2+}$ release was as significant as that released by $InsP_3$ from the microsomal fraction. We've tried to repeat that, but have been unable to see any $InsP_3$ sensitivity in bovine adrenal chromaffin granules.

*Putney:* Exocrine cell granules are different from chromaffin granules. $Ca^{2+}$ in exocrine granules has been 'condensed'. The $Ca^{2+}$ pump present in mature zymogen granules is probably vestigial. The $Ca^{2+}$ stored in exocrine gland granules is not, for example, releasable by ionophores, so the majority of it at least is not diffusible. It seems unlikely that that $Ca^{2+}$ could be dynamically regulated.

*Nathanson:* The ultimate proof of whether or not there is ER in this region of the cell will rely on electron microscopic studies. We have been unable so far to use the anti-$InsP_3$ receptor antibody for electron microscopic localization. I could only make guesses at this stage.

*Berridge:* Have you tested antibodies to other $InsP_3$ receptors in the more basal zone?

*Nathanson:* We've used antibodies that recognize the type 1 receptor (which may also recognize the type 2 receptor) and haven't been able to find evidence for $InsP_3$ receptors anywhere else in the acinar cell. It is possible that these receptors are present but in very small numbers, or there may be other subtypes localized there, or there may not be $InsP_3$-sensitive stores there at all.

*Irvine:* Your work has suggested that although the type 3 $InsP_3$ receptors are in the same region as the secretory granules, they cover only part of it, and there are regions of the secretory granule area that do not contain any receptors. That suggests to me that the secretory granules aren't the organelles containing the type 3 $InsP_3$ receptors.

*Nathanson:* I would agree with that. We are sure that the area labelled is at the extreme apex, almost more apical than the secretory granules.

*Thomas:* One hypothesis is that high-affinity and low-affinity $InsP_3$ receptors are present in different parts of the cell. Is anything known about the relative affinity of these two types of $InsP_3$ receptor?

*Nathanson:* I'm not sure that's been characterized for the type 3 receptor.

*Meyer:* A different density should give the same result as different $InsP_3$ affinities.

*Nathanson:* We've also made measurements of $Ca^{2+}$ waves in rat pancreatic acinar cells by confocal microscopy. We've looked at these waves in acinar cells within intact acini, as well as in single acinar cells. We've noticed two differences in the responses in these two preparations. First, the wave speed is about three times higher in cells within intact acini. Second, even at very low doses of agonist stimulation, waves in cells in intact acini always seem to propagate all the way across the cell in a non-diminishing fashion. In single acinar cells Dr Kasai has shown that the waves don't propagate all the way across the cell, and even at higher agonist concentrations your results suggested that the wave diminishes towards the basal area.

*Kasai:* That difference may result from the activity of phospholipase being lower in the isolated cell so that less $InsP_3$ can be made.

*Nathanson:* With reference to the functional implications of these observations, do you think that secretion is regulated the same way in single cells and cells within intact acini?

*Kasai:* I assume so.

*Berridge:* Dr Thomas found differences in wave kinetics between single hepatocytes and cells in the intact liver.

*Thomas:* The rates of rise and fall of spikes in single cells are significantly slower than in connected cells. We know these waves are passing from cell to cell, so in a three-dimensional structure you may get a re-entrant stimulation that would prolong the spike, analogous to what can occur in the heart. I don't know whether that could happen in a pancreatic acinar cell.

*Berridge:* That's quite different from what Michael Nathanson was saying.

*Thomas:* He noted that $Ca^{2+}$ waves travel faster in cells within intact acini. I don't think our waves go faster in the intact liver, because they can be resolved with the image time resolution of two seconds as taking about two images to cross each cell.

*Jaffe:* But there is a difference in the comparison. You're (Thomas) talking about the slower rate in the whole tissue, which includes cell-to-cell communication, and he (Nathanson) is talking about a faster rate within a single cell.

*Thomas:* No. The measurements we were discussing were within a single cell and are based on our preliminary confocal measurements where we have two-second resolution and we see a wave cross a cell in between one and two images. Dr Nathanson, in your experiments measuring the rates of wave propagation across cells within intact acini, do the waves slow down as they go into the basal pole, as in isolated cells?

*Nathanson:* No, they don't.

*Thomas:* Dr Kasai, it was somewhat surprising to me that your waves slowed down so much (Fig. 4E). Are the $Ca^{2+}$ waves really going slower once they get going, or is there a delay as they move from one propagation mechanism to the next?

*Kasai:* Sometimes there is a delay and sometimes not. I would agree that the $Ca^{2+}$ wave in my preparation is slower than in the intact acini. This helped us to identify the mechanism of $Ca^{2+}$ release.

*Goldbeter:* Dr Nathanson, have you studied wave propagation at different stimulus intensities? If a decrease in phospholipase activity is responsible for the slow propagation, you might observe a similar rate at a low intensity of stimulation.

*Nathanson:* Even with low concentrations of agonist we observe a non-attenuated wave in cells within intact acini. The speed is slower than that observed with maximal stimulation, about two-thirds of the maximal speed, but is still not as slow as in single dispersed cells.

*Berridge:* The acetylcholine-induced spikes Dr Kasai showed seemed to be much longer than those described by Professor Petersen. Is this just a species difference?

*Petersen:* There is a slight difference between the rat and mouse. Dr Kasai's spikes last 5–10 s whereas ours last 1–5 s.

In most exocrine cells that secrete a lot of fluid there are $Ca^{2+}$-activated $K^+$ channels. For the push phase to be maximally effective, when $Cl^-$ channels are activated, $K^+$ channels should also be activated. The ideal situation would then be for $K^+$ channels to be clustered either in the luminal membrane or in the lateral membrane close to the lumen. The model would then function well. I agree with Dr Kasai that this pushing force would by itself be an effective way of secreting fluid.

*Miyazaki:* Dr Kasai, could you tell me at what level the $Cl^-$ equilibrium potential is?

*Kasai:* It's about −40 mV.

*Miyazaki:* That means that the resting potential is more negative than the $Cl^-$ equilibrium potential, and that the membrane potential goes up following activation of cation channels. Does the membrane potential go up and down across the $Cl^-$ equilibrium potential?

*Kasai:* I assume this happens in intact tissue. Unfortunately, isolated acinar cells do not have a resting membrane potential.

*Petersen:* The $Cl^-$ equilibrium potential is actually midway between the acetylcholine equilibrium potential and the resting potential. The resting potential, measured with microelectrodes, is about −40 mV, the equilibrium potential of acetylcholine is about −15 to −20 mV, and the $Cl^-$ equilibrium is somewhere in between (Petersen et al 1981). This is the beauty of Dr Kasai's model. If you activate only the $Cl^-$ channels, $Cl^-$ will of course have to move out of the cell. If you activate them together with non-selective cation channels, you move the membrane potential across the $Cl^-$ equilibrium potential and then the gradient reverses, so that $Cl^-$ will move through open channels into the cell rather than out of the cell. This is completely consistent with all the measurements we have made with microelectrodes in intact cells. There is a problem specifically in patch clamp experiments (whole-cell configuration) in which the resting potential sometimes seems to be low.

## References


Fasolato C, Zottini M, Clementi E, Lacchetti D, Meldolesi J, Pozzan T 1991 Intracellular $Ca^{2+}$ pools in PC12 cells. Three intracellular pools are distinguished by their turnover and mechanisms of $Ca^{2+}$ accumulation, storage, and release. J Biol Chem 266:20159–20167

Petersen OH, Maruyama Y, Graf J, Laugier R, Nichiyama A, Pearson GT 1981 Ionic currents across pancreatic acinar cell membranes and their role in fluid secretion. Philos Trans R Soc Lond B Biol Sci 296:151–166

Yoo S-H, Albanesi JP 1990 Inositol 1,4,5-trisphosphate-triggered $Ca^{2+}$ release from bovine adrenal medulary secretory vesicles. J Biol Chem 265:13446–13448

# Calcium signalling during chemotaxis

Fredric S. Fay, Susan H. Gilbert and Rodney A. Brundage*

*Department of Physiology Program in Molecular Medicine, University of Massachusetts Medical School, 373 Plantation Street, Worcester, MA 01605, USA*

*Abstract.* The role of $Ca^{2+}$ in chemotaxis of eosinophils from the newt *Taricha granulosa* was investigated using fluorescent indicators and digital imaging microscopy. In response to serum chemoattractant, cytoplasmic $Ca^{2+}$ concentration ($[Ca^{2+}]_i$) rises prior to polarization. In polarized locomoting cells $[Ca^{2+}]_i$ gradients (tail-high–front-low) are always seen, and when cells turn $[Ca^{2+}]_i$ rises transiently and falls fastest and furthest in the new direction of cell motion. These $Ca^{2+}$ signals, which are required for polarization and locomotion, arise from $Ca^{2+}$ derived from internal stores released in response to inositol 1,4,5-trisphosphate ($InsP_3$) (because microinjected heparin fully blocks them). 1,2-Diacyl-*sn*-glycerol (DAG), which is co-produced with $InsP_3$, has an inhibitory effect on $Ca^{2+}$ signals, an effect apparently mediated by protein kinase C. Studies with caged $InsP_3$ reveal that $InsP_3$-responsive stores appear to be concentrated in the nuclear and microtubule-organizing centre regions and that $InsP_3$ moves so rapidly within the cell that it is effectively a global secondary messenger. Thus, stable $[Ca^{2+}]$ gradients observed during unidirectional migration appear to result from the concentration of $InsP_3$-responsive $Ca^{2+}$ stores in the rear of the cell. By contrast, we propose that reorientation of the $[Ca^{2+}]$ gradient prior to a change in direction of motion results from the joint actions of $InsP_3$ and DAG, with $InsP_3$ acting as a global secondary messenger stimulating $Ca^{2+}$ release and DAG, through protein kinase C, acting as a spatially restricted secondary messenger inhibiting $[Ca^{2+}]$ increases locally near the site of chemotactic stimulation.

*1995 Calcium waves, gradients and oscillations. Wiley, Chichester (Ciba Foundation Symposium 188) p 121–140*

Virtually all cells at some stage polarize and migrate in response to chemical gradients in their environment. Embryonic development, wound healing and chemotaxis by granular leukocytes all depend critically on this cellular capability. Cells respond to gradients of chemoattractant with remarkable sensitivity. Human neutrophils, perhaps the best studied system, appear able to detect a difference in occupancy of receptors of less than 2% between front and rear (Devreotes & Zigmond 1988). Polarization and chemotaxis are persistent; even after the stimulus has been removed, the cell continues for some time up the

*Present address: Fox-Chase Cancer Center, 7701 Burholme Avenue, Philadelphia, PA 19111, USA.

gradient. However, when the orientation of the gradient is suddenly changed, such cells respond either by turning their front toward the new source of chemoattractant or by forming a new head and tail, thereby effecting a complete reorganization of the cytoskeleton and cytoplasmic organelles before moving in the new direction.

The chemotactic response of all cells involves two basic steps. First, the cell goes from a symmetrical shape in the absence of chemoattractant to a polarized configuration, with a broad lamellipod closest to the chemoattractant and the region of the cell most distant from the chemoattractant source drawn into a uropod, and the nucleus located posteriorly and the microtubule-organizing centre just in front of it. Second, the cell begins to move towards the source of chemoattractant in a highly coordinated fashion, involving continuous extension of the leading lamellipod and constriction at the uropod. Locomotion also requires continuous formation of adhesions to the substratum near the front and release at the rear.

These regional differences in morphology and behaviour must result from regional differences in the distribution and activities of proteins responsible for locomotion and cell adhesion. Myosin II, which interacts with actin filaments, has been shown to be concentrated in the cell cortex at the rear, probably producing a constrictive force in the uropod that squeezes the cell's contents towards the source of chemoattractant (Fukui et al 1989). Myosin I in locomotory cells is localized in the leading lamellipod (Fukui et al 1989) and may be responsible for generating force between actin filaments and membrane structures; it may also be involved in the rearward movement of actin filaments that form continuously at the advancing edge of the lamellipod (Forscher & Smith 1988). Many other proteins thought to be involved either directly or indirectly in cell motility are asymmetrically distributed along the axis of motion, presumably as part of the coordinated set of chemical activities that drive the cell forward. This pattern of molecular organization is dictated in some way by the external stimulus gradient. What are the steps in signal transduction that link this gradient to the particular intracellular organization that ensures polarization and migration in the appropriate direction?

Because $Ca^{2+}$ is known to affect the activities of many of the proteins thought to control cell shape and motion, it has long been suspected that spatial differences in intracellular $Ca^{2+}$ concentration ($[Ca^{2+}]_i$) might play an important role in translating the stimulus to effect the response. Early experiments to test this possibility focused on the effects of removing $Ca^{2+}$ from the medium. Although the chemotactic response of many cells is blocked at least temporarily by this treatment (Scanlon 1987), such results are difficult to interpret because inhibition of chemotaxis may be due to other effects of removing extracellular $Ca^{2+}$ that are unrelated to those on intracellular $Ca^{2+}$ signals (e.g., effects of cell adhesion or ligand–receptor binding), and they provided no direct means of linking $Ca^{2+}$ signals to locomotion. It has also been argued that the persistence of chemotaxis in other cells even in the absence of extracellular $Ca^{2+}$ shows that $Ca^{2+}$ signals play no role in cell polarization and directed locomotion (Zigmond et al 1988). Because most cells have a considerable amount of $Ca^{2+}$ sequestered in intracellular stores, however, it is not at all certain that removal of extracellular $Ca^{2+}$ necessarily disrupts normal $Ca^{2+}$ signalling, at least acutely. The correct conclusion can be ensured only by directly measuring cytoplasmic $[Ca^{2+}]_i$. Finally, it should be noted that there is likely to be considerable redundancy in the control of this fundamental cellular process. Thus, a more or less normal response to a chemotactic stimulus after inhibition of a single signalling system should not be taken to mean that this system plays no role under physiological conditions.

The earliest work in which $[Ca^{2+}]_i$ was measured directly in locomotory cells was done in the early 1980s in the giant free-living *Amoeba proteus* with the $Ca^{2+}$-sensitive photoprotein aequorin, the only intracellular $Ca^{2+}$ indicator available at that time (Taylor et al 1980). Aequorin luminescence was highest in the tail and lowest in the front, although there were bursts of light emission near the front as well that were possibly associated with phagocytosis. These experiments produced images with rather poor signal:noise characteristics, and their interpretation was complicated by the fact that light emission by aequorin in cells may be influenced by other ions (Blinks & Moore 1986). Also, and perhaps most importantly, because aequorin is not a ratiometric indicator, the intensity of light emission at a given point in the cell is affected by local variations in cell thickness and the fraction of the cell volume accessible to the protein at any location.

The development of several new classes of brightly fluorescent, cell-permeant ratiometric $Ca^{2+}$ indicators allowed measurements in smaller cells (Tsien 1989). Although several of these studies revealed that $[Ca^{2+}]_i$ in polarized chemotactic cells was distributed non-uniformly, the spatial patterns of $[Ca^{2+}]$ reported were highly variable. In some cells the highest $[Ca^{2+}]_i$ was reported to be in the rear (Poenie et al 1987), in others in the front (Sawyer et al 1985), and in some cells no $[Ca^{2+}]_i$ gradients were observed (Marks & Maxfield 1990). Furthermore, the patterns of $[Ca^{2+}]_i$ distribution were not usually measured in conjunction with an assessment of cell behaviour, which they should be if changes in $[Ca^{2+}]_i$ distribution are to be related to cell functioning. Finally, because the proteins presumed to be responsible for cell polarization and locomotion are all cytoplasmic, it is imperative that the indicators report cytoplasmic $[Ca^{2+}]$. This cannot always be assured because the loading of indicators into cells depends on endogenous esterases trapping exogenous esterified membrane-permeant indicators, and the indicators often end up in intracellular, membrane-bound compartments under these conditions (Moore et al 1990). Incompletely de-esterified dye, which is fluorescent but $Ca^{2+}$-insensitive and often compartmentalized, further complicates the interpretation of such studies (Scanlon et al 1987). Hence, to investigate the spatial aspects of cytoplasmic $Ca^{2+}$ signalling one needs *cytoplasmic* $Ca^{2+}$ indicators; one possibility is to microinject the free acid form of ratiometric indicators (ideally linked to a relatively high molecular mass dextran) into single chemotactic cells.

To satisfy these criteria, we have used large eosinophils from the newt (*Taricha granulosa*) as a model system (Fig. 1). These cells show a brisk chemotactic response to a chemoattractant (Brundage et al 1993). They polarize within a few seconds of exposure to a gradient of chemoattractant, with the basic morphological and functional specialization shown by other polarized cells. They

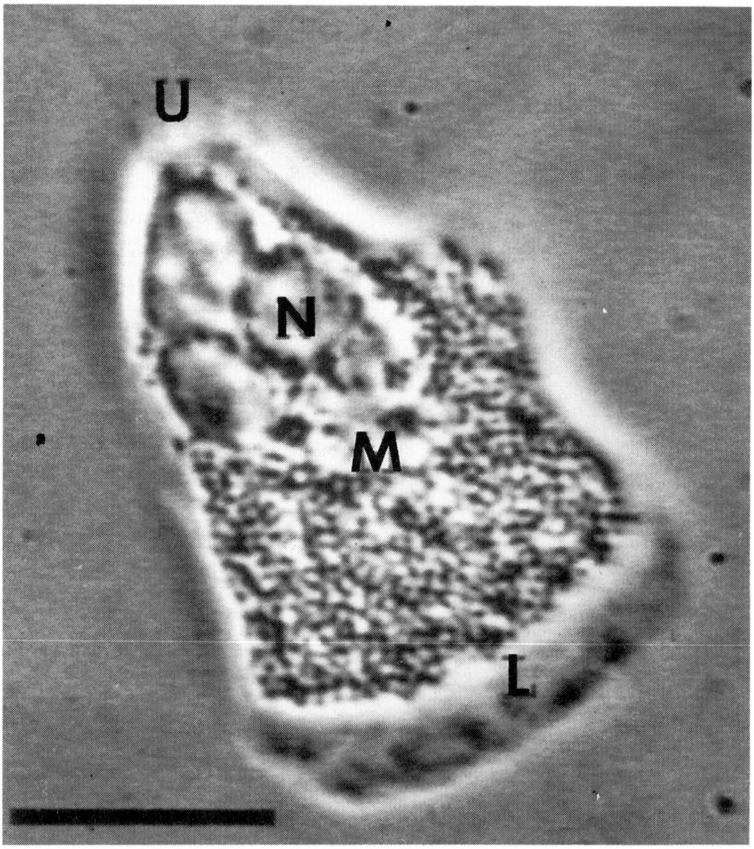


FIG. 1. Phase-contrast image of a typical newt (*Taricha granulosa*) eosinophil locomoting in 10% newt serum. The cell is moving towards the lower right of the figure. Note the lamellipod (L), the granule-filled cytoplasm behind it which has a clear zone containing the microtubule-organizing centre (M). The posterior of the cell contains a multi-lobed nucleus (N) and a constricted uropod (U). Scale bar, 20 μm.

respond to shifts in the direction of the stimulus and migrate towards the source at maximum speeds of 30 μm/min. Their principal advantage is that they are relatively large (70 μm long and 30 μm wide), which facilitates both microinjection of indicators and other agents and the subsequent resolution by optical methods of local differences in the concentration of $Ca^{2+}$ and other molecular species. To determine whether $Ca^{2+}$ plays a role in chemotaxis, in preliminary studies we tested the effect on the chemotactic response of agents expected to interfere with $Ca^{2+}$ signalling. The eosinophils lost their polarity and stopped moving in response to prolonged (>10 min) exposure to a $Ca^{2+}$-free medium, agents that should block $Ca^{2+}$ entry through the plasma membrane, $Ca^{2+}$ ionophores, caffeine and neomycin (Brundage et al 1991). The common feature of these treatments is that they interfere with $Ca^{2+}$ homeostasis, which we therefore concluded is likely to be necessary for polarization and locomotion in these cells.

We measured $[Ca^{2+}]_i$ by microinjecting fura-2 as either the free acid or the dextran conjugate into single eosinophils. The two forms of the indicator yielded essentially identical information. Fluorescent images were acquired at the same time as phase-contrast images with a novel digital imaging microscope equipped with a cooled, slow-scan charge-coupled device (CCD) camera (Brundage et al 1993). We found that when a cell is stimulated with a chemoattractant applied uniformly in the medium, $[Ca^{2+}]_i$ rises transiently to at least 400 nM within a few seconds, as shown in Fig. 2. Thereafter, the formerly spherical cell flattens onto the substrate and produces a lamellipod. The lamellipod continues to expand away from the centre of the cell and the nucleus, which ultimately comes to lie in the rear of the polarized cell in a constricted uropod. As the cell continues to move, $[Ca^{2+}]_i$ is often seen to increase periodically. These $[Ca^{2+}]_i$ spikes, which occur at a rate of about three a minute, are usually followed either by a change in direction (by the formation of a new lamellipod and the cessation of movement and retraction of the pre-existing lamellipod) or by an increase in velocity (Brundage et al 1991). Cytoplasmic $[Ca^{2+}]_i$ is uniform in the resting spherical cell, but as the cell flattens and becomes polarized, $[Ca^{2+}]_i$ is highest at the tail and lowest in the actively advancing lamellipod, as shown in Fig. 3.

A similar relationship between the spatial pattern of $[Ca^{2+}]_i$ and cell behaviour and organization is observed when cells are stimulated with a chemoattractant applied as a point source from a pipette (Brundage et al 1991, 1993). The initial stimulation produces a $[Ca^{2+}]_i$ spike and the cell subsequently polarizes and moves towards the source of chemoattractant. Relocation of the pipette causes a transient rise in $[Ca^{2+}]_i$ followed by the formation of a new lamellipod closest to the pipette. As the cell persistently moves toward the chemoattractant source, $[Ca^{2+}]_i$ remains higher than in the resting cell but usually lower than the peak level observed when the chemoattractant is first applied or when the direction of the chemoattractant source is suddenly changed.

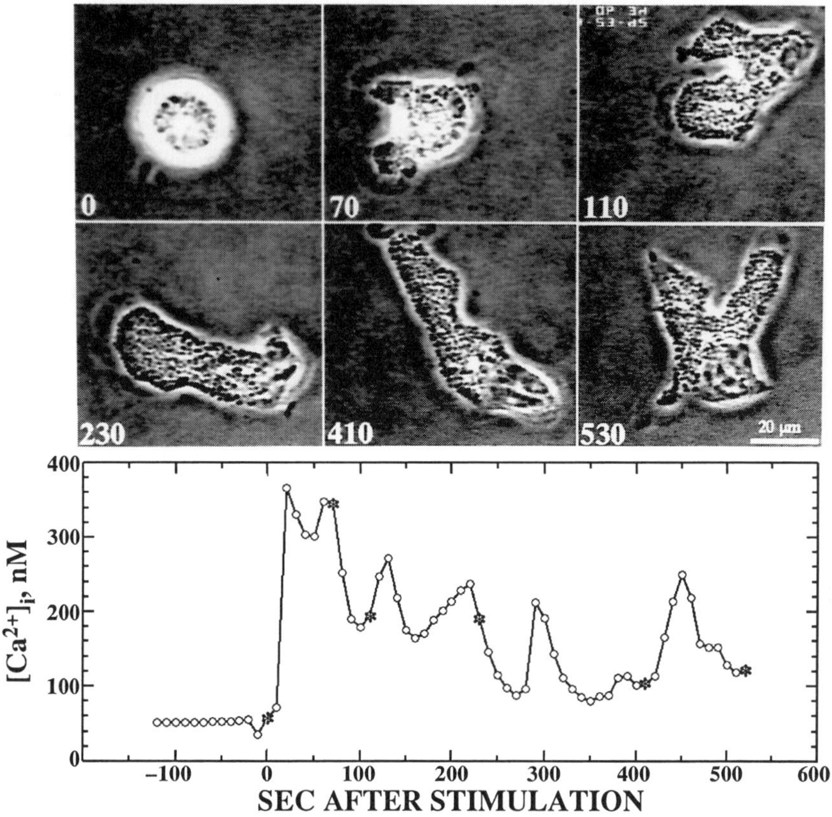


FIG. 2. Time course of the $Ca^{2+}$ response of a newt eosinophil to stimulation with 10% newt serum. Intracellular $Ca^{2+}$ concentration ($[Ca^{2+}]_i$) values of all pixels within the outline of the cell shown in the top panel were averaged at each time point to construct the graph below, in which the asterisks correspond to the images shown in A. $[Ca^{2+}]$ was calculated from the ratio of fluorescence excited at 340 and 380 nm of fura-2, as described in Gilbert et al (1994). Serum was added at $t = 0$.

In active cells, $[Ca^{2+}]_i$ is almost never spatially uniform. In spherical resting cells, however, $[Ca^{2+}]_i$ appears to be spatially uniform; this may be artefactual, a result of the large depth of focus of the fluorescence microscope and the consequent superposition of information from regions of the cell that may in fact have a quite different $[Ca^{2+}]_i$. Further work using methods that enhance axial resolution will help to resolve this issue.

The rear-to-front gradient of $[Ca^{2+}]$ seems essential for polarization and locomotion. Although removing external $[Ca^{2+}]$ has no initial effect on either $[Ca^{2+}]_i$ or polarity and motility, within 5–10 min $[Ca^{2+}]_i$ does fall and this is always followed by cessation of locomotion and loss of polarity, with the nucleus

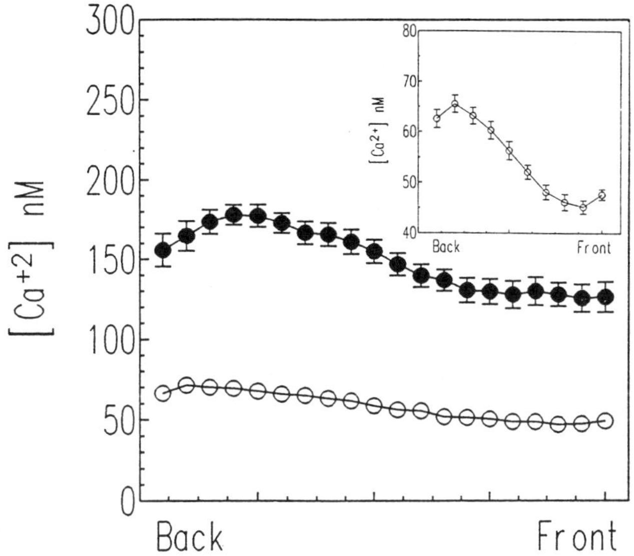


FIG. 3. Average intracellular $Ca^{2+}$ concentration ($[Ca^{2+}]_i$) gradients in cells moving straight (○, 13 images of five cells) and cells turning (●, 16 images of five cells) in the presence of 10% newt serum. Average $Ca^{2+}$ (concentrations ± SE at 20 points) along the long axis of cells which were turning or moving persistently in one direction are shown. The inset shows a coarser sampling of the average $[Ca^{2+}]_i$ gradient in cells moving persistently in one direction, drawn on a different scale to show the gradient more clearly.

moving to the centre of the cell and several lamellipods spreading radially in all directions (Gilbert et al 1994). Microinjection of excessive amounts of $Ca^{2+}$ indicator ($[\text{fura-2}]_i > 50\,\mu M$), which increases intracellular $Ca^{2+}$ buffering, has a similar effect on cell morphology. Under these conditions the cell flattens and forms lamellipods on stimulation, and the rate of rise of $[Ca^{2+}]_i$ is slowed severalfold.

The observation that cells are able to generate several $[Ca^{2+}]_i$ spikes and maintain a $[Ca^{2+}]_i$ gradient for several minutes in $Ca^{2+}$-free medium indicates that $Ca^{2+}$ discharged from internal stores is sufficient for normal cytoplasmic $Ca^{2+}$ signals for several minutes (Gilbert et al 1994). Furthermore, injecting cells with heparin, which is widely used to inhibit binding of inositol 1,4,5-trisphosphate ($InsP_3$) to its receptor, completely blocked both the $Ca^{2+}$ response and the morphological changes usually seen on stimulation, whereas de-*N*-sulphated heparin, which does not bind to the receptor, had no effect (Gilbert et al 1994), as shown in Fig. 4. Hence, the rise in $[Ca^{2+}]_i$ that follows stimulation appears to be mediated by an increase in intracellular $[InsP_3]$. It is quite likely that

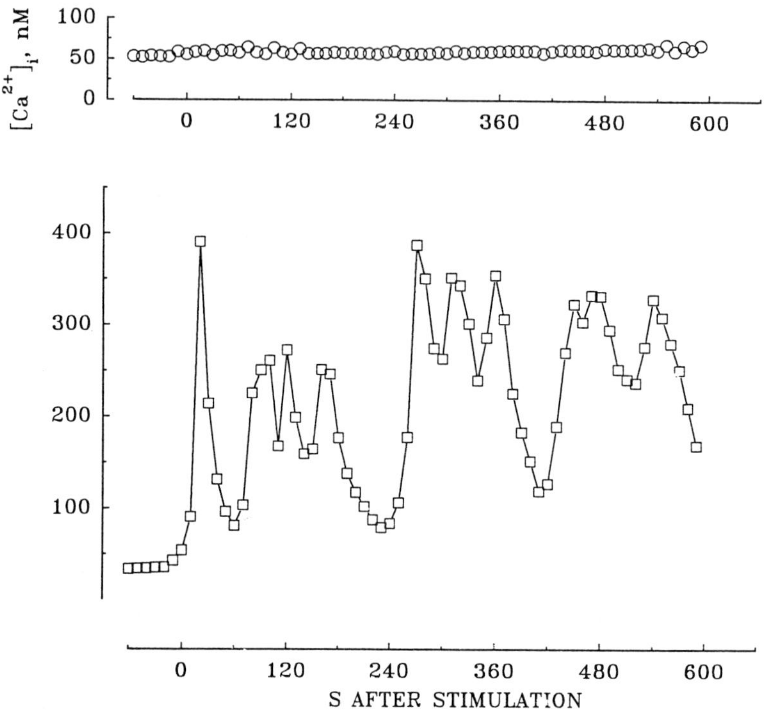


FIG. 4. Effect of injection of heparin sulphate (*top*) of de-*N*-sulphated heparin (*bottom*) on the time course of the $Ca^{2+}$ response to stimulation with 10% newt serum. The intracellular concentrations, 108 $\mu$g/ml heparin sulphate and 104 $\mu$g/ml de-*N*-sulphated heparin, were calculated as described in Gilbert et al (1994). $[Ca^{2+}]_i$, intracellular $Ca^{2+}$ concentration.

subsequent $[Ca^{2+}]_i$ spikes, also resulting from release of $Ca^{2+}$ from internal stores, are also triggered by $InsP_3$. To demonstrate this definitively, however, it will be necesssary to microinject caged heparin or some other agent whose ability to block $InsP_3$ binding to its receptor can be suddenly changed.

Production of $InsP_3$ is often associated with the co-production of 1,2-diacyl-*sn*-glycerol (DAG), which in turn activates many isoforms of protein kinase C (Berridge 1993). Hence, we investigated whether the DAG–protein kinase C pathway might also play some role in the changes in $[Ca^{2+}]_i$ or perhaps some other process involving the cytoskeleton, many components of which are known to be targets of protein kinase C. We found that when cells were injected with a peptide corresponding to the auto-inhibitory domain of protein kinase C, the $[Ca^{2+}]_i$ increase normally observed following stimulation with chemoattractant was significantly increased (Gilbert et al 1994). In contrast, when cells were micro-injected with the control peptide, identical except for a single amino acid which renders it ineffective as an inhibitor of protein kinase C, the changes in $[Ca^{2+}]_i$ observed on stimulation were similar to those cells injected with fura-2 alone. These results indicate that activated protein kinase C acts as a negative regulator of some process or processses underlying the increase in $[Ca^{2+}]_i$ that follows stimulation with a chemoattractant. Further support for this proposition comes from the finding that protein kinase C activators, including both phorbol esters and diacylglycerol analogues, supress the $Ca^{2+}$ response to stimulation. Application of these agents in the absence of chemoattractant caused cells to flatten onto the substrate and to form lamellipods in several directions. The cells failed, however, to develop the typical polarized shape and were incapable of significant motion under these conditions.

Although these observations indicate that the $Ca^{2+}$ signals generated in response to chemoattractant are the result, at least in part, of the antagonistic actions of $InsP_3$ and protein kinase C on $Ca^{2+}$ signalling, the mechanisms underlying the observed spatial patterns remained elusive. We therefore undertook experiments to probe directly the spatial organization of elements of the $Ca^{2+}$ signalling system in these cells. We focused initially on $InsP_3$-responsive elements, using caged $InsP_3$ to reveal regional differences in $InsP_3$-sensitive $Ca^{2+}$release. The caged compound was co-injected with the 70 kDa dextran conjugate of calcium green, to ensure that only cytosolic $Ca^{2+}$ changes would be detected and to avoid overlap between the excitation spectrum of calcium green and that of the caged compound in the near UV. Because $Ca^{2+}$ binding causes an increase in brightness rather than a spectral shift, images acquired after photolysis were normalized using the fluorescence image obtained just before photolysis. In this way, images of spatial patterns of changes in $[Ca^{2+}]_i$ response to $InsP_3$ could be obtained and effects on fluorescence intensity of regional variations in cell thickness eliminated. The interval between the pre-photolysis reference image and all post-photolysis test images was always $\leqslant$800 ms, to ensure that rapid movement of organelles or translation of the cell itself did not produce artefacts in the result. We verified that this was the case in the image series acquired without photolysis, which proved to have an intensity ratio of 1.0 distributed uniformly throughout the cell.

To probe for regional differences in $InsP_3$ sensitivity, we generated $InsP_3$ uniformly throughout the cell by brief (5 ms) exposure of the cell to UV light (the 350 and 363 nm lines of an argon/ion laser). There was an increase in $[Ca^{2+}]_i$ throughout the cell within 50–100 ms, the magnitude of which was graded with light intensity. The $[Ca^{2+}]_i$ increase was not spatially uniform, however, with the largest increases routinely being seen in the perinuclear and microtubule-organizing centre (MTOC) regions of the cell, as shown in Fig. 5.

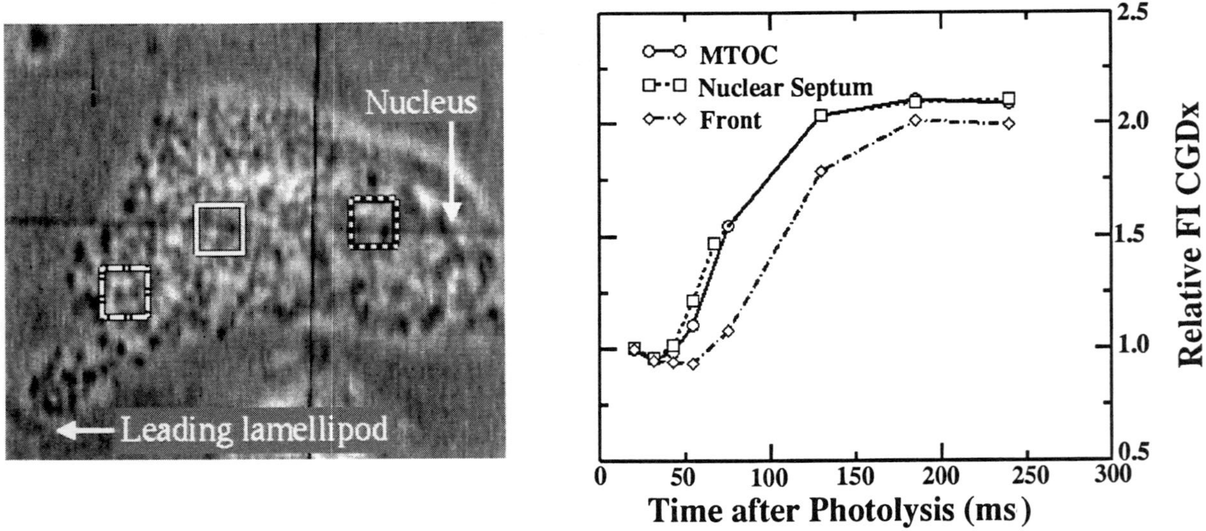


FIG. 5. Relative time course and magnitude of $[Ca^{2+}]$ changes in a newt eosinophil following a uniform increase in $[InsP_3]$ produced by photolytics of caged $InsP_3$. $[Ca^{2+}]$ changes were determined by comparing the fluorescence intensity (FI) of 70 kDa dextran–calcium green (CGDx) point-by-point within the cell just before and then at the times indicated after photolysis. The relative changes in $[Ca^{2+}]$ were calculated for three separate $5 \times 5$ $\mu$m boxes positioned at the microtubule organizing centre (MTOC, ○), over the nuclear septum (□), and at the front of the cell (◇) just behind an active lamellipod, as indicated in the phase-contrast image of the cell. Note that in response to this uniform increase in $InsP_3$, $[Ca^{2+}]_i$ rose faster and to higher levels at the MTOC and over the nucleus than in the front. The cell was injected with solution corresponding to about 5% of its initial volume, containing 12.5 $\mu$M 70 kDa dextran–calcium green (dextran : fluorophore = 1 : 9) and 100 $\mu$M caged $InsP_3$ (D-*myo*-inositol 1,4,5-trisphosphate, P4[5]-1-[2-Nitrophenyl]ethyl ester; Cal. Biochem.).

To determine whether this spatial pattern of $InsP_3$ sensitivity could result from a high density of $Ca^{2+}$ stores in these regions, we incubated cells with the acetoxymethyl ester of either fura-2 or magfura-2 (a variant of fura-2 with similar spectral properties but low affinity for $Ca^{2+}$). With this method, indicators are loaded into both the cytoplasm and the organelles of many cells. With both indicators, the fluorescence intensity ratio (340:380 nm excitation) was highest over the perinuclear and MTOC regions, a pattern not observed in cells injected with the impermeant free-acid forms of these probes, indicating that the high ratios observed over the nuclear and MTOC regions were coming from membrane-enclosed $Ca^{2+}$ stores. Furthermore, exposure of these cells to the internal membrane dye $DiOC_{6(3)}$ revealed that the perinuclear region, and especially the MTOC region, contained a high density of membranes, which presumably are the membranes surrounding regions of high $[Ca^{2+}]$ revealed by loading with the acetoxymethyl ester form of fura-2 (Brundage et al 1993). Hence, the large $Ca^{2+}$ release in the perinuclear/MTOC regions following a photolytically induced increase in $[InsP_3]$ would appear to result, at least in part, from a high concentration of $InsP_3$ -sensitive $Ca^{2+}$ stores in that region.

It is quite likely that during chemotaxis occupancy of chemoattractant receptors, and hence activation of phospholipase C, is non-uniform. The next step was therefore to determine how the cell responded to localized photolytic generation of $InsP_3$ from its caged precursor. We found that when $InsP_3$ was generated over the MTOC region there was a prompt, large rise in $[Ca^{2+}]_i$, which, although largest over the MTOC, also occurred throughout the rest of the cell. The rise in $[Ca^{2+}]_i$ was virtually complete within 500 ms of photolysis. When $InsP_3$ was produced near the advancing lamellipod, some 20–30 $\mu$m distant from the MTOC/perinuclear region, the resulting $[Ca^{2+}]_i$ increase had a distinctly different spatial and temporal pattern. $[Ca^{2+}]_i$ rose in the immediate vicinity of $InsP_3$ production within 150–300 ns, and a smaller increase also occurred simultaneously near the MTOC. A further $[Ca^{2+}]_i$ increase was observed in the MTOC/perinuclear region in the next few hundred milliseconds, and little further increase in $[Ca^{2+}]_i$ was noted thereafter.

These results might be explained if $InsP_3$ produced at the front moved rapidly through the cell, causing a diffusion-dependent delayed rise in $[Ca^{2+}]_i$ at the perinuclear/MTOC region. Alternatively, $Ca^{2+}$ released at the site of generation of $InsP_3$ might diffuse to the perinuclear/MTOC region, to release additional $Ca^{2+}$ by a $Ca^{2+}$-induced $Ca^{2+}$ release mechanism. This, however, does not appear likely, because release of $Ca^{2+}$ from the photosensitive $Ca^{2+}$ chelator, Nitr 7 (Adams et al 1988), into the cytoplasm near, but not directly at, the MTOC caused no secondary release of $Ca^{2+}$. Furthermore, the decay in the local $[Ca^{2+}]_i$ increase appeared to be dominated by reuptake and binding to exogenous and endogenous buffers, because there was no detectable radial spread of the local $[Ca^{2+}]_i$ increase. Hence, the spreading of the $Ca^{2+}$ response in the first few hundred milliseconds following local generation of $InsP_3$ must reflect rapid diffusion of $InsP_3$ from its site of generation within the eosinophil. This conclusion is consistent with recent measurements of the diffusion coefficient for $InsP_3$ in *Xenopus* oocyte cytosol showing it to be more than 30-fold greater than that of $Ca^{2+}$ (Allbritton et al 1992). Our results indicate that, although there may be local heterogeneities in the generation of $InsP_3$ during physiological stimulation, because of its apparently high rate of diffusion in relation to the duration of typical $Ca^{2+}$ spikes, $InsP_3$ is acting as a global second messenger in these cells. Hence, the pattern of $Ca^{2+}$ release resulting from the effects of $InsP_3$ will be determined largely by the distribution of $InsP_3$-sensitive $Ca^{2+}$ stores.

These observations can explain why $Ca^{2+}$ spikes in polarized cells that are persistently migrating in a given direction are always highest at the tail and lowest in the front. However, additional mechanism(s) are required to account for the rapid reversal of the $[Ca^{2+}]_i$ gradient when a cell experiences a shift in the orientation of the stimulus gradient during chemotaxis. We believe that the early reorientation of the $[Ca^{2+}]_i$ gradient seen under those conditions may be explained by the inhibitory actions of DAG–protein kinase C on the $Ca^{2+}$ signalling pathways in these cells. The lipophilic nature of DAG and some isoforms of activated protein kinase C could confine these molecules to membranes in the immediate vicinity of the site of generation, where diffusion is two to three orders of magnitude slower than for hydrophilic molecules such as $InsP_3$ diffusing in cytosol (Wolf & Edidin 1981). Thus, the inhibitory action of DAG–protein kinase C on $Ca^{2+}$ signalling would be expected to be highly local in nature, confined principally to the regions of the cell closest to the stimulus source where chemoattractant receptor occupancy, and consequently the activity of phospholipase C, are presumably highest. The shift in the $[Ca^{2+}]_i$ gradient would be expected to cause a reorganization of the cytoskeleton (see below). We propose that this, in turn, causes the $InsP_3$-sensitive $Ca^{2+}$ stores to be moved with/by the cytoskeleton to the presumptive rear end of the cell. This reorganization of $Ca^{2+}$ stores would then stabilize and reinforce the direction of the new $[Ca^{2+}]_i$ gradient within the cell.

How might the gradients of $[Ca^{2+}]_i$ and the presumed underlying gradients of protein kinase C and phospholipase C activities be involved in organizing and controlling the activity of cytoskeletal components that effect polarization and directed locomotion of these cells? Myosin II, whose interaction with actin is controlled by $Ca^{2+}$ via a $Ca^{2+}$-dependent phosphorylation of myosin light chains, would be expected to be maximally activated at the rear of the cell. Because myosin II is found principally in the cell cortex, the tail-high–front-low gradient of force generation it might produce would tend to squeeze the cytosol from the rear to the front of the cell (Taylor & Fechheimer 1982). Myosin I, whose activity is highest at low $[Ca^{2+}]$ (Collins & Matsudaira 1991), may be activated mainly at the front of the cell. One role postulated for myosin I is that of producing forces between actin filaments and membranes. Its preferential activation at the front of the cell could therefore contribute to the membrane and actin flows underlying the expansion of the leading lamellipod. Actin polymerization, which is thought to be the principal factor underlying the forward expansion of the leading lamellipod, should be enhanced at the front of the cell. This follows, because actin-severing activities of proteins such as gelsolin (Yin & Stossel 1979) would be the lowest at the front because their activities are dependent on $Ca^{2+}$, the concentration of which is lowest at the front of the cell. Attachment of the cell to some substrates via integrins would be expected to be weakest at the rear where the higher $[Ca^{2+}]$ would be expected to favour dephosphorylation of integrins by the $Ca^{2+}$-dependent phosphatase calcineurin. Binding of cellular integrins to extracellular matrix proteins such as vitronectin is believed to be inhibited following dephosphorylation (Hendey et al 1992). Hence, the gradients of $[Ca^{2+}]$ should cause gradients in the activities of $Ca^{2+}$-dependent processes that would favour the polarization and directed locomotion of cells toward a source of chemoattractant.

Changes in the direction of the gradient of chemoattractant produce a large, relatively uniform $[Ca^{2+}]_i$ increase; the $[Ca^{2+}]_i$ appears to fall fastest and, ultimately, to its lowest level in the part of the cell that will become the front with the new direction of motion. This pattern of $[Ca^{2+}]_i$ changes should facilitate the reorganization of the cell that enables it to move in the new direction towards the chemoattractant. Specifically, the large, relatively uniform increase in $[Ca^{2+}]_i$ that immediately follows the change in the direction of chemoattractant would be expected to cause global actin depolymerization, the detachment of integrins from extracellular matrix sites and overall activation of myosin II, resulting in a 'cringing' retraction of the cell from the substrate. As $[Ca^{2+}]_i$ falls and a new $[Ca^{2+}]_i$ gradient forms, this new gradient is likely to produce gradients of activity of elements underlying the polarization and directed locomotion of these cells. Furthermore, if the force-producing molecules are not locked in place, they will become concentrated in regions of highest activity/force (Bray & White 1988), thereby amplifying gradients of activity of the cytoskeleton expected to result from gradients in $[Ca^{2+}]_i$. Finally, the reorganization of elements of the cytoskeleton is likely to cause a reorganization of calcium storage sites within the cell in view of the known association of internal membrane systems with elements of the cytoskeleton, especially the microtubule system (Dabora & Sheetz 1988).

It is worth briefly considering the functional significance of the concentration of $InsP_3$-sensitive $Ca^{2+}$ stores around the MTOC. It is possible this concentration provides a means whereby $InsP_3$ generated at the plasma membrane could control the chemistry of the MTOC and in that way control functions associated with the microtubule system. Certainly, $Ca^{2+}$ is known to have profound effects on the structure of the centrosome, especially the pericentriolar material that controls microtubule nucleation (Paintrand et al 1992). Although the mechanisms of the effects of $Ca^{2+}$ on the system are not understood, $Ca^{2+}$-sensitive proteins such as centrin and calmodulin, and various protein kinases including $p34^{cdc2}$ protein kinase, are known to be associated with the MTOC in cells of various types (Kalnins 1992). The role of microtubules in cell polarization and locomotion is controversial, but there is considerable evidence indicating that disruption of microtubules with chemical agents interferes with normal polarization and locomotion of some cells (Bornens et al 1989, Malech et al 1977). Furthermore, studies on newt eosinophils have shown that the activity of the MTOC, presumably through its role in regulating microtubule functioning, is essential for coordinated migration in response to chemoattractant; laser micro-irradiation of the MTOC interferes rather specifically with the polarization and locomotion of these cells (Koonce et al 1984). The specific manner in which $InsP_3$-generated signals at the MTOC exert control over microtubules and functions associated with them remains a subject for future investigation. The use of caged compounds in conjunction with fluorescent indicators and analogues should help to unravel the links between $InsP_3$-sensitive $Ca^{2+}$ signals at the MTOC and other regions of the cell and the functioning of this and other components of the cytoskeleton.

*Acknowledgements*

We thank Kristine Perry for technical assistance, Richard Tuft for designing and constructing optical systems, Kevin Fogarty and Doug Bowman for software development and Karen Cawrse for help in the preparation of the manuscript. This work was supported in part by grants from the NIH (HL-14523 and HL-47530) and the NSF (BIR-9200027).

## References

Adams SR, Kao JPY, Grynkiewicz G, Minta A, Tsien RY 1988 Biologically useful chelators that release $Ca^{2+}$ upon illumination. J Am Chem Soc 110:3212–3220

Allbritton NL, Meyer T, Stryer L 1992 Range of messenger action of calcium ion and inositol 1,4,5-trisphosphate. Science 258:1812–1815

Berridge MJ 1993 Inositol trisphosphate and calcium signalling. Nature 361:315–325

Blinks JR, Moore ED 1986 Practical aspects of the use of photoproteins as biological calcium indicators. Soc Gen Physiol Ser 40:229–238

Bornens M, Paintrand M, Celati C 1989 The cortical microfilament system of lymphoblasts displays a periodic oscillatory activity in the absence of microtubules: implications for cell polarity. J Cell Biol 109:1071–1083

Bray D, White JG 1988 Cortical flow in animal cells. Science 239:883–888

Brundage RA, Fogarty KE, Tuft RA, Fay FS 1991 Calcium gradients underlying polarization and chemotaxis of eosinophils. Science 254:703–706

Brundage RA, Fogarty KE, Tuft RA, Fay FS 1993 Chemotaxis of newt eosinophils: calcium regulation of chemotactic response. Am J Physiol 265:C1527–C1543

Collins K, Matsudaira P 1991 Differential regulation of vertebrate myosin-I and myosin-II. J Cell Sci Suppl 14:11–16

Dabora SL, Sheetz MP 1988 The microtubule-dependent formation of a tubulovesicular network with characteristics of the ER from cultured cell extracts. Cell 54:27–35

Devreotes PN, Zigmond SH 1988 Chemotaxis in eukaryotic cells: a focus on leukocytes and *Dictyostelium*. Annu Rev Cell Biol 4:649–686

Forscher P, Smith SJ 1988 Actions of cytochalasins on the organization of actin filaments and microtubules in a neuronal growth cone. J Cell Biol 107:1505–1516

Fukui Y, Lynch TJ, Brzeska H, Korn RY 1989 Myosin I is located at the leading edges of locomoting *Dictyostelium* amoebae. Nature 341:328–331

Gilbert SH, Perry K, Fay FS 1994 Mediation of chemoattractant-induced changes in $[Ca^{2+}]_i$ and cell shape, polarity and locomotion by $InsP_3$, DAG and protein kinase C. J Cell Biol 127:489–503

Hendey B, Klee CB, Maxfield FR 1992 Inhibition of neutrophil chemokinesis on vitronectin by inhibitors of calcineurin. Science 258:296–299

Kalnins VI (ed) 1992 The centrosome. Academic Press, San Diego, CA

Koonce MP, Cloney RA, Berns MW 1984 Laser irradiation of centrosomes in newt eosinophils: evidence of centriole role in motility. J Cell Biol 98:1999–2010

Malech HL, Root RK, Gallin JI 1977 Structural analysis of human neutrophil migration—centriole, microtubule, and microfilament orientation and function during chemotaxis. J Cell Biol 75:666–693

Marks PW, Maxfield FR 1990 Transient increases in cytosolic free calcium appear to be required for the migration of adherent human neutrophils. J Cell Biol 110: 43–52

Moore EDW, Becker PL, Fogarty KE, Williams DA, Fay FS 1990 $Ca^{2+}$ imaging in single living cells: theoretical and practical issues. Cell Calcium 11:157–179

Paintrand M, Moudjou M, Delacroix H, Bornens M 1992 Centrosome organization and centriole architecture: their sensitivity to divalent cations. J Struct Biol 108: 107–128

Poenie M, Tsien RY, Schmitt-Verhulst AM 1987 Sequential activation and lethal hit measured by $[Ca^{2+}]_i$ in individual cytolytic T cells and targets. EMBO (Eur Mol Biol Organ) J 6:2223–2232

Sawyer DW, Sullivan JA, Mandell GL 1985 Intracellular free calcium localization in neutrophils during phagocytosis. Science 230:663–666

Scanlon M 1987 Cellular mechanism of neutrophil chemotaxis: the role of $Ca^{2+}$ as viewed with the fluorescent dye, fura-2, in the polarization of human polymorphonuclear leukocytes following stimulation with the chemoattractant, f-methionyl-leucyl-phenylalanine. PhD thesis, University of Massachusetts Medical School, Worcester, MA, USA

Scanlon M, Williams DA, Fay FS 1987 A $Ca^{2+}$-insensitive form of fura 2 associated with polymorphonuclear leukocytes. J Biol Chem 262:6308–6312

Taylor DL, Fechheimer M 1982 Cytoplasmic structure and contractility: the solation–contraction coupling hypothesis. Philos Trans R Soc Lond Ser B Biol Sci 299: 185–197

Taylor DL, Binks JR, Reynolds G 1980 Contractile basis of ameboid movement. 8. Aequorin luminescence during ameboid movement, endocytosis, and capping. J Cell Biol 86:599–607

Tsien RY 1989 Fluorescent indicators of ion concentrations. Methods Cell Biol 30: 127–156

Wolf DE, Edidin M 1981 Methods of measuring diffusion and mobility of molecules in surface membranes. In: Baker P (ed) Techniques in cellular physiology. Elsevier Biomedical, Amsterdam, vol 1:1–14

Yin HL, Stossel TP 1979 Control of cytoplasmic actin gel–sol transformation by gelsolin, a calcium-dependent regulatory protein. Nature 281:583–586

Zigmond SH, Slonczewski JL, White MW, Carson M 1988 Polymorphonuclear leukocyte locomotion is insensitive to lowered cytoplasmic calcium levels. Cell Motil Cytoskeleton 9:184–189

## DISCUSSION

*Michell:* As the cell changed from being non-polarized to polarized and began to move, the microtubule-organizing centre (MTOC) seemed to be ahead of the nucleus, underlying the granular part of the lamellipod. The $Ca^{2+}$ gradient flows right back through the cell from front to back, so why is it that the very back end ends up with the highest $[Ca^{2+}]$?

*Fay:* That's a good point. The presumed gradients of DAG and protein kinase C, which have a negative influence on $Ca^{2+}$ signal generation, would tend to provide an additional bias that might cause the regions of the cell closest to the stimulus to exhibit the smallest increase in $[Ca^{2+}]_i$ in response to a global increase in inositol 1,4,5-trisphosphate ($InsP_3$) concentration.

*Michell:* Do you then assume that DAG modulates the $Ca^{2+}$ stores rather than generating the signal controlling the $Ca^{2+}$ stores?

*Fay:* Yes; we are suggesting that there's an effect on the responsiveness of the stores to $InsP_3$, rather than on the $InsP_3$ signal responsible for causing release from the stores.

*Irvine:* You must be implying that protein kinase C controls the behaviour of $InsP_3$ receptors; the control can't be prior to $InsP_3$ generation. What is known about the influence of protein kinase C on $InsP_3$ receptors?

*Berridge:* As I understand it, protein kinase C doesn't have much effect on $InsP_3$ receptors.

*Fay:* Although such an effect would readily explain what we have seen, there are other possibilities. Protein kinase C might be acting through an effect on $Ca^{2+}$ pumping.

*Michell:* Have you tested the effect of protein kinase C regulatory molecules with caged $InsP_3$?

*Fay:* We haven't done that yet.

*Putney:* Do you think that protein kinase C and DAG impart directionality by virtue of their ability to impart a $Ca^{2+}$ gradient, or is it just as likely that protein kinase C inhibits the $Ca^{2+}$ signal but imparts directionality through some other mechanism?

*Fay:* My response is diplomatic but truthful. I think both probably occur. The $Ca^{2+}$ gradient is significantly influenced by what we assume to be a gradient of DAG and protein kinase C activity in the cell. The $Ca^{2+}$ gradient that we observe would be expected preferentially to activate myosin II at the tail and cause a squeezing action in that region, whereas the low $[Ca^{2+}]$ at the front would favour activation of myosin I motor activity and actin polymerization at the front, which would favour lamellipod extension. Many proteins regulating actin polymerization and depolymerization are known to be influenced by $[Ca^{2+}]$; hence the polarity of actin filament assembly could well be explained by the $Ca^{2+}$ gradients. It is also true that protein kinase C and various phospholipids have significant effects on actin polymerization through interaction with proteins such as profilin and the MARCKS (myristoylated alanine-rich C kinase substrate) protein that are presumed to be involved in controlling locally assembly of actin into filaments in cells. Thus, gradients of generation of phospholipids could also have a more direct action not mediated by $Ca^{2+}$.

*Meyer:* Your model suggests a simple experiment. What effect does DAG have when applied locally with a micropipette? A simple prediction would be that the cell would turn around to the tip.

*Fay:* We have begun experiments with caged DAG but the results are preliminary. In some cells, when we photolyse caged DAG, we see a local drop in $[Ca^{2+}]_i$ that persists for some time. This is consistent with our line of thinking, but we need to know whether the elements responding to DAG are localized in some way, which might explain some of the variability in the results that we have seen.

*Thomas:* When you applied serum to the cells globally, you got a transient global $[Ca^{2+}]_i$ increase, with the gradient becoming evident only later when the cell polarized. Is it possible that there is a pre-existing polarization, which does not depend on the $Ca^{2+}$ gradient?

*Fay:* As $[Ca^{2+}]_i$ goes down after the global increase there is in fact a slight 'tilt' to the seemingly flat $[Ca^{2+}]_i$ image. $[Ca^{2+}]_i$ is lower in the region that becomes the new leading edge. The system is inherently poised to polarize in response to a stimulus. This inherent polarity is perhaps created by something associated with the nuclear–MTOC axis. The external stimulus ultimately orients that internal axis appropriately, which, in turn, further stabilizes the polarity and directionality of the cell.

*Sanderson:* If the cell is already polarized in one direction, and you suddenly change the chemical gradient, the cell responds with a large pulse of increased $[Ca^{2+}]_i$. You implied that this response is a form of resetting signal. What's the function of the reset? Also, if the system is already polarized, how does it suddenly respond with a global change in $[Ca^{2+}]_i$ when the stimulus comes from a different direction?

*Fay:* It's easier to answer your first question than the second. The large rise in $[Ca^{2+}]_i$ that proceeds repolarization could be a scrambling signal for the actin. It would tend to break the actin cytoskeleton up into smaller bits, allowing it to re-form along a different axis. The fact that the system responds to a change in direction with a larger rise in $[Ca^{2+}]_i$ suggests that there is some complex spatial adaptation in response to a stimulus graded in a particular direction that gives rise to increased sensitivity to a change in direction of the chemoattractant. We clearly don't understand the nature of that mechanism.

*Berridge:* Is that true in the case of the killer cells that Roger Tsien has worked on? From what I can remember, the $Ca^{2+}$ signal occurs on the side opposite to the initial site of contact.

*Fay:* In other polarized cells, killer T cells (Poenie et al 1987) and *Amoeba proteus* (Taylor et al 1980), it has been found that the tail has a high [$Ca^{2+}$] and the front a low [$Ca^{2+}$].

*Jaffe:* Some years ago we did a vibrating probe study on amoebae and found what we interpreted as a large, steady flow of $Ca^{2+}$ ions into the tail, but we also saw pulses of current entering nascent pseudopods (Nuccitelli et al 1977). We thought that meant that there was a different kind of $Ca^{2+}$ control, brief rises at the front which somehow induce the pseudopods. Do you see any hint of little flashes of $Ca^{2+}$ at the front?

*Fay:* Not at the rate at which we've been able to sample and with the number of images that we've collected so far. If these $Ca^{2+}$ pulses were truly brief and rare, we may not have looked long and hard enough yet.

*Smith:* Even the low [$Ca^{2+}$] in the lamellipod in the locomoting cell appeared always to be higher than the low [$Ca^{2+}$] in the quiescent cell. Is that correct?

*Fay:* When a round cell polarizes, the [$Ca^{2+}$] does go up overall, but it goes up less at the front.

*Smith:* That could be related to Lionel Jaffe's observation of little pulses at the front.

*Fay:* The overall [$Ca^{2+}$]$_i$ is higher in cells moving persistently in one direction in a uniform chemoattractant than in those being led with a pipette. The higher [$Ca^{2+}$] is associated with a higher velocity of travel. The average [$Ca^{2+}$] of the cell appears to be related to velocity. We could again speculate about connections between $Ca^{2+}$ and $Ca^{2+}$-sensitive cytoskeletal proteins such as gelsolin that might explain these observations.

*Thomas:* Do you have any evidence about whether the $Ca^{2+}$ stores, which you have shown to be located where the [$Ca^{2+}$] is high, reorganize in advance of the change in direction of cell migration?

*Fay:* As the cell reorients, the MTOC changes its position; one might expect this to be associated with a change in the distribution of the stores. One is looking for some mechanism by which the directionality of movement of the cell would be stabilized. An effect of the cytoskeleton on the $Ca^{2+}$ stores tending to reorganize the stores in a manner similar to what is seen in a cell that is persistently migrating towards a chemoattractant would tend to further stabilize the direction of travel. It is interesting that the MTOC appears to contain such a high density of $InsP_3$-sensitive stores, and that the microtubule system itself is known to attract and interact with $Ca^{2+}$ storage vesicles.

*Williams:* When a cell changes its direction there is an input of $Ca^{2+}$ to the system. The cell then begins to move in its new direction after the reorganization. At that reorganization stage a low [$Ca^{2+}$] gradient is maintained from one end of the cell to the other, from about 50 to 150 nM, as I understand it. Is it known if there is a very strong binding centre for $Ca^{2+}$? Do you know where the $Ca^{2+}$ inlet is? I don't think the gradient would be a standing gradient, but rather a flow, a $Ca^{2+}$ current running from one end of the cell to the other. Are the $Ca^{2+}$ pumps placed or moved to one end and the $Ca^{2+}$ channels to the other end of the cell? You would then have a $Ca^{2+}$ current, like that seen by Dr Jaffe with his vibrating probe.

*Fay:* Actually, we went to Dr Jaffe's laboratory to try to do exactly that experiment. Unfortunately, these cells move so quickly that we had difficulty keeping the probe in the right position for long enough. It would be interesting if we could see a net flow of $Ca^{2+}$ from the rear to the front.

*Williams:* The system would also need high-binding, slow-reacting $Ca^{2+}$ centres, because to be effective at 50 nM, the $Ca^{2+}$-binding constants must be way above what is normal for something like calmodulin.

*Fay:* Unless for some reason calmodulin is being trapped.

*Williams:* Of course; calmodulin, if it's bound to a special centre, can have totally different binding constants from free calmodulin, and totally different kinetics.

*Thomas:* One way to get a flow of $Ca^{2+}$ through the cytosol is to transfer it back to the release sites through the lumen of the endoplasmic reticulum. Our results indicate there is continuity within the endoplasmic reticulum. Perhaps the $Ca^{2+}$ is released at one end and taken back up by pumps at the other end.

*Fay:* That's entirely possible. There are $InsP_3$-responsive elements at the front—they are just less sensitive or perhaps less concentrated than at the rear.

*Thomas:* Or there could be more pumps at the front.

*Berridge:* Some time ago, Francis Crick considered the processes involved in setting up gradients. Initially, he came up with the idea of a source and a sink but soon realized that this was not the best way of setting up a gradient. He concluded that all the elements in the system have to contribute to the gradient. At one end you have more source than sink and at the other end you have the opposite. The gradient is still based on pump and leak, but it is just biased differently at each end in order to set up the gradient. Every element within the system has to contribute to produce a dynamic gradient. Such a mechanism would explain how the gradient in neutrophils can change so rapidly.

*Thomas:* In the absence of extracellular $Ca^{2+}$, there must be internal recycling, from one end of the cell to the other.

*Michell:* Dr Fay, when you look at the phase images of the cells as they go through the reorientation, can you make any judgement about which parts of the cell reorient first when you ask the cells to reverse their direction?

*Fay:* We've tried to answer this question but it has been a somewhat unrewarding task. One of the earliest things that we see upon reorientation is snapping back of the MTOC towards the nucleus, and then it heads towards the new direction of motion. Another early response to a reorientation of a gradient of the chemoattractant is cessation of the forward motion of the active lamellipod. In many cells we initially see bilateral lamellipods in response to a change in the position of the chemoattractant gradient. The posterior lamellipod is then inhibited and the lamellipod closest to the chemoattractant source continues to extend toward the source of chemoattractant. Kinetically, the response can be divided into two stages: establishment of a new axis and a polarization decision.

*Nathanson:* What is the factor in the serum to which the eosinophils respond?

*Fay:* All of the experiments are done with serum. The active factor in serum is a protein of about 70 kDa, but it is not yet well characterized. It would be nice to have a well-defined molecule that the cell responded to.

## References

Nuccitelli R, Poo M-M, Jaffe LF 1977 Relations between ameboid movement and membrane-controlled electrical currents. J Gen Physiol 69:743–763

Poenie M, Tsien RY, Schmitt-Verhulst AM 1987 Sequential activation and lethal hit measured by $[Ca^{2+}]_i$ in individual cytolytic T cells and targets. EMBO (Eur Mol Biol Organ) J 6:2223–2232

Taylor DL, Blinks JR, Reynolds G 1980 Contractile basis of ameboid movement. 8. Aequorin luminescence during ameboid movement, endocytosis, and capping. J Cell Biol 86:599–607

# General discussion I

## Calcium signalling and localization of endoplasmic reticulum in ascidian embryos

*Speksnijder:* Ascidian embryos show a series of 12–25 repetitive $Ca^{2+}$ transients during the completion of meiosis following fertilization (Speksnijder et al 1989a). Imaging studies with the $Ca^{2+}$-specific photoprotein aequorin show that these transients are initiated in distict regions of the zygote, from which they travel as waves (Speksnijder et al 1990a). Each $Ca^{2+}$ wave is followed by a contractile wave in the same direction. The first wave starts in the region where the sperm triggered the large fertilization wave of $Ca^{2+}$ that precedes the shorter repetitive waves, usually in the animal hemisphere. However, the initiation site of the subsequent waves becomes progressively localized in the vegetal–contraction pole region (see below) of the egg over a period of about five minutes (Speksnijder et al 1990a,b). These observations have recently been confirmed with confocal ratio imaging of $Ca^{2+}$ (A. McDougal & C. Sardet, personal communication 1994). In collaboration with Lionel Jaffe and Christian Sardet I have investigated the basis of this progressive localization of the wave initiation point.

Before fertilization, ascidian eggs show a distinct polarity in their organization. The vegetal hemisphere is enriched in cortical endoplasmic reticulum (ER) and actin filaments and contains a mitochondrion-rich subcortical domain called the myoplasm (Sardet et al 1992). Sperm entry occurs preferentially in the animal hemisphere (Speksnijder et al 1989b). Studies with cytoskeleton-disrupting agents have shown that actin filaments, but not microtubules, are instrumental in maintaining this polarity (Sardet et al 1992).

Fertilization results in a wave of $Ca^{2+}$ release which, in turn, triggers the first phase of ooplasmic segregation. A cortical contraction wave originating from the animal pole propagates towards the vegetal pole region in about two to three minutes, ending in a constricted region close to but not necessarily coincident with the vegetal pole. We have named this region the vegetal–contraction pole area (Speksnijder et al 1990b). The driving force for this contraction wave is provided by actin filaments, because it can be blocked by treatment with cytochalasin but not colchicine.

During this contraction wave, the mitochondrial domain originally present in the entire vegetal hemisphere becomes concentrated in the vegetal–contraction pole region. Concomitantly, a thick layer of ER accumulates between the plasma membrane and the mitochondrial domain at the vegetal–contraction pole (Gualtieri & Sardet 1989). To explore the role of this ER domain, we investigated its organization and fate by labelling the ER network *in vivo* and observing its distribution before and after fertilization in the living egg using laser-scanning confocal microscopy (Speksnijder et al 1993). Using this method, we confirmed our previous observation that in the unfertilized egg the cortical ER network is denser in the vegetal hemisphere than in the animal hemisphere. Furthermore, we showed that the ER network is an extremely dynamic structure, as demonstrated by the striking rearrangements that occur after fertilization. During the wave of cortical contraction that sweeps over the egg's surface after fertilization, the ER becomes progressively concentrated into a cortical location in the vegetal–contraction pole region. This progressive concentration of the ER corresponds closely with the progressive localization of the $Ca^{2+}$ wave initiation site.

Several experimental observations suggest that this ER domain at the vegetal–contraction pole may be involved in initiating the repetitive $Ca^{2+}$ waves that sweep across the zygote after fertilization. First, removal of external $Ca^{2+}$ does not affect the frequency, amplitude or spatial characteristics of these waves, suggesting that they are supported by release from internal stores, probably the ER. Second, when the cortical contraction that concentrates the ER network and the mitochondrial domain at the vegetal–contraction pole is blocked with cytochalasin, the repetitive $Ca^{2+}$ waves no longer start in this region, tending instead to originate from the region of sperm entry (Speksnijder et al 1990a). This suggests that during the cortical contraction some component necessary for the initiation of the repetitive $Ca^{2+}$ waves becomes localized in the vegetal–contraction pole area. The nature of this component was investigated by centrifuging unfertilized eggs such that their cytoplasmic components, including the mitochondrial domain, are displaced. This treatment, which does not affect the polar distribution of cortical components such as the ER network, does not alter the starting position of the $Ca^{2+}$ waves (Speksnijder 1992). This suggests that the mitochondrial domain is not instrumental in wave initiation, and that a component that is initially localized in the cortex of the unfertilized egg and subsequently becomes concentrated in the vegetal–contraction pole region after fertilization is involved in initiating the repetitive $Ca^{2+}$ waves. We suggest that this component is the ER domain at the vegetal–contraction pole.

What is the function of these repetitive waves? They occur in embryos that continue to develop into normal tadpole larvae, which suggests that they are natural events. Their timing initially suggested a role in the completion of meiosis but recent experiments by A. McDougall & C. Sardet (unpublished work 1994) indicate that this is not the case. It is noteworthy, however, that the waves appear to be attenuated, which results in an oscillating gradient of $Ca^{2+}$ in the egg, with $Ca^{2+}$ periodically increased in the vegetal–contraction pole region (Speksnijder et al 1990a). This area is transiently endowed with axial information (Jeffery 1990), so perhaps these waves are somehow involved in pattern development. Alternatively, they may be involved in creating the conditions

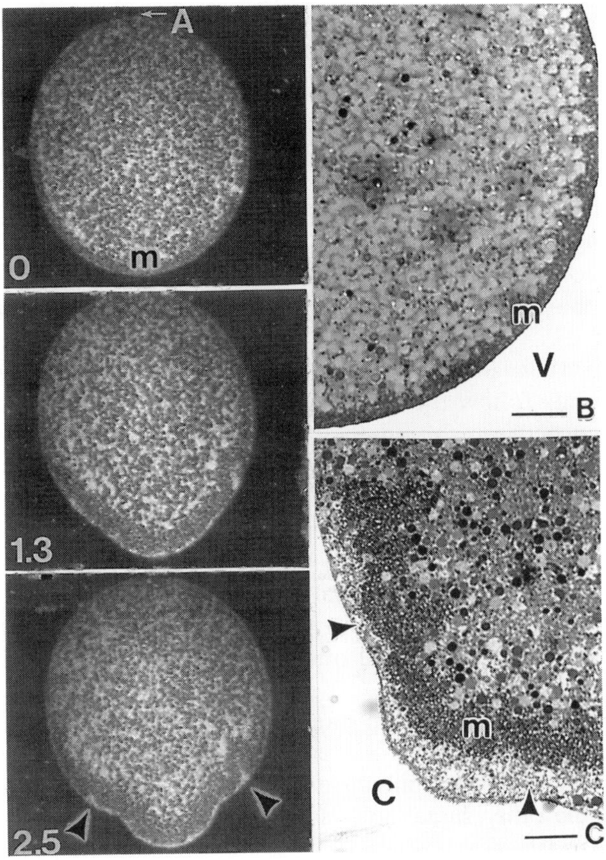


FIG. 1. (*Speksnijder*) Reorganization of the endoplasmic reticulum (ER) network after fertilization of the ascidian egg. (A) Time series of a median confocal section showing the ER network after injection with the carbocyanine dye DiI. The egg is oriented with its animal (A)–vegetal axis parallel to the focal plane. The numbers at the bottom left (0, 1.3, 2.5) indicate time in minutes after fertilization. The series starts with an unfertilized egg, showing the subcortical, ER-poor mitochondrial domain (m) in the vegetal hemisphere. After fertilization, the myoplasm thickens and ER accumulates in the vegetal–contraction pole area (arrowheads). Scale bar, 20 $\mu$m. (B) Section of an unfertilized egg showing the mitochondrial domain (m) closely apposed to the plasma membrane. Scale bar, 10 $\mu$m. (C) Section of the vegetal–contraction pole (C) five minutes after fertilization. A 2–6 $\mu$m thick layer (arrowheads) appears between the folded mitochondrial domain (m) and the plasma membrane of the vegetal–contraction pole constriction in which the ER has accumulated. Scale bar, 10 $\mu$m. Reproduced from Speksnijder et al (1993) by copyright permission of the Rockefeller University Press.

required for the second phase of ooplasmic segregation, which involves a microtubule-based movement of the mitochondrial domain from the vegetal–contraction pole to a more equatorial position. During subsequent divisions, this mitochondrial domain (also called the myoplasm) is inherited by the major precursors of the primary muscle cell lineage. As the ER domain moves simultaneously with the mitochondrial domain, the muscle precursors will become endowed with a large intracellular store of releasable $Ca^{2+}$. This prelocalization might be relevant for fast-developing organisms such as ascidians.

*Berridge:* Do you think there's any correlation between the contractile events you see and the spiking activity? Do you think the spikes are triggering the contractions?

*Speksnijder:* Yes; in dual-imaging studies we have clearly seen that the contractile events follow the $Ca^{2+}$ waves after a delay of 30–40 s. A similar delay was seen between the onset of the fertilization wave of $Ca^{2+}$ and the cortical contraction wave that concentrates the ER at the vegetal–contraction pole.

*Williams:* Do you know anything about where the $Ca^{2+}$ finally goes? A lot of it may be bound somewhere and not in the stores.

*Speksnijder:* Most of it is probably pumped back into the ER, because these $Ca^{2+}$ waves do not require external $Ca^{2+}$. The mitochondria may also be involved in pumping some of the $Ca^{2+}$ back because at the end of the fertilization wave the $[Ca^{2+}]$ drops faster in the area that is dense with mitochondria.

*Thomas:* In the prefertilization stage there is already a greater density of ER at the vegetal pole, yet the $Ca^{2+}$ waves are initiated at the animal pole and then migrate. Do you think that migration is due to the concentration of the ER or is there some other signal such as puff sites or concentration of inositol 1,4,5-trisphosphate?

*Speksnijder:* I don't know. So far we have only morphological observations. We were actually surprised to find that there was a higher density of cortical ER in the vegetal hemisphere of the unfertilized egg. The sperm has a preference for entering the animal hemisphere, so we expected a greater density of ER in the animal hemisphere.

*Putney:* Whatever it is that changes the initiation site from one pole to the other, it probably migrates around the periphery of the cell rather than simply reorganizing at the opposite pole, because I gather that you have seen intermediate cases where the waves were initiated half way around.

*Speksnijder:* I agree, but I was reluctant to make that point because the process is rather complex. The cortical actin network accumulates with the ER in the vegetal–contraction pole region and, since cortical actin filaments interact extensively with the ER, this would suggest that the ER accumulation is the result of an actomyosin-driven migration of the cortical ER network. However, not only do the cortical and subcortical organelles move, but also particles stuck to the outside of the egg plasma membrane. The whole transmembrane complex must be moving and considerable forces must be involved.

*Berridge:* Is this analogous to the rotation in *Xenopus* eggs at fertilization?

*Speksnijder:* It's a little different. In *Xenopus* the cytoplasm rotates along the cortex. In the ascidian egg both membrane cortical and subcortical components move unipolarly to the vegetal pole. Furthermore, the rotation in *Xenopus* is microtubule dependent, whereas this movement in the ascidian egg is dependent on actin filaments.

## References

Gualtieri R, Sardet C 1989 The endoplasmic reticulum network in the ascidian egg: localization and calcium content. Biol Cell 65:301–304

Jeffery WR 1990 Ultraviolet irradiation during ooplasmic segregation prevents gastrulation, sensory cell induction, and axis formation. Dev Biol 140:388–400

Sardet C, Speksnijder JE, Terasaki M, Chang P 1992 Polarity of the ascidian egg cortex before fertilization. Development 115:221–237

Speksnijder JE 1992 The repetitive calcium waves in the fertilized ascidian egg are initiated near the vegetal pole by a cortical pacemaker. Dev Biol 153:259–271

Speksnijder JE, Corson DW, Sardet C, Jaffe LF 1989a Free calcium pulses following fertilization in the ascidian egg. Dev Biol 135:182–190

Speksnijder JE, Jaffe LF, Sardet C 1989b Polarity of sperm entry in the ascidian egg. Dev Biol 133:180–184

Speksnijder JE, Sardet C, Jaffe LF 1990a Periodic calcium waves cross ascidian eggs after fertilization. Dev Biol 142:246–249

Speksnijder JE, Sardet C, Jaffe LF 1990b The activation wave of calcium in the ascidian egg and its role in ooplasmic segregation. J Cell Biol 110:1589–1598

Speksnijder JE, Terasaki M, Hage WJ, Jaffe LF, Sardet C 1993 Polarity and reorganization of the endoplasmic reticulum during fertilization and ooplasmic segregation in the ascidian egg. J Cell Biol 120:1337–1346

# Calcium signalling in cardiac muscle cells

Withrow Gil Wier*, José Ramón López-López*, Philip S. Shacklock† and C. William Balke*†

*Department of Physiology* and Department of Medicine†, University of Maryland School of Medicine, 655 West Baltimore Street, Baltimore, MD 21201, USA*

*Abstract.* In heart cells, several distinct kinds of transient spatial patterns of cytoplasmic calcium ion concentration ($[Ca^{2+}]_i$) can be observed: (1) $[Ca^{2+}]_i$ waves, in which regions of spontaneously increased $[Ca^{2+}]_i$ propagate at high velocity (100 $\mu$m/s) through the cell; (2) $Ca^{2+}$ 'sparks', which are spontaneous, non-propagating changes in $[Ca^{2+}]_i$ that are localized in small ($\approx 2\,\mu$m) subcellular regions; and (3) evoked $[Ca^{2+}]_i$ transients that are elicited by electrical depolarization, in association with normal excitation–contraction (E–C) coupling. In confocal $[Ca^{2+}]_i$ images, evoked $[Ca^{2+}]_i$ transients appear to be nearly spatially uniform throughout the cell, except during their rising phase or during small depolarizations. In contrast to $[Ca^{2+}]_i$ waves and spontaneous $Ca^{2+}$ sparks, evoked $[Ca^{2+}]_i$ transients are triggered by L-type $Ca^{2+}$ channel current and they are 'controlled', in the sense that stopping the L-type $Ca^{2+}$ current stops them. Despite their different characteristics, all three types of $Ca^{2+}$ transient involve $Ca^{2+}$-induced release of $Ca^{2+}$ from the sarcoplasmic reticulum. Here, we address the question of how the autocatalytic process of $Ca^{2+}$-induced $Ca^{2+}$ release, which can easily be understood to underlie spontaneous regenerative ('uncontrolled'), propagating $[Ca^{2+}]_i$ waves, might be 'harnessed', under other circumstances, to produce controlled changes in $[Ca^{2+}]_i$, as during normal excitation–contraction coupling, or changes in $[Ca^{2+}]_i$ that do not propagate. We discuss our observations of $Ca^{2+}$ waves, $Ca^{2+}$ sparks and normal $Ca^{2+}$ transients in heart cells and review our results on the 'gain' of $Ca^{2+}$-induced $Ca^{2+}$ release. We discuss a model involving $Ca^{2+}$ microdomains beneath L-type $Ca^{2+}$ channels, and clusters of $Ca^{2+}$-activated $Ca^{2+}$ release channels in the sarcoplasmic reticulum which may form the basis of the answer to this question.

*1995 Calcium waves, gradients and oscillations. Wiley, Chichester (Ciba Foundation Symposium 188) p 146–164*

The normal signal for contraction in cardiac muscle is a brief rise in the cytoplasmic free calcium ion concentration ($[Ca^{2+}]_i$). Here, we shall be concerned with the spatiotemporal pattern of that brief rise in $[Ca^{2+}]_i$ during normal excitation–contraction (E–C) coupling and with the cellular processes that produce that pattern. Other spatiotemporal patterns of $[Ca^{2+}]_i$ in heart cells, such as $Ca^{2+}$ waves (Takamatsu & Wier 1990a,b, Lipp & Niggli 1993, see also Wier 1993) and $Ca^{2+}$ sparks (Cheng et al 1993) are produced by the same cellular processes, but in modes of operation that are not, strictly speaking, physiological. Nevertheless, these other spatiotemporal patterns of $[Ca^{2+}]_i$ can reveal information about the cellular processes involved in normal E–C coupling and it is from this point of view that all the observed types of $[Ca^{2+}]_i$ transient are reviewed and discussed here.

During normal E–C coupling (for reviews see Wier 1990, Bers 1991, Stern 1992a) most of the $Ca^{2+}$ appearing in the cytoplasm is released from the sarcoplasmic reticulum (SR). The amount of $Ca^{2+}$ released from the SR is controlled by L-type $Ca^{2+}$ channel current in the surface membrane. $Ca^{2+}$ entering via L-type $Ca^{2+}$ channels activates release of $Ca^{2+}$ from the SR by binding to $Ca^{2+}$ release channels (or ryanodine receptors) on the SR. This process is known as $Ca^{2+}$-induced release of $Ca^{2+}$ (Fabiato 1985). The major uncertainty in this scheme is how uncontrolled positive feedback of $Ca^{2+}$-induced $Ca^{2+}$ release is avoided during the normal $[Ca^{2+}]_i$ transient (as might occur if $Ca^{2+}$ released from the SR activated its own further release). Here, we present our findings relevant to this issue from our observations of $Ca^{2+}$ waves, $Ca^{2+}$ sparks and normal $[Ca^{2+}]_i$ transients in heart cells. We also review our earlier results on the 'gain' of $Ca^{2+}$-induced $Ca^{2+}$ release which provide an important clue to the way in which $Ca^{2+}$-induced release of $Ca^{2+}$ might be controlled. Finally, we discuss a model, in terms of these results and current concepts in the field, which involves $[Ca^{2+}]$ domains beneath L-type $Ca^{2+}$ channels, clusters of $Ca^{2+}$-activated SR $Ca^{2+}$ release channels and variability in the amount of $Ca^{2+}$ available to be released. Such a model may be capable of explaining phenomena as apparently diverse as propagating $Ca^{2+}$ waves and tight control of $[Ca^{2+}]_i$ transients by $Ca^{2+}$ influx during normal E–C coupling.

## Calcium waves

Under certain conditions, spontaneous release of intracellular $Ca^{2+}$ in mammalian heart takes the form of propagating changes in $[Ca^{2+}]_i$ known as $[Ca^{2+}]_i$ waves (Wier et al 1987, Lipp & Niggli 1993, Takamatsu & Wier 1990a). For our present purposes, 'spontaneous' release of $Ca^{2+}$ in mammalian heart muscle can be defined as a release of $Ca^{2+}$ from the SR that is not elicited by an action potential (or voltage-clamp pulse). The $[Ca^{2+}]_i$ transient elicited by an action potential in the heart does not normally propagate and is much more spatially uniform within the cell (Takamatsu & Wier 1990b, Lipp & Niggli 1994). In contrast to the situation in other tissues (see Jaffe 1993), $[Ca^{2+}]_i$ waves in the mammalian heart are probably not involved either in normal signal transduction or in normal E–C coupling. Spontaneous $Ca^{2+}$ release seems to represent a response of the cell to a state of high $Ca^{2+}$ loading (Stern et al 1988).

Of particular interest here is the fact that these waves almost certainly propagate by a reaction–diffusion mechanism involving $Ca^{2+}$-induced release of $Ca^{2+}$ and diffusion of $Ca^{2+}$ released from the SR. Under these conditions, $Ca^{2+}$-induced release of $Ca^{2+}$ is clearly not 'controlled' because, once they have started, $[Ca^{2+}]_i$ waves may propagate from one end of the cell to the other. As already mentioned, this differs from the operation of $Ca^{2+}$-induced release of $Ca^{2+}$ during normal E–C coupling, when the release of $Ca^{2+}$ from the SR seems to be under the control of $Ca^{2+}$ entering via L-type $Ca^{2+}$ channels and not of $Ca^{2+}$ already released from the SR.

**Calcium sparks**

Recently, confocal microscopy has been used to observe, in unstimulated heart cells, spontaneous, non-propagating local changes in $[Ca^{2+}]_i$, termed $Ca^{2+}$ sparks (Cheng et al 1993). It was shown clearly in that study that $Ca^{2+}$ sparks result from release of $Ca^{2+}$ through $Ca^{2+}$ release channel(s) in the SR and that $Ca^{2+}$ sparks can trigger propagating $Ca^{2+}$ waves, probably when the SR is highly loaded with $Ca^{2+}$. $Ca^{2+}$ sparks are of substantial interest because they represent release of $Ca^{2+}$ from the SR but under conditions, or in a mode of operation, in which this release does not feed back to cause further $Ca^{2+}$ release and the propagation of a $Ca^{2+}$ wave. An earlier study (O'Neill et al 1990) had also shown that local changes in $[Ca^{2+}]_i$ could fail to propagate. Examples of $Ca^{2+}$ sparks recorded with our confocal imaging system (MRC-600, Bio-Rad Microscience Division, Herts, UK) are presented in Fig. 1. As discussed below, the fact that $Ca^{2+}$ sparks do not propagate or spread could be relevant to the 'control' of $Ca^{2+}$ release from the SR by $Ca^{2+}$ entering via L-type $Ca^{2+}$ channels. Two situations might be imagined that could account for the failure of the spark to spread through the cell via $Ca^{2+}$-induced release of $Ca^{2+}$. If, as has been postulated (Cheng et al 1993), the $Ca^{2+}$ spark is the result of the opening of a single ryanodine receptor, it must be concluded that, for reasons as yet unknown, other ryanodine receptors in the vicinity of the $Ca^{2+}$ spark simply fail to be activated. Alternatively, the $Ca^{2+}$ spark could be the result of the concerted opening of several ryanodine receptors all activated together in a cluster (Stern 1992a), perhaps at an individual transverse-tubule SR junction. In this case, regenerative $Ca^{2+}$ release would occur *within* the cluster, but propagation throughout the cell would not ensue because clusters of ryanodine receptors are physically isolated from each other, allowing them to function independently. (These ideas have also been discussed recently by Wier et al 1994). In either case, the failure of the $Ca^{2+}$ spark to propagate is an important indication that local changes in $[Ca^{2+}]_i$ need not result in uncontrolled positive feedback via $Ca^{2+}$-induced release of $Ca^{2+}$.

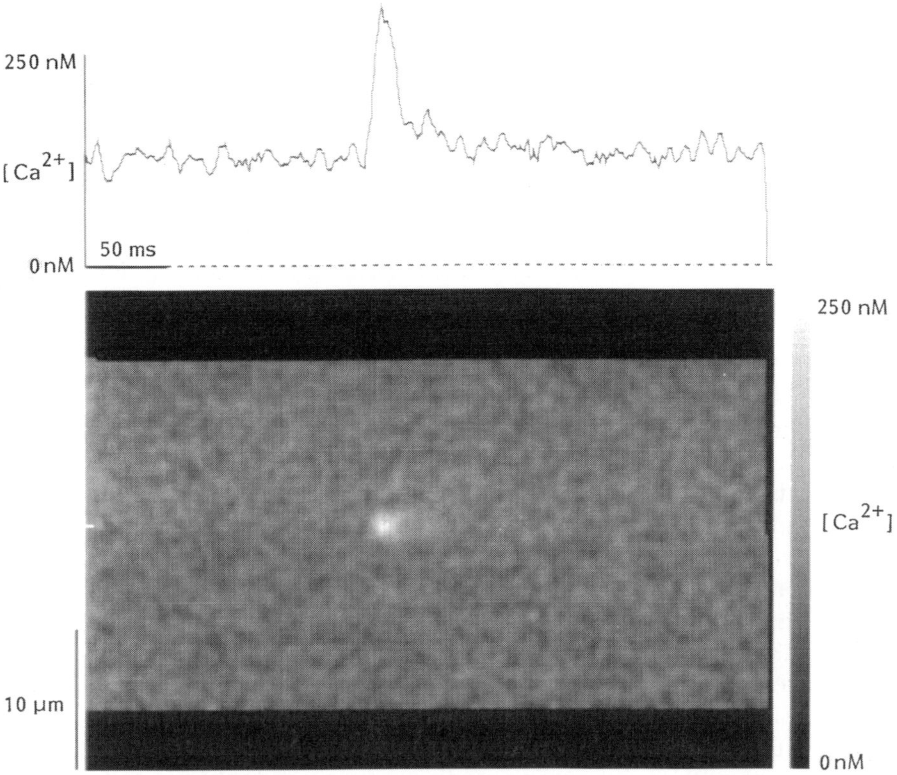


FIG. 1. $Ca^{2+}$ spark recorded by confocal microscopy from a single, voltage-clamped rat ventricular myocyte perfused internally with fluo-3 (100 $\mu$M). The holding potential was −40 mV. The image is a confocal line-scan image of $[Ca^{2+}]_i$ obtained by methods published previously (López-López et al 1994). The recording above the image represents the time course of the spatial average $[Ca^{2+}]_i$ in a volume of about 2 $\mu m^3$, in which the $Ca^{2+}$ spark occurred. The dashed line indicates $[Ca^{2+}]_i$ of 0 nM; the vertical line extends to 250 nM. The horizontal line indicates 50 ms for both the line-scan image and the spatial average $[Ca^{2+}]_i$ recording. Line-scan image calibration, 10 $\mu$m; grey scale on right ranges from 80 nM $[Ca^{2+}]_i$ (black) to 250 nM $[Ca^{2+}]_i$ (white).

## Excitation–contraction coupling calcium transients

As indicated above, our major goal is to understand the normal $[Ca^{2+}]_i$ transients accompanying E–C coupling, as these are the 'normal' $Ca^{2+}$ signals in heart cells. Until recently, the available results indicated that, in contrast to $[Ca^{2+}]_i$ waves and $Ca^{2+}$ sparks, these $[Ca^{2+}]_i$ transients were spatially uniform, although the existence of $[Ca^{2+}]_i$ gradients within the cell during E–C coupling had been predicted from theory (Cannell & Allen 1984, Wier & Yue 1986). $[Ca^{2+}]_i$ transients during E–C coupling have now been studied with confocal microscopy (Cheng et al 1993, Lipp & Niggli 1994, López-López et al 1994), particularly in the line-scanning mode, as discussed below. These studies have revealed some spatial heterogeneity within the $[Ca^{2+}]_i$ transients that are evoked during an action potential (Cheng et al 1993) and during voltage-clamp depolarization (López-López et al 1994). This heterogeneity is possibly the result of the evocation of local $[Ca^{2+}]_i$ transients that are similar to the spontaneous $Ca^{2+}$ sparks discussed above. Although confocal microscopy has the potential to reveal information on the spatial heterogeneity of $[Ca^{2+}]_i$ transients that may be very useful in understanding E–C coupling, the bulk of the studies to date have been done on spatially averaged (i.e., whole-cell) $[Ca^{2+}]_i$ transients, and we therefore discuss these findings first.

Figure 2 illustrates spatially averaged $[Ca^{2+}]_i$ transients observed during normal E–C coupling in a typical rat ventricular myocyte (Balke et al 1994). In this study, cells exhibiting spontaneous or 'uncontrolled' $Ca^{2+}$ release from the SR, manifested as $Ca^{2+}$ waves, were deliberately excluded from consideration. In such (normal) cells, the bell-shaped dependence of the $[Ca^{2+}]_i$ transient on clamp-pulse potential and the existence of $[Ca^{2+}]_i$ transients upon repolarization are indicative of release of $Ca^{2+}$ from the SR that is controlled by L-type $Ca^{2+}$ current via $Ca^{2+}$-induced release of $Ca^{2+}$ (Beuckelmann & Wier 1988). Also in such cells, it has been shown that the rising phase of the $[Ca^{2+}]_i$ transient may be terminated prematurely by stopping the $Ca^{2+}$ current through depolarization to a membrane potential at which the driving force for $Ca^{2+}$ entry is small or zero (Cleemann & Morad 1991, Wier et al 1994). Taken together, these results indicate unequivocally that release of $Ca^{2+}$ from the SR of mammalian cardiac muscle is indeed $Ca^{2+}$ induced and that this release is controlled by the $Ca^{2+}$ entering via L-type $Ca^{2+}$ channels.

## The mechanism of excitation–contraction coupling

A full understanding of the mechanisms that might be involved in controlling $Ca^{2+}$ release from the SR during E–C coupling requires knowledge of the actual SR $Ca^{2+}$ release flux ($F_{SR,rel}$). An important clue to the mechanism of E–C coupling has emerged from studies on spatially averaged $[Ca^{2+}]_i$ transients. It has been shown that the 'gain' of $Ca^{2+}$ release from the SR varies with membrane voltage. We have previously (Wier et al 1994) advanced the explanation that this is because of local events (namely, L-type $Ca^{2+}$ channel currents and evoked local $[Ca^{2+}]_i$ transients) that differ from the macroscopically observable events discussed above. As mentioned above and discussed later, the nature of these putative local events can now be investigated directly, to some extent, with confocal microscopy. First, however, we discuss the results from spatially averaged $[Ca^{2+}]_i$ transients, that led us to postulate the involvement of such local events in order to explain E–C coupling, particularly the phenomenon of variable gain in $Ca^{2+}$-induced release

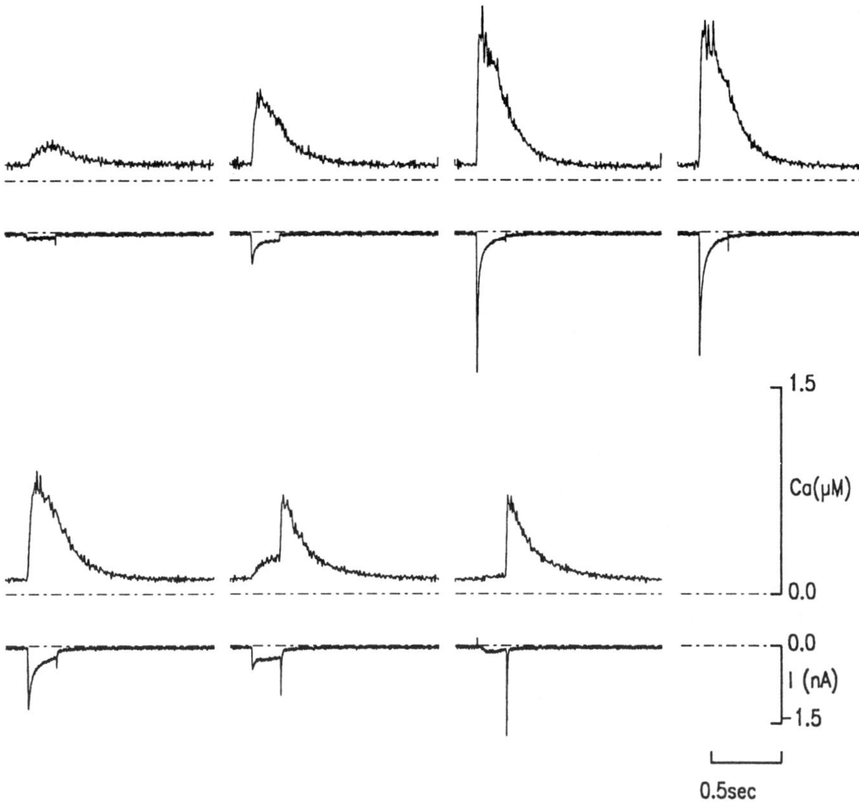


FIG. 2. Representative $[Ca^{2+}]_i$ transients and $Cd^{2+}$-sensitive membrane currents in a rat cardiac ventricular cell. ($Cd^{2+}$-sensitive current is L-type $Ca^{2+}$ channel current.) The holding potential was −40 mV and clamp-pulse depolarizations of 200 ms duration were given to −30, −20, 0, +20, +40, +60 and +80 mV (pulse voltage increasing from left to right and top to bottom). Membrane currents are $Ca^{2+}$ currents through L-type $Ca^{2+}$ channels. Temperature, 21 °C; cell capacity, 149 pF; $K^+$ was replaced by $Cs^+$ and tetraethylammonium, inside and out. From Balke et al (1994).

of $Ca^{2+}$. This necessitates a brief discussion of the measurement of SR $Ca^{2+}$ release fluxes.

### *Sarcoplasmic reticulum calcium fluxes*

The E–C coupling $[Ca^{2+}]_i$ transient is the result of many cellular processes that influence $[Ca^{2+}]_i$ and not simply of release of $Ca^{2+}$ from the SR (Wier 1990). Because this release of $Ca^{2+}$ from the SR cannot be measured directly in intact cells, it must be computed, according to a theory of the $[Ca^{2+}]_i$ transient. This computation is based on experimental measurements of the $Ca^{2+}$ current and the free $Ca^{2+}$ transient. The theory of the $[Ca^{2+}]_i$ transient in cardiac muscle used for the computation is similar to that used for the same purpose in skeletal muscle (Melzer et al 1987) and has been published already (Sipido & Wier 1991, Wier et al 1994). Developed before spatial resolution of $[Ca^{2+}]_i$ transients became possible through confocal microscopy, this theory produces the *total* whole-cell $F_{SR,rel}$. The theory relies on the assumption that the evoked $[Ca^{2+}]_i$ transients are spatially uniform. The first step in computing $F_{SR,rel}$ is to account quantitatively for the uptake of $Ca^{2+}$ by the SR $Ca^{2+}$-pumping ATPase, so that the flux through it ($F_{SR,pump}$) can be subtracted from the net SR $Ca^{2+}$ flux (Sipido & Wier 1991) to yield $F_{SR,rel}$. The methods for characterizing the functioning of the SR $Ca^{2+}$-pumping ATPase, during the declining phase of the $[Ca^{2+}]_i$ transient, have been described in detail previously (Balke et al 1994). Briefly, analysis of the declining phase of the $[Ca^{2+}]_i$ transient results in estimations of the pump parameters ($K_m$ and $V_{max}$), and the leak of $Ca^{2+}$ from the SR ($F_{SR,leak}$) that is needed to maintain a steady-state $[Ca^{2+}]_i$ at rest. The smooth line through the declining phase of the $[Ca^{2+}]_i$ transient illustrated in Fig. 3 has been calculated according to these methods, with $V_{max} = 0.14$ mM/s, $K_m = 0.24\,\mu$M and $F_{SR,leak} = 0.018$ mM/s, with the assumption that two $Ca^{2+}$ ions bind to each ATPase and that two ATPases cooperate in the transport of four $Ca^{2+}$ ions (Klein et al 1991). To calculate $F_{SR,rel}$, however, we then assume that this characterization of the processes removing $Ca^{2+}$ from the cytoplasm applies, with the same parameters, throughout the entire $[Ca^{2+}]_i$ transient, and not just its declining phase. $F_{SR,rel}$ is then calculated as shown in **1**:

$$F_{Sr,rel} = \frac{d[Ca^{2+}]}{dt} - F_{SR,pump} - F_{SR,leak} - F_{ICa} + \sum_{n=1}^{N} \frac{d[CaL]_n}{dt} \qquad \mathbf{1}$$

where $F_{ICa}$ is calculated from the measured $Ca^{2+}$ current and other measured parameters according to methods described by Sipido & Wier (1991) and, in rat cells, by Balke et al (1994). It is assumed that there are five ($N$) species of $Ca^{2+}$-binding ligands (L): calmodulin, troponin C, indo-1, SR binding sites and plasmalemmal binding sites. The $Ca^{2+}$ bound to the $n^{th}$ $Ca^{2+}$-binding ligand (L) is thus indicated by $[CaL]_n$. The flux due to the $Na^+/Ca^{2+}$ exchanger is not considered because the experiments are done in the total absence of $Na^+$. The $[Ca^{2+}]_i$ transient, $F_{ICa}$, $F_{SR,rel}$, $F_{SR,leak}$ and $F_{SR,pump}$ shown in Fig. 3 are representative of those obtained in more than 25 different cells.

A notable feature of the result illustrated in Fig. 3 is that the peak flux of $Ca^{2+}$ from the SR is about 12-fold greater than the peak influx through L-type $Ca^{2+}$ channels. This fact is the central issue in $Ca^{2+}$-induced release of $Ca^{2+}$ from the SR: how can SR $Ca^{2+}$ release be controlled by the relatively small influx of $Ca^{2+}$, given that so much more $Ca^{2+}$ is released from the SR

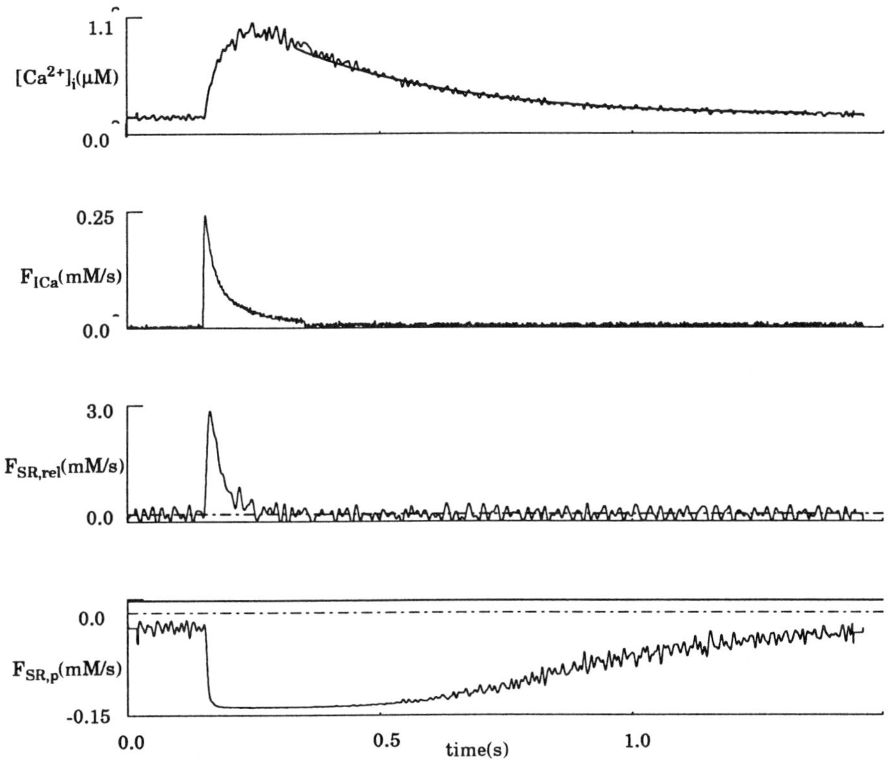


FIG. 3. Computation of sarcoplasmic reticulum (SR) $Ca^{2+}$ fluxes from measured intracellular $Ca^{2+}$ concentration ($[Ca^{2+}]_i$) transients and $Ca^{2+}$ currents. (Data from Wier et al 1994.) From top to bottom, the records are: a $[Ca^{2+}]_i$ transient; the flux through L-type $Ca^{2+}$ channels ($F_{ICa}$); the calculated SR release flux ($F_{SR,rel}$); and the calculated SR pump flux ($F_{SR,p}$). Dot-dash lines are at zero. The solid line in the lowest record indicates the level of the constant leak into the cytoplasm, $F_{SR,leak}$. The smooth line superimposed on the $[Ca^{2+}]_i$ transient (top trace) is the $Ca^{2+}$ removal function, as described in the text.

and that it might activate its own further release? This question can be addressed further by considering the 'gain' of SR $Ca^{2+}$ release, as discussed next.

### *The theory of gain in sarcoplasmic reticulum calcium release*

The operation of $Ca^{2+}$-induced release of $Ca^{2+}$ from the SR has been quantified through the use of the concept of 'gain' (Niggli & Lederer 1990, Stern 1992a, Wier et al 1994). The concept of gain is essential to the issues being considered here because uncontrolled $[Ca^{2+}]_i$ transients, such as $[Ca^{2+}]_i$ waves, could be considered to represent the system operating in a high-gain mode. Yet, high gain is inconsistent with control of SR

$Ca^{2+}$ release during normal E–C coupling, as described above and discussed extensively by Stern (1992a).

Gain has been defined recently (Wier et al 1994) as the ratio of the peak $F_{SR,rel}$ to the peak $F_{ICa}$, a definition similar to that advanced by Stern (1992a). When defined in this way, gain is a function of membrane potential (i.e., it is variable). Gain may also vary in time, but, at present, its measurement has been restricted to the peaks of $F_{SR,rel}$ and $F_{ICa}$. Figure 4 shows the voltage dependence of $F_{SR,rel,peak}$, $F_{ICa,peak}$ (A), and gain (B). Clearly, gain is not related to the whole-cell $Ca^{2+}$ current; the voltage dependencies of these two quantities are fundamentally different. Nevertheless, as discussed above, SR $Ca^{2+}$release is controlled by the $Ca^{2+}$ current (Beuckelmann & Wier 1988). However, the voltage dependence of gain is similar to that of the unitary L-type $Ca^{2+}$ current (compare Figs 4B and 4C). Therefore, we could hypothesize that, during normal E–C coupling, gain is related to the *size of the unitary L-type* $Ca^{2+}$ *channel current*. This led us to postulate that the essential signal for SR $Ca^{2+}$ release during E–C coupling is actually a transient 'microdomain' of $[Ca^{2+}]_i$, produced solely by influx through an L-type $Ca^{2+}$ channel, in the region of a cluster of SR $Ca^{2+}$ release channels. Direct experimental examination of this hypothesis is impossible at present; therefore, we shall consider this hypothesis theoretically below.

The first step in considering how unitary $Ca^{2+}$ currents might control SR $Ca^{2+}$ release is to consider the voltage dependence of their magnitude. In previous publications, we have measured single L-type $Ca^{2+}$ channel currents (Rose et al 1992) in elevated external $[Ca^{2+}]$ (10 mM), but under otherwise physiological conditions, such as in the absence of $Ca^{2+}$-channel agonists (which increase channel open times). Clearly, as stated above, the dependence of single-channel current on voltage, shown in Fig. 4C, is very similar to that of gain.

From published computations, it is known already that the $Ca^{2+}$ transient in a semi-infinite medium bounded by a membrane containing a $Ca^{2+}$ channel (Chad & Eckert 1984, Simon & Llinás 1985, Bers & Peskoff 1991, Stern 1992b) rises extremely rapidly to very high concentrations ($\approx 0.1$ mM) within a fraction of a millisecond after opening of the channel, that a steady-state gradient of $[Ca^{2+}]_i$ is reached quickly, and that the gradient of $[Ca^{2+}]_i$ collapses very rapidly upon closure of the channel. In cardiac muscle, L-type $Ca^{2+}$ channels involved in E–C coupling are not facing an infinite space, but are facing a restricted space (i.e., 'junctional region') of about 20 nm in width (Frank 1990) between the transverse (T) tubule and the terminal cistern of the SR. The theoretical microdomain of $[Ca^{2+}]_i$ in this junctional region, or cleft, during and after a brief (0.2 ms) opening of a hypothetical L-type $Ca^{2+}$ channel can be computed (Fig. 5). It can be seen that the $[Ca^{2+}]$ in the junction rapidly reaches a steady-state distribution, with

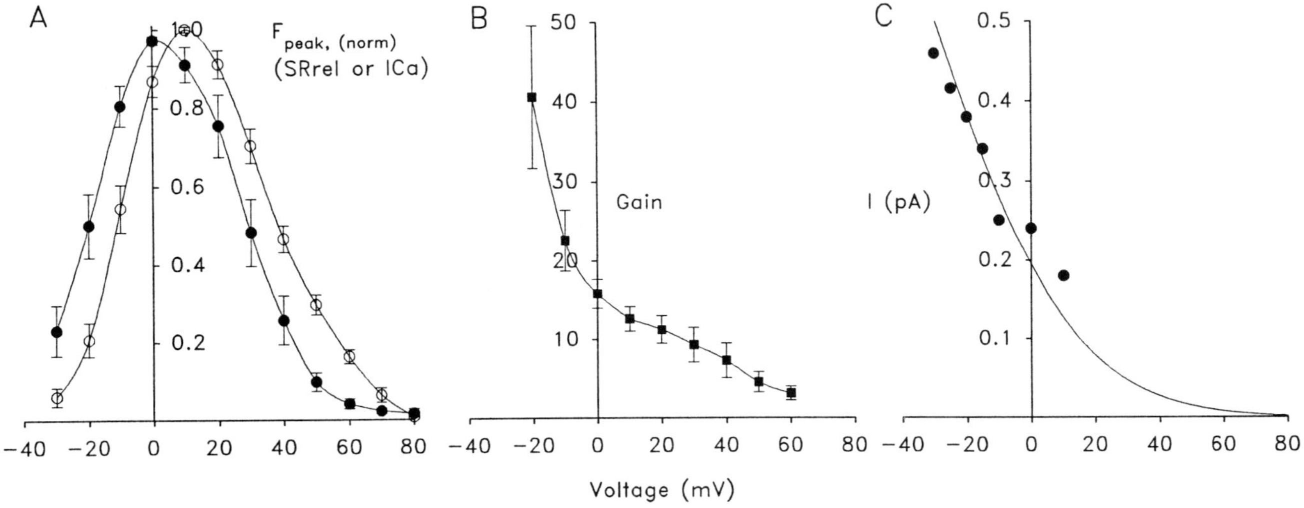


FIG. 4. Voltage dependence of the sarcoplasmic reticulum release flux ($F_{SR,rel}$), the flux through L-type $Ca^{2+}$ channels ($F_{ICa}$), gain and unitary current through L-type $Ca^{2+}$ channels. Mean ± SEM values ($n = 9$) of the normalized peak fluxes $F_{SR,rel,peak}$ (●) and $F_{ICa,peak}$ (○) (A) and gain (B) are plotted against membrane potential. (Data from Wier et al 1994.) (C) shows the single-channel current–voltage relationship for L-type $Ca^{2+}$ channels. The data were taken from Rose et al (1992) and were fitted to the Goldman–Hodgkin–Katz equation (Hodgkin & Katz 1949) (extracellular $[Ca^{2+}]$, 10 nM; intracellular $[Ca^{2+}]$, 0.1 $\mu$M; permeability (P), $1.0 \times 10^{-4}$ cm/s).

a steep gradient of concentration along and across the cleft. This gradient collapses in less than 50 $\mu$s when the channel closes. The microdomains produced in the steady-state at different membrane depolarizations, varying from those arising from relatively large single-channel currents (i.e., $-30$ mV) to those arising from relatively small single-channel currents ($+30$ mV) are shown in Fig. 5 (column labelled steady-state). The essential feature of the $[Ca^{2+}]_i$ microdomain for our purposes is that its amplitude and spatial extent are determined mainly by the amplitude of single-channel currents. The possible significance of this is discussed below.

### *Calcium transients recorded with confocal microscopy*

Confocal microscopy of $[Ca^{2+}]_i$ in heart cells has the potential to reveal spatial heterogeneity of $[Ca^{2+}]_i$ during E–C coupling, should it exist, although it would not be capable of detecting microdomains of $[Ca^{2+}]_i$ of the type predicted above. Figure 6 illustrates $[Ca^{2+}]_i$ transients recorded with confocal microscopy. The results illustrate that $[Ca^{2+}]_i$ transients studied in this way exhibit the graded (i.e., controlled) release of $Ca^{2+}$ in response to $Ca^{2+}$ current and other phenomena established previously using whole-cell recording techniques in mammalian cardiac cells (Barcenas-Ruiz & Wier 1987, Cannell et al 1987, Beuckelmann & Wier 1988, Balke et al 1994). The phenomenon of variable gain described above (Wier et al 1994) is evident from a comparison of the responses illustrated in Figs 6A and 7C (i.e., similar currents, at different membrane potentials, produced very different transients). It is this phenomenon that we have postulated previously is accounted for by local events different from the macroscopic events (local $[Ca^{2+}]_i$ transients and unitary L-type $Ca^{2+}$ channel currents) during E–C coupling. Such local events may be evident in the $[Ca^{2+}]_i$ transients evoked by small clamp-pulse depolarizations (Fig. 6A). Clearly, these images show a heterogeneity of $[Ca^{2+}]_i$ that appears similar to the spontaneous $Ca^{2+}$ sparks. These observations lend support to the idea that the evoked $[Ca^{2+}]_i$ transient during E–C coupling is composed of local events similar to the spontaneous $Ca^{2+}$ sparks (Fig. 1).

## **Summary**

The results indicate that release of $Ca^{2+}$ from the SR is not highly regenerative, unless the SR is highly or abnormally loaded with $Ca^{2+}$, which is usually the condition under which $[Ca^{2+}]_i$ waves occur. Two observations support this conclusion. (1) Under normal conditions, spontaneous release of $Ca^{2+}$ from the SR occurs rarely, in the form of $Ca^{2+}$ sparks, but such changes in $[Ca^{2+}]_i$ do not propagate. (2) The 'gain' of $Ca^{2+}$ release from the SR has a voltage dependence similar to that of the single-channel $Ca^{2+}$ current. This could come

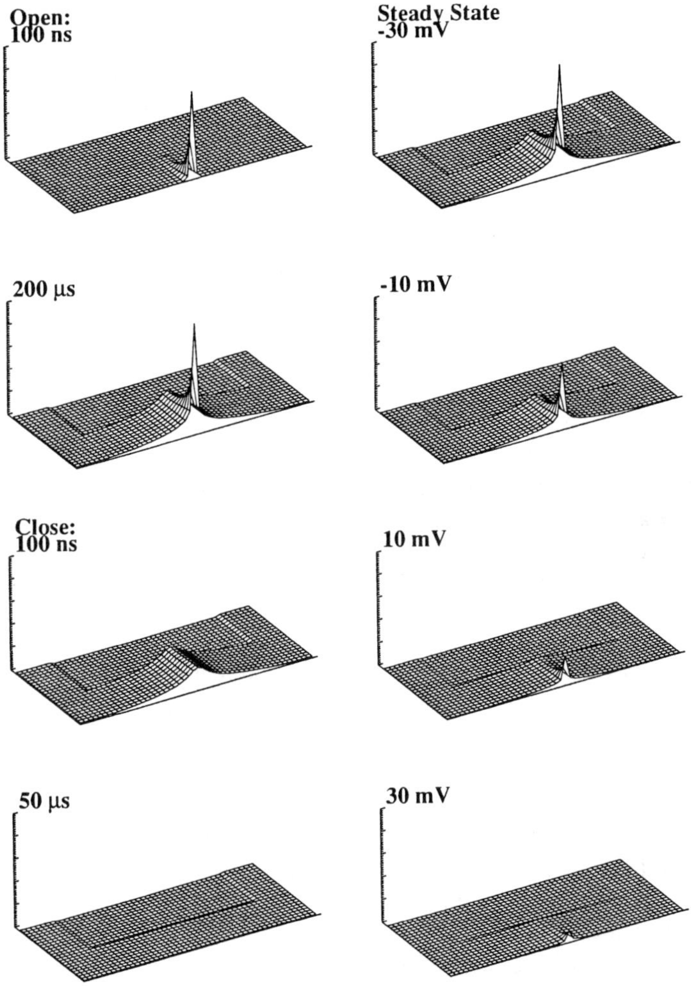


FIG. 5. Computed $[Ca^{2+}]$ in the junctional space between the transverse (T) tubule and the SR. The rectangular space showing intracellular $[Ca^{2+}]$ of 0.0 $\mu$M corresponds to the SR. The L-type channel is placed in the T tubule membrane in the centre of the junction. $[Ca^{2+}]$, on the vertical axis, ranges from 0 to 50 $\mu$M. (Resting $[Ca^{2+}]$, 100 nM.) The *left* part shows the time course of the $Ca^{2+}$ transient elicited by the opening of an L-type $Ca^{2+}$ channel at $-30$ mV for 0.2 ms. The *right* part shows the steady-state intracellular $[Ca^{2+}]$ obtained when the channel is open at the indicated voltages.

about only if, on the macroscopic level, $Ca^{2+}$ release from the SR is not, in fact, highly regenerative. A possible explanation for the fact that $Ca^{2+}$ sparks do not normally propagate is that although regenerative $Ca^{2+}$ release may occur within a cluster of ryanodine receptors, the clusters are sufficiently far

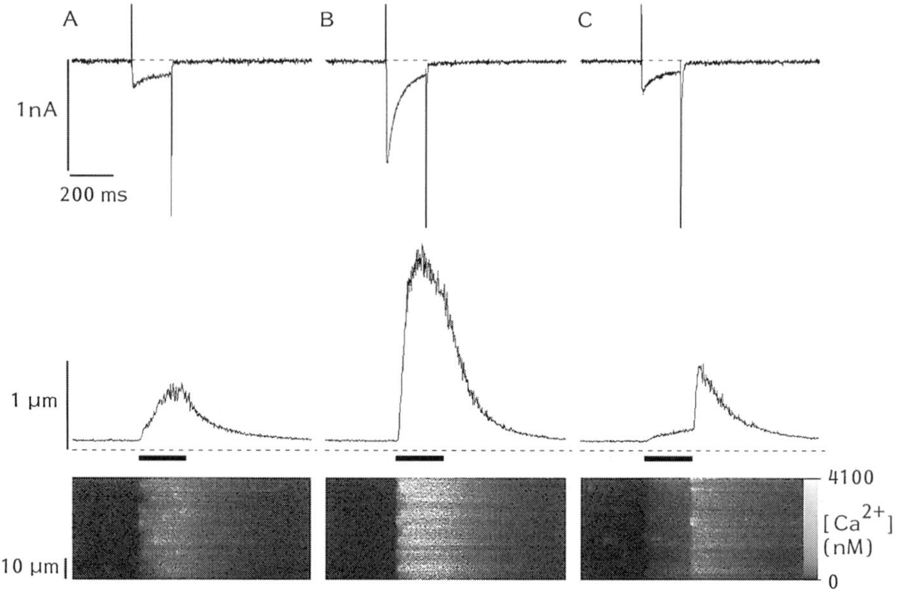


FIG. 6. Confocal $[Ca^{2+}]_i$ transients during excitation–contraction coupling. (Data from López-López et al 1994.) Confocal $[Ca^{2+}]_i$ transients and $Ca^{2+}$ currents in a voltage-clamped rat ventricular myocyte perfused internally with fluo-3 (100 $\mu$M). The cell was depolarized for 200 ms, from a holding potential of −40 mV, to −30 mV (A), +10 mV (B) and +45 mV (C). The upper traces in each case are L-type $Ca^{2+}$ channel current, obtained as the $Cd^{2+}$-sensitive current. $Cs^+$ replaced $K^+$. The lower traces in each case are the 'line $[Ca^{2+}]_i$ transient', obtained from the line-scan images (bottom) as described in López-López et al (1994). The $Ca^{2+}$ current, the plot of average $[Ca^{2+}]_i$ (the line $[Ca^{2+}]_i$ transient) and the successive 'line scans' (which constitute a 'line-scan image') have been aligned temporally. At all clamp-pulse potentials, the line $[Ca^{2+}]_i$ transient is similar to the spatially averaged $[Ca^{2+}]_i$ transient recorded previously in voltage-clamped rat ventricular cells.

apart for neighbouring clusters not to be activated. During normal E–C coupling, a similar situation may also exist. In this case, all the ryanodine receptors within a cluster are activated and the probability of the entire cluster being activated is determined initially by a 'microdomain' of $[Ca^{2+}]_i$, established by influx of $Ca^{2+}$ through an L-type $Ca^{2+}$ channel.

An alternative explanation for these phenomena would be that $Ca^{2+}$ sparks are, in fact, the result of the opening of a single ryanodine receptor. Evidently, neighbouring ryanodine receptors are not activated, although the mechanism preventing them from being activated by the $Ca^{2+}$ released in their vicinity is not known. In this case, the very high $[Ca^{2+}]_i$ within the microdomain established by influx through a single L-type $Ca^{2+}$ channel would determine the number of ryanodine receptors activated, and hence the dependence of the gain of $Ca^{2+}$ release from the SR on membrane voltage would be similar to that of the unitary current amplitude.

$[Ca^{2+}]_i$ waves would be produced when the SR became 'overloaded' with $Ca^{2+}$, such that gain would become very high. In this case, as suggested previously (Wier et al 1994), $Ca^{2+}$-induced release of $Ca^{2+}$ may propagate via a reaction–diffusion mechanism from cluster to cluster. An additional factor may be the modulation of release of $Ca^{2+}$ from the SR by $Ca^{2+}$ within the lumen of the SR (Ikemoto et al 1991).

## References

Balke CW, Egan TM, Wier WG 1994 Processes that remove calcium from the cytoplasm during excitation–contraction coupling in intact rat heart cells. J Physiol 474:447–462

Barcenas-Ruiz L, Wier WG 1987 Voltage dependence of intracellular $[Ca^{2+}]$-transients in guinea-pig ventricular myocytes. Circ Res 61:148–155

Bers DM 1991 Excitation–contraction coupling and cardiac contractile force. Kluwer Academic, Dordrecht

Bers DM, Peskoff A 1991 Diffusion around a cardiac calcium channel and the role of surface bound calcium. Biophys J 59:703–721

Beuckelmann DJ, Wier WG 1988 Mechanism of release of calcium from sarcoplasmic reticulum of guinea-pig cardiac cells. J Physiol 405:233–255

Cleemann L, Morad M 1991 Role of $Ca^{2+}$ channel in cardiac excitation–contraction coupling in the rat: evidence from $Ca^{2+}$ transients and contraction. J Physiol 432:283–312

Cannell MB, Allen DG 1984 Model of calcium movements during activation in the sarcomere of frog skeletal muscle. Biophys J 45:913–925

Cannell MB, Berlin JR, Lederer WJ 1987 Effect of membrane potential changes on the calcium transient in single rat cardiac muscle cells. Science 238:1419–1423

Chad JE, Eckert R 1984 Calcium domains associated with individual channels can account for anomalous voltage relations of Ca-dependent responses. Biophys J 45:993–999

Cheng H, Lederer WJ, Cannell MB 1993 Calcium sparks: elementary events underlying excitation–contraction coupling in heart muscle. Science 262:740–744

Fabiato A 1985 Time and calcium dependence of activation and inactivation of calcium-induced release of calcium from the sarcoplasmic reticulum of a skinned canine cardiac Purkinje cell. J Gen Physiol 85:247–289

Frank JS 1990 Ultrastructure of the unfixed myocardial sarcolemma and cell surface. In: Langer GA (ed) Calcium and the heart. Raven Press, New York, p 1–25

Hodgkin AL, Katz B 1949 The effect of sodium ions on the electrical activity of the giant axon of the squid. J Physiol 108:37–77

Ikemoto N, Antoniu B, Kang JJ, Meszaros LG, Ronjat M 1991 Intravesicular calcium transient during calcium release from sarcoplasmic reticulum. Biochemistry 30:5230–5237

Jaffe LF 1993 Classes and mechanisms of calcium waves. Cell Calcium 14:736–745

Klein MG, Kovacs L, Simon BJ, Schneider MF 1991 Decline of myoplasmic $Ca^{2+}$, recovery of $Ca^{2+}$ release and sarcoplasmic $Ca^{2+}$ pump properties in frog skeletal muscle. J Physiol 414:639–671

Lipp P, Niggli E 1993 Microscopic spiral waves reveal positive feedback in subcellular calcium signalling. Biophys J 65:2272–2276

Lipp P, Niggli E 1994 Sodium current-induced calcium signals in isolated guinea-pig ventricular myocytes. J Physiol 474:439–446
López-López JR, Shacklock PS, Balke CW, Wier WG 1994 Local, stochastic release of $Ca^{2+}$ in voltage-clamped rat heart cells: visualization with confocal microscopy. J Physiol 480:21–29
Melzer W, Rios E, Schneider MF 1987 A general procedure for determining the rate of calcium release from the sarcoplasmic reticulum in skeletal muscle fibers. Biophys J 51:849–863
Niggli E, Lederer WJ 1990 Voltage-independent calcium release in heart muscle. Science 250:565–568
O'Neill SC, Mill JG, Eisner DA 1990 Local activation of contraction in isolated rat ventricular myoctes. Am J Physiol 258:C1165–C1168
Rose WC, Balke CW, Wier WG, Marban E 1992 Macroscopic and unitary properties of physiological ion flux through L-type $Ca^{2+}$ channels in guinea-pig heart cells. J Physiol 456:267–284
Simon SM, Llinás RR 1985 Compartmentalization of the submembrane calcium activity during calcium influx and its significance in transmitter release. Biophys J 48:485–498
Sipido KR, Wier WG 1991 Flux of $Ca^{2+}$ across the sarcoplasmic reticulum of guinea-pig cardiac cells during excitation–contraction coupling. J Physiol 435:605–630
Stern MD 1992a Theory of excitation–contraction coupling in cardiac muscle. Biophys J 63:497–517
Stern MD 1992b Buffering of calcium in the vicinity of a channel pore. Cell Calcium 13:183–192
Stern MD, Capogrossi MC, Lakatta EG 1988 Spontaneous calcium release from the sarcoplasmic reticulum in myocardial cells: mechanisms and consequences. Cell Calcium 9:247–256
Takamatsu T, Wier WG 1990a Calcium waves in mammalian heart: quantification of origin, magnitude, wave-form, and velocity. FASEB (Fed Am Soc Exp Biol) J 4:1519–1525
Takamatsu T, Wier WG 1990b High temporal resolution video imaging of intracellular calcium. Cell Calcium 11:111–120
Wier WG 1990 Dynamics of control of cytosolic calcium ion concentration. Annu Rev Physiol 52:467–485
Wier WG 1993 $[Ca^{2+}]_i$-waves in heart: more than a passing fancy. Biophys J 65:2270–2271
Wier WG, Yue DT 1986 Intracellular Ca-transients underlying the short-term force-interval relationship in ferret ventricular myocardium. J Physiol 376:507–530
Wier WG, Cannell MB, Berlin JR, Marban E, Lederer WJ 1987 Cellular and subcellular heterogeneity of $[Ca^{2+}]_i$ in single heart cells revealed by fura-2. Science 235:325–328
Wier WG, Egan TM, López-López JR, Balke CW 1994 Local control of excitation–contraction coupling in rat heart cells. J Physiol 474:463–471

## DISCUSSION

*Berridge:* This was really exciting and tied in nicely with what Ian Parker talked about. There seems to be a kind of spatial locking between the trigger system that's providing the signal and the amplification system. Your little microdomain functions as a 'chemical synapse' to lock the two processes together. Basically, this was what Ian Parker was saying with regard to $Ca^{2+}$ puffs, which represent the trigger $Ca^{2+}$ that activates the intracellular stores. If the puff fails to lock into the regenerative $Ca^{2+}$ stores, the puff never spreads and it decays away.

*Parker:* Dr Wier, you say that a local microdomain of $Ca^{2+}$ triggers the internal release, but to explain the graded relationship, must the conductance of the sarcoplasmic reticulum (SR) channels be lower than that of the L-type $Ca^{2+}$ channels? Otherwise, I would expect a similar microdomain which should also give rise to an all-or-none release characteristic.

*Wier:* Yes, that is an unresolved issue. If the flux of $Ca^{2+}$ from an individual ryanodine receptor is similar or larger, as might be the case, than that through an L-type $Ca^{2+}$ channel, then that might establish a domain which might activate others.

*Berridge:* Doesn't this depend on the state of loading of the stores? If the SR is overloaded, then the regenerative process clearly takes place, but if the stores are not sufficiently loaded, the sensitivity of the receptors will be reduced such that they fail to lock in on the initial signal.

*Wier:* That's what would be concluded from the fact that one doesn't observe the out-of-control, regenerative phenomenona except under conditions where the SR is overloaded with $Ca^{2+}$. The final explanation will be rather more complicated than this, involving the geometry of the ryanodine receptor, the exact way in which it releases $Ca^{2+}$, and the fact that these microdomains can be very steep and very brief. The geometry of the relationship between the L-type $Ca^{2+}$ channel and the ryanodine receptor may well be such that the activating site at which $Ca^{2+}$ acts on the ryanodine receptor just doesn't 'see' the $Ca^{2+}$ released from the ryanodine receptor. There is other evidence for that. The inactivation of the L-type $Ca^{2+}$ channel is $Ca^{2+}$ dependent, in part, and that inactivation is much more sensitive to $Ca^{2+}$ flux through the L-type $Ca^{2+}$ channel itself than it is to $Ca^{2+}$ release through the SR. So there are other lines of evidence suggesting that for some reason the released $Ca^{2+}$ doesn't get back to this site.

*Putney:* Some time ago, Fabiato (1985a) showed that the ability of $Ca^{2+}$ applied directly to the SR to induce $Ca^{2+}$-induced $Ca^{2+}$ release depended more on the rate at which it was applied than on the actual amount. In terms of your model, does that mean that you have to apply $Ca^{2+}$ at a rapid rate to raise the [$Ca^{2+}$] before it is buffered quickly, for example, by the SR? What's the relationship between that phenomenon and the ability to get a graded response?

*Wier:* In Fabiato's model (1985a,b,c) the graded response and the phenomenon you mentioned were both the result of the fact that there was supposed to be an activating site and an inactivating site, with $Ca^{2+}$ binding more rapidly to the activating site than to the inactivating site, although the affinity of the inactivating site was higher. That could explain how with a very high [$Ca^{2+}$] release didn't occur at all, because the inactivating site predominated. This is a critical issue, which bears on what turns off the release. The results in the literature are equivocal. To my knowledge, there is no good evidence that there is $Ca^{2+}$-dependent inactivation of release in these cells.

*Putney:* Is this not a property of the ryanodine receptor?

*Wier:* You can show inactivation of the ryanodine receptor in bilayers or reconstituted channels, but physiological, functional, evidence is lacking. One can also show slow inactivation after flash photolysis of caged $Ca^{2+}$ around SR release channels in a bilayer. That's slow but it might still have the characteristics appropriate for turning off $Ca^{2+}$ release from the SR.

I'd like to raise the notion that an inactivation mechanism may not be needed, that the activity might just die out by itself as a result of the stochastic properties of the channels. Mike Stern (1992) published this first. He noted that with a small cluster of channels, even if $[Ca^{2+}]$ is high so that the channels are activated, there is still a finite probability that the channels will close. In bilayers, with a high $[Ca^{2+}]$, the open probability may still be only 0.5. Then, half the time the channel would be closed. With several of these channels together, there is a finite probability of them all being closed at the same time. The local $Ca^{2+}$ would then dissipate in microseconds, so that by the time the channels opened again or became available for activation by $Ca^{2+}$, the $Ca^{2+}$ would be gone. This is called stochastic attrition.

*Putney:* But this is a $Ca^{2+}$-independent process. In bilayers, the inactivation is $Ca^{2+}$ dependent.

Do you know of any other examples, other than the heart, where $Ca^{2+}$-induced $Ca^{2+}$ release is physiologically graded?

*Wier:* What we observe seems to contrast with most, if not all, of what's been presented so far, where the assumption, at least, is that the release is not graded.

*Putney:* Is there gradation in other cells? How often does $Ca^{2+}$-induced $Ca^{2+}$ release happen in a all-or-none way?

*Berridge:* Graded $Ca^{2+}$ release in response to caffeine has been demonstrated in chromaffin cells (Cheek et al 1993). Low doses of caffeine give small signals which gradually fade if the cells lack external $Ca^{2+}$. However, if such cells are given a higher dose of caffeine, further $Ca^{2+}$ is released. The cells seem to have a population of stores with variable sensitivities to caffeine. At any dose of caffeine, therefore, a proportion of the stores are activated in an all-or-none manner without recruiting the rest of the stores. A similar phenomenon might explain graded release in cardiac cells. The number of ryanodine receptors that are recruited during each depolarization may depend on their sensitivity. A graded response occurs even though release at each channel is ultimately all-or-none.

*Wier:* Ultimately, the release becomes all-or-none. We excluded from our experiments those cells that exhibit all-or-none behaviour because we think such behaviour is abnormal for a mammalian cardiac cell.

*Cobbold:* Are your cells electrically paced? Rat myocytes would normally show a negative staircase over the first several contractions because they release less $Ca^{2+}$ later on in the sequence of contractions.

*Wier:* We were careful to address that point. They are paced. We give them electrical stimuli at regular intervals.

*Cobbold:* Do you see the sparks in electrically paced cells?

*Wier:* No; we've looked for sparks only in resting cells.

*Fay:* There's a possibility that you don't see all-or-none release of $Ca^{2+}$ because you have geometrically separated release from the stimulus. To get out of that dilemma, you talk about stochastic attrition accounting for channel behaviour. Is there a possibility that the more limited nature of $Ca^{2+}$ release is due to limits imposed by the amount of $Ca^{2+}$ that's cued up ready to go through the channel, perhaps $Ca^{2+}$ held by calsequestrin that is closely associated with each channel? That is another way of potentially limiting the amount that could be released at any point in time.

*Berridge:* Your idea is that the calsequestrin near the ryanodine receptor that's loaded with $Ca^{2+}$ would enhance the sensitivity.

*Fay:* Also that the $Ca^{2+}$ bound to calsequestrin that is closely associated with the channel would perhaps be the only deliverable $Ca^{2+}$ that channel opening could release before significant slower movements of $Ca^{2+}$ would be required.

*Friel:* The ability of $Ca^{2+}$-induced $Ca^{2+}$ release to generate all-or-nothing $Ca^{2+}$ spikes is expected to depend not only on the rate of $Ca^{2+}$ release from the SR, but also on the rate at which $Ca^{2+}$ is transported from and buffered within the cytosol. Only if $Ca^{2+}$-induced $Ca^{2+}$ overwhelms the restorative systems that are present is it expected to produce all-or-nothing $Ca^{2+}$ spikes.

*Jaffe:* The results of Ikemoto et al (1991) suggest that the fundamental mechanism for initiating the $Ca^{2+}$ wave is a conformational change inside the lumen of the endoplasmic reticulum (ER). That conformational change depends on the $[Ca^{2+}]$ in the ER. To me, that furnishes a plausible hypothesis to explain your results.

*Wier:* You are suggesting that the $[Ca^{2+}]$ in the SR sets the sensitivity of the system.

*Jaffe:* I'm suggesting that there's a second mode of $Ca^{2+}$-induced $Ca^{2+}$ release which depends not on the opening of channels, but on a conformational change inside the ER.

*Berridge:* Dr Wier has pointed out that such effects operating from the SR represent an abnormal situation.

*Jaffe:* But there was a discussion of why it doesn't happen normally.

*Fay:* To what extent is the gain factor influenced by different kinds of buffers with different kinetics and different $Ca^{2+}$ affinities in the space between the plasma membrane and the SR? Knowing this might provide a way of differentiating mechanisms that are operating inside the lumen from those operating on the outside.

*Wier:* That is a good idea but a difficult experiment. Part of the problem is that we don't have fast buffers. Stern (1992) has shown that it would be difficult to buffer the $Ca^{2+}$ around the mouth of a channel. It would be hard to interpret that kind of experiment, because when you abolished the $Ca^{2+}$ transient you couldn't be confident that you were truly abolishing or affecting $Ca^{2+}$ transients in spaces that you can't really observe.

*Cannell:* I've done an unpublished experiment which bears on this question. We pulse-injected indo-1 into a cardiac cell, raising the buffering power, while trying desperately to keep the $Ca^{2+}$ current absolutely constant. The free $[Ca^{2+}]$ went down linearly with the buffering power, such that the total amount released always seemed to be essentially the same. We don't see any particular, odd properties of the buffers, at least not by perturbing them. The system seems to be programmed by the $Ca^{2+}$ current.

*Wier:* That suggests there is little feedback of released $Ca^{2+}$ onto SR $Ca^{2+}$ release, because you always get the same amount released whether or not a change in free $[Ca^{2+}]$ results from the release.

*Fay:* Or does it mean that it is the $Ca^{2+}$ in the SR (that you're not able to influence) that is determining the gain?

*Cannell:* That's another possibility.

*Berridge:* This might be an appropriate moment for you to tell us about $Ca^{2+}$ sparks.

## References

Cheek TR, Moreton RB, Berridge MJ, Stauderman KA, Murawsky MM, Bootman MD 1993 Quantal $Ca^{2+}$ release from caffeine-sensitive stores in adrenal chromaffin cells. J Biol Chem 268:27076–27083

Fabiato A 1985a Time and calcium dependence of activation and inactivation of calcium-induced release of calcium from the sarcoplasmic reticulum of a skinned canine Purkinje cell. J Gen Physiol 85:247–289

Fabiato A 1985b Rapid ionic modifications during the aequorin-detected calcium transient in a skinned canine cardiac Purkinje cell. J Gen Physiol 85:189–246

Fabiato A 1985c Simulated calcium current can both cause calcium loading in and trigger calcium release from the sarcoplasmic reticulum of a skinned canine cardiac Purkinje cell. J Gen Physiol 85:291–320

Ikemoto N, Mészáros LG, Ronjat M 1991 Intravesicular calcium transient during calcium release from sarcoplasmic reticulum. Biochemistry 30:5230–5237

Stern MD 1992 Theory of excitation–contraction coupling in cardiac muscle. Biophys J 63:497–517

# General discussion II

### Calcium sparks in cardiac muscle

*Cannell:* The first application of video-rate calcium imaging to cardiac muscle indicated that intracellular free calcium ($[Ca^{2+}]_i$) was uniformly distributed, both at rest and at the peak of the electrically evoked $[Ca^{2+}]_i$ transient (Cannell et al 1987). However, the conventional (wide-field) fluorescence microscope has a large depth of focus, which compromises the spatial resolution of the instrument. The confocal microscope reduces the contribution of out-of-focus fluorescence to in(focal)-plane fluorescence, which results in images with a greatly improved spatial resolution. In collaboration with W. J. Lederer and H. Cheng (whose contribution I would like to acknowledge and who have been intimately involved in the development of the ideas I shall present), we decided to re-examine the uniformity of $[Ca^{2+}]_i$ in isolated rat cardiac myocytes. To our surprise, we found that spontaneous local changes in $[Ca^{2+}]_i$ occurred in completely quiescent cells (Cheng et al 1993). I would like to describe how our new observations of spontaneous and electrically evoked local non-uniformities in $[Ca^{2+}]_i$ support the hypothesis that excitation–contraction (E–C) coupling in the heart is a local stochastic process. $[Ca^{2+}]_i$ images of quiescent rat ventricular cardiac myocytes taken with the confocal microscope and fluo-3 (Minta et al 1989) revealed discrete local increases in fluo-3 fluorescence that varied in position between images (Fig. 1A). The peak fluorescence associated with these events was approximately double the resting fluorescence level, a difference that could not be explained by photon noise in the image. Because these events were never observed outside the cell boundary, they could not be due to artefacts arising within the imaging system. We therefore concluded that these increases in fluorescence were due to spontaneous local increases in $[Ca^{2+}]_i$, which we called '$Ca^{2+}$ sparks'. These $Ca^{2+}$ sparks occupied a region of the cell about 1.5 $\mu$m in diameter and occurred at a frequency of about 1.3 $s^{-1}$.

Because the $Ca^{2+}$ sparks persisted for less than the image acquisition time (as shown by the sparks changing position between images), we used the microscope in line-scan mode, in which a single confocal line was scanned repeatedly. A line-scan image is built from these scans by placing successive lines beneath each other to form an image in which time increases from the top to the bottom of the image, and fluorescence intensity along the scan line is displayed horizontally (see Fig. 1B,C). We therefore discard one dimension

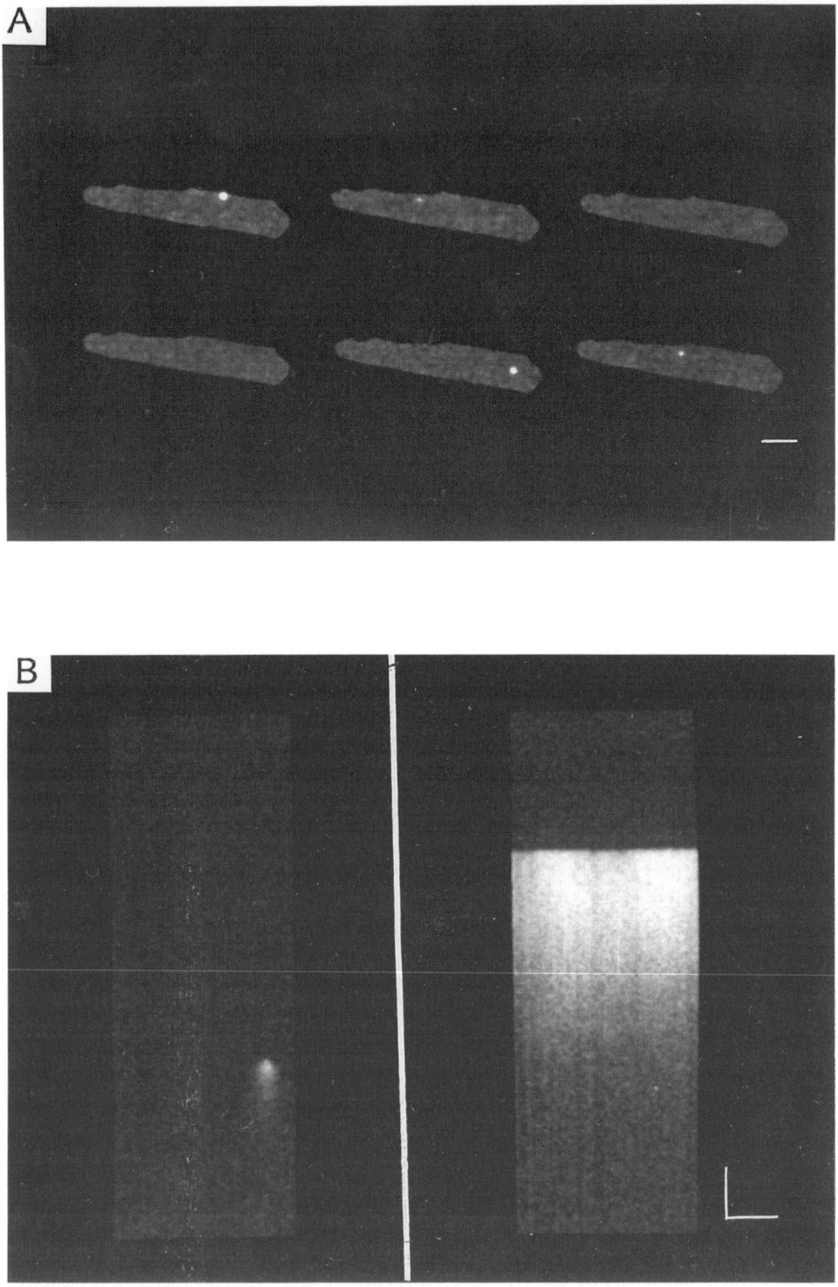


FIG. 1A and B (*caption opposite*)

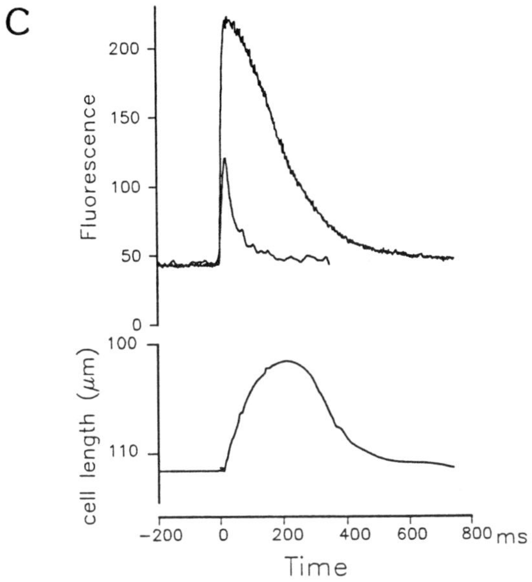


FIG. 1. (*Cannell*) (A) Confocal images of a quiescent rat cardiac myocyte loaded with fluo-3. Note the presence of bright patches of fluorescence that vary in position between the images (taken 1 s apart). These changes in fluorescence are due to '$Ca^{2+}$ sparks'. Scale bar 14 $\mu$m. (B) Line-scan images of a $Ca^{2+}$ spark (*left*) and an electrically evoked intracellular free calcium ($[Ca^{2+}]_i$) transient (*right*). Note the discrete local nature of the $Ca^{2+}$ spark compared with the evoked transient. (C) Time course of the $Ca^{2+}$ spark and electrically evoked $[Ca^{2+}]_i$ transient. Note that the spark is smaller and faster than the transient. Reproduced from Cheng et al 1993, with permission, ©1993 by the AAAS.

of spatial information to increase the temporal resolution of the microscope to 2 ms, the maximum scan rate of our Biorad MRC600 confocal microscope. As we were scanning only a tiny fraction of the cell, a large number of images had to be taken before we could catch the site of initiation of a calcium spark. In successful experiments, the calcium spark appeared to originate in a small region of the cell that was about 1 $\mu$m in diameter, and was confined to a small region rather than spreading across the cell (Fig. 1B). Using a self-ratio method, we estimated that the peak of the $Ca^{2+}$ sparks corresponded to a $[Ca^{2+}]_i$ of about 270 nM (assuming a resting $[Ca^{2+}]_i$ of 100 nM) and the $Ca^{2+}$ spark declined with a half-time of about 25 ms. These data can be compared with the time course of an electrically evoked $[Ca^{2+}]_i$ transient in the same cell, which reached a peak of about 1.5 $\mu$M and declined with a half-time of about 164 ms (Fig. 1C). In addition, the electrically evoked $[Ca^{2+}]_i$ transient arises almost synchronously throughout the cell, as shown by the step-like nature of the increase in fluorescence in line-scan images.

The local increase in $[Ca^{2+}]_i$ associated with the $Ca^{2+}$ spark is probably removed by the same mechanisms that reduce $[Ca^{2+}]_i$ after an electrically evoked $[Ca^{2+}]_i$ transient. The increased rate of decline in the $Ca^{2+}$ spark can probably be explained by three factors: (1) intracellular $Ca^{2+}$ buffers will be less saturated during the $Ca^{2+}$ spark (because the peak $[Ca^{2+}]_i$ is lower) and may be able to take up $Ca^{2+}$ during the decline in the $Ca^{2+}$ spark, thereby accelerating its decline; (2) alternatively or additionally, intrinsic $Ca^{2+}$ buffers will be more saturated with $Ca^{2+}$ during the larger electrically evoked $[Ca^{2+}]_i$ transient and will slow its decline by releasing $Ca^{2+}$; (3) the diffusion of calcium from its site of release will also help reduce $[Ca^{2+}]_i$ (during the electrically evoked $[Ca^{2+}]_i$ transient $[Ca^{2+}]$ increases everywhere, so the contribution of diffusion to the decline in $[Ca^{2+}]_i$ would be reduced).

An estimate of the amount of $Ca^{2+}$ required to generate a $Ca^{2+}$ spark can be made from the $Ca^{2+}$-buffering power of the cell, the peak change in $[Ca^{2+}]_i$ during a spark and the volume occupied by the spark. We estimated that the buffering power of the cell was about 100 $\mu$M (free + bound) per $\mu$M (free), whereas the volume of the spark was about 10 fl and the peak change in $[Ca^{2+}]_i$ was 0.2 $\mu$M. The spark would have resulted from the appearance of about $2 \times 10^{-19}$ M $Ca^{2+}$ or 30 000 $Ca^{2+}$ ions in the myoplasm. Because the peak of the $Ca^{2+}$ spark occurred within about 10 ms, this would require a $Ca^{2+}$ flux corresponding to an ionic current of 4 pA (for 10 ms).

The source of $Ca^{2+}$ for the $Ca^{2+}$ spark was not extracellular, because removal of external $Ca^{2+}$ (for brief periods) did not prevent the occurrence of sparks. In addition, $Ca^{2+}$ sparks were not prevented by the addition of $Cd^{2+}$ or D600, inhibitors of $Ca^{2+}$ channels. In marked contrast to the lack of immediate effect of interventions that modify $Ca^{2+}$ influx, ryanodine, which affects the sarcoplasmic reticulum (SR) $Ca^{2+}$ release channel, dramatically altered spark behaviour. There were no $Ca^{2+}$ sparks in most cells (14 out of 15) exposed to ryanodine concentrations greater than 1 $\mu$M. After exposure to intermediate concentrations of ryanodine (100–300 nM) for one to three minutes, spark frequency increased severalfold. However, some regions showed an even greater increase in spark frequency and in those regions the kinetics of the spark appeared unaltered, suggesting that ryanodine had increased only the probability of their occurrence. Ryanodine also evoked long-lasting sparks of lower amplitude, as shown in Fig. 2.

These effects of ryanodine closely mimic its effects on isolated SR $Ca^{2+}$ release channels in bilayer experiments. At low doses, ryanodine promotes channel opening, whereas at intermediate doses it induces the appearance of long-lasting subconductance states. At high doses ryanodine blocks the SR $Ca^{2+}$ release channel. Given that the ionic current underlying the $Ca^{2+}$ spark was similar to that expected from bilayer experiments on isolated SR $Ca^{2+}$ release channels, these results raised the possibility that we were observing the spontaneous opening of single SR $Ca^{2+}$ release channels. However, it is also

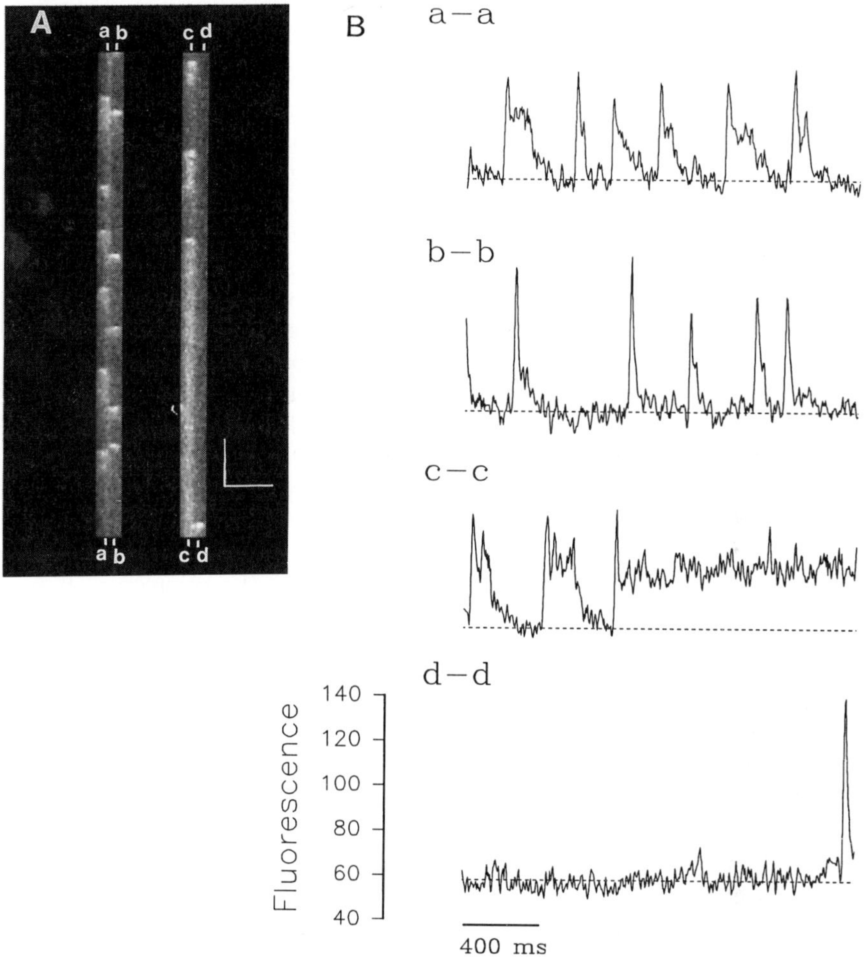


FIG. 2. (*Cannell*) Calcium sparks in the presence of ryanodine. (A) Line-scan images of a sparking region. Note the increased frequency of $Ca^{2+}$ sparks (compared with Fig. 1B) and that the two regions appear to behave independently, despite their proximity (scale bar 10 $\mu$m horizontally, 200 ms vertically). (B) Time course of fluorescence change in the regions indicated in (A). Both normal sparks and long-lasting sparks are evident. Reproduced from Cheng et al 1993, with permission, © 1993 by the AAAS.

possible that the $Ca^{2+}$ spark arises from a number of SR $Ca^{2+}$ release channels acting in concert. If the SR $Ca^{2+}$ release channels are clustered into functional release units, their gating will mimic a single channel because the activation of one SR $Ca^{2+}$ release channel will cause the others to open. However, the functional unit will eventually close because the intrinsic gating of the channels will eventually allow the local $[Ca^{2+}]$ to decline to a level that does not support reactivation. This process has been termed 'stochastic attrition' (Stern 1992) and can explain why SR $Ca^{2+}$ release can have high gain as well as stability.

From the results described above, the open time of the activated functional unit would be about 10 ms, and it is notable that this is similar to the rise time of the normal $[Ca^{2+}]_i$ transient. Thus, the release of $Ca^{2+}$ from the SR during the normal $Ca^{2+}$ transient is triggered by the $Ca^{2+}$ influx mediated by the opening of L-type $Ca^{2+}$ channels, which causes a local increase in $[Ca^{2+}]_i$ around the functional SR $Ca^{2+}$ release units. This causes the opening of at least one ryanodine receptor, which, in turn, activates others. This is the regenerative phase of the $[Ca^{2+}]_i$ transient. However, after about 10 ms the functional units close because the finite mean open time of the SR release channels results in the simultaneous closure of a number of the SR channels, and the local $[Ca^{2+}]_i$ rapidly declines (mainly by diffusion). In addition, it is possible that local depletion of the SR luminal $Ca^{2+}$ contributes to this process because SR $Ca^{2+}$ depletion will result in a reduction in the flux through the activated unit, which will lead to a decline in the local $[Ca^{2+}]_i$ and an increase in the probability of release unit deactivation.

An analogy to how E–C coupling works might be the behaviour observed at parties. People don't like to be overheard at parties so that when there is lots of noise they are encouraged to talk. However, if no one is talking, conversation is unlikely to be initiated. In this analogy, the people at the party represent individual SR $Ca^{2+}$ release channels, whereas the noise they produce represents the SR $Ca^{2+}$ release. If people are already talking, provided there are not too many people at the party, some of them will eventually stop talking at the same time and the party will suddenly become quiet—a phenomenon that is probably a common experience. However, if the doorbell rings and someone arrives from outside (the sarcolemmal $Ca^{2+}$ channels open), the resulting greetings rapidly activate new conversation, which encourages everyone to start talking and the party noise rapidly increases (until the next embarrassing silence). Such a system is very responsive to activation by $Ca^{2+}$ from outside the functional unit but does not remain permanently activated, despite positive feedback, because stochastic gating ensures the eventual cessation of activation. Another factor nicely demonstrated by the analogy is the importance of the number of SR release channels in a functional unit. If the functional unit is small it is not very responsive (a dead party), wherease if it is very large it will exhibit uncontrolled regenerative behaviour (the party is permanently noisy). The number of SR release channels in a functional unit is critical in determining the behaviour of E–C coupling.

Conditions that increase resting $[Ca^{2+}]_i$ can lead to the development of spontaneous waves of $[Ca^{2+}]_i$ that traverse the cell. However, the initiating event for the $Ca^{2+}$ wave is unclear. We increased resting $[Ca^{2+}]_i$ by increasing the $[Ca^{2+}]$ of the bathing solution to 10 mM; under these conditions the rate

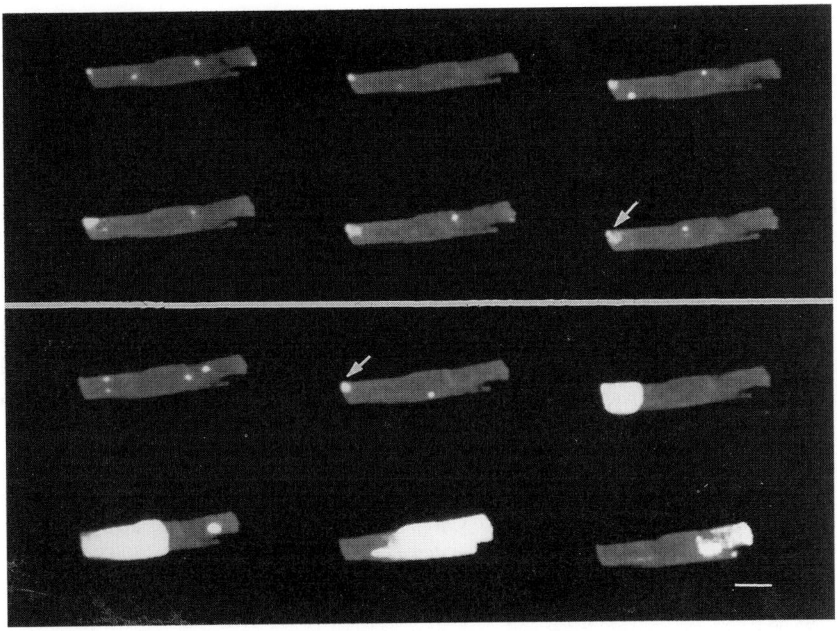

FIG. 3. (*Cannell*) Calcium overload increases the frequency of $Ca^{2+}$ sparks and produces waves of $[Ca^{2+}]_i$. Note that at the top left of the cell, a long lasting 'macro spark,' which fails to generate a regenerative wave in the top panel but initiates a wave 4 min later in the lower panel (at the arrows). Scale bar 20 $\mu$m. Reproduced from Cheng et al 1993, with permission, ©1993 by the AAAS.

of $Ca^{2+}$ sparking increased about four-fold. Interestingly, $Ca^{2+}$ sparks were often observed at the site of initiation of the $Ca^{2+}$ wave, suggesting that the $Ca^{2+}$ spark could be the trigger (Fig. 3). However, it is difficult to explain the appearance of waves simply on the basis of this moderate increase in spark frequency. It is more likely that the amount of $Ca^{2+}$ in the SR plays a critical role in determining whether or not a spontaneous $Ca^{2+}$ spark can spread. In $Ca^{2+}$ overload we often observed local increases in $[Ca^{2+}]_i$ that are larger, last longer and spread further before decaying than the usual $Ca^{2+}$ spark (see arrows in Fig. 3). We termed such events 'macro sparks'. They probably result from the recruitment of additional $Ca^{2+}$ release units and their failure to propagate may be due the surrounding SR content being insufficient to support regeneration. It is also possible that the probability of SR release unit activation by the local $[Ca^{2+}]_i$ depends on the SR luminal $Ca^{2+}$ content (Thedford et al 1994, Tripathy & Meissner 1994).

If the $Ca^{2+}$ spark reflects the activation of functional SR $Ca^{2+}$ release units, $Ca^{2+}$ sparks should be present during normal E–C coupling. However, the global increase in $[Ca^{2+}]_i$ associated with electrical stimulation decreases spatial contrast in the image and makes spark detection difficult. An alternative approach is to reduce the probability of $Ca^{2+}$ spark production by inhibiting sarcolemmal $Ca^{2+}$ channel opening. In cells exposed to submaximal $Ca^{2+}$ channel antagonists such as verapamil, D600 or $Cd^{2+}$, electrical stimulation resulted in a very non-uniform $[Ca^{2+}]_i$ transient (Cannell et al 1994). In regions where $Ca^{2+}$ release occurred, the time course and amplitude of the change in $[Ca^{2+}]_i$ were similar to those of the spontaneous $Ca^{2+}$ sparks described above, except that the production of these sparks was linked to the stimulus. In regions between the evoked $Ca^{2+}$ sparks, there was a slower and smaller increase in $[Ca^{2+}]_i$, as might be expected from the diffusion of $Ca^{2+}$ from outside the confocal plane. $Ca^{2+}$ sparks can therefore be activated by the $Ca^{2+}$ current, supporting the idea that they represent the activation of functional SR $Ca^{2+}$ release units whose summation underlies the $[Ca^{2+}]_i$ transient. These results are the first demonstration that $Ca^{2+}$ release from the SR is restricted to functional units that are controlled by the local $[Ca^{2+}]$. We have also demonstrated that SR $Ca^{2+}$ release during E–C coupling is a stochastic event whose probability of occurrence depends on the local activation of L-type $Ca^{2+}$ channels. Thus, interventions that reduce the amplitude of the current (either by decreasing the mean open time of the $Ca^{2+}$ channel or by decreasing the gradient for $Ca^{2+}$ entry) will reduce the spatial uniformity of $[Ca^{2+}]_i$. Future models of E–C coupling should include the stochastic behaviour of functional SR $Ca^{2+}$ release units whose activation underlies $Ca^{2+}$ sparks and the myocardial $[Ca^{2+}]_i$ transient.

*Putney:* In line-scan mode, you are likely to be seeing a tangential cut through the spark. Does that influence the kinetics?

*Cannell:* We can't detect any anisotropy in the spark, so that the orientation of the scan with regard to the axis of the cell doesn't seem to influence the time course of the spark.

*Putney:* But if you section tangentially, if you didn't catch the centre, would you still get the same kinetics?

*Cannell:* You can simulate that by looking at the time course of the spark in line-scan data just off its centre. There you see a slightly less rapid peak, i.e., slower kinetics. The change in time course is what you would expect from diffusion from a point source into an absorbing medium.

*Jaffe:* How far apart are the spark release centres?

*Cannell:* All I can tell you is the closest together that we have observed discrete, defined sparks occurring. The closest we've seen, in spontaneous events, is about 2 $\mu$m, but it's really difficult to catch these events because the normal spark rate is about one per line-scan image, which takes about 1 s to acquire. We have tried very hard to image spatial non-uniformity during evoked transients, but there you lose contrast in the image owing to the global increase in $[Ca^{2+}]$. If we turn down the strength of the E–C coupling by inhibiting the $Ca^{2+}$

current, then the pattern of release is very similar to that of spontaneous sparks. I think the spark is an elementary event underlying E–C coupling, and normal E–C coupling is simply the summation of a large number of sparks.

*Jaffe:* How far apart are the sparks in the presence of $Ca^{2+}$ channel inhibitors?

*Cannell:* The sarcomere spacing is about 2 μm longitudinally, and the sarcomeres are about 1 μm across and separated by about 0.5 μm, giving a centre-to-centre distance of about 1.5 μm. That's just about the closest that we've ever observed isolated sparks.

*Petersen:* The cardiac ryanodine receptor is sensitive to cyclic ADP (cADP)-ribose. Are there circumstances under which [cADP-ribose] is known to be increased, or is it possible to inject cADP-ribose and see sparks?

*Cannell:* Regulation of the cardiac sarcoplasmic reticulum $Ca^{2+}$ channel *in situ* is really unexplored. If we can measure the probability of the cardiac release channel opening by measuring spark frequency, that would open up whole new fields for studying regulation. I would do such experiments if I could get access to the right equipment.

*Berridge:* You are giving the impression that each spark results from the opening of only one channel, whereas Dr Parker suggested that puffs result from an ensemble of channels.

*Cannell:* The flux was about an order of magnitude smaller but the single-channel conductance (of the ryanodine receptor) is about an order of magnitude greater. So the flux associated with a spark is much closer to that expected for a single channel than is the case for Dr Parker's experiments. Unfortunately, we can't be sure whether or not it is a single channel, because in single-channel records the only way you can determine whether there is more than one channel is by observing multiple opening levels. In the experiments with ryanodine, we can see long-lasting events that have fairly constant levels, but that doesn't exclude the possibility of more than one channel being involved. There could be several channels activated together, perhaps flickering, so that we can't see changes in fluorescence resulting from sequential openings. In this case, we could be seeing multiple channels acting in concert as if they were a single channel.

*Berridge:* So there's never any hint of steps in the rising phase.

*Cannell:* Our data aren't good enough to be sure.

*Thomas:* You suggested that ryanodine increases the frequency of sparking by pushing the channel into a subconductance with a more sustained open state. Are there any differences in the kinetics of the rise of the spark in ryanodine-treated cells?

*Cannell:* No.

*Thomas:* Is there a sufficient difference in the ryanodine-activated conductance compared with the normal conductance of a single channel on the basis of bilayer studies?

*Cannell:* The ryanodine-modified conductance is about half the normal conductance. One problem is that although the steady level of [$Ca^{2+}$] that we saw during a long-lasting opening was smaller than the peak level of fluorescence in an individual spark—which would be consistent with a subconductance state—it is also possible that we're seeing a dynamic redistribution of $Ca^{2+}$.

## References


Cannell MB, Berlin JR, Lederer WJ 1987 Intracellular calcium in cardiac myocytes: calcium transients measured using fluorescence imaging. In: Mandel LJ, Eaton DC (eds) Cell calcium and control of membrane transport. Rockefeller University Press, New York, p 201–214

Cannell MB, Cheng H, Lederer WJ 1994 Nifedipine decreases the spatial uniformity of the depolarization-evoked $Ca^{2+}$ transient in isolated rat cardiac myocytes. J Physiol 477:P25(abstr)

Cheng H, Lederer WJ, Cannell MB 1993 Calcium sparks: elementary events underlying excitation–contraction coupling in heart muscle. Science 262:740–744

Minta A, Kao JP, Tsien RY 1989 Fluorescent indicators for cytosolic calcium based on rhodamine and fluorescence chromophores. J Biol Chem 264:8171–8178

Stern MD 1992 Theory of excitation–contraction coupling in cardiac muscle. Biophys J 63:497–517

Thedford SE, Lederer WJ, Valdivia HH 1994 Activation of sarcoplasmic reticulum calcium release channels by intraluminal $Ca^{++}$. Biophys J 66:A20(abstr)

Tripathy A, Meissner G 1994 Effect of SR lumenal $Ca^{++}$ on the open probability of the skeletal muscle ryanodine receptor/$Ca^{++}$ release channel. Biophys J 66:A416(abstr)

# Intercellular calcium waves mediated by inositol trisphosphate

Michael J. Sanderson*

*Department of Anatomy and Cell Biology, UCLA School of Medicine, University of California, Los Angeles, CA 90024-1763, USA*

*Abstract.* Intercellular calcium waves occur in diverse cells. Those that are induced by mechanical stimulation have been extensively investigated in epithelial and glial cells. Mechanical stimulation of an individual cell initiates an increase in the intracellular free calcium concentration, $[Ca^{2+}]_i$, that spreads across the cell. At the cell border this intracellular $Ca^{2+}$ wave is arrested but, after a brief delay, similar $Ca^{2+}$ waves occur in adjacent cells. The repetition of this process results in the propagation of an intercellular $Ca^{2+}$ wave through a limited number of cells. The propagation of intercellular $Ca^{2+}$ waves correlates with the presence of functional gap junctions and occurs in the absence of extracellular $Ca^{2+}$ or following the microinjection of inositol 1,4,5-trisphosphate ($InsP_3$). The propagation of intercellular $Ca^{2+}$ waves is inhibited by heparin (an $InsP_3$ receptor antagonist) and by U73122 (a phospholipase C inhibitor) or when intracellular $Ca^{2+}$ stores are depleted with thapsigargin. These characteristics suggest that mechanical stimulation initiates $InsP_3$ production and that intercellular $Ca^{2+}$ waves are propagated through the movement of $InsP_3$ through gap junctions. Mathematical modelling supports the idea that diffusion of $InsP_3$ is a viable hypothesis for the generation of intercellular $Ca^{2+}$ waves. The ability of cells to display changes in $[Ca^{2+}]_i$ that are independent of neighbouring cells (i.e., asynchronous $Ca^{2+}$ oscillations) and the low diffusion constant of $Ca^{2+}$ suggest that $Ca^{2+}$ itself is not a major messenger moving between cells to propagate $Ca^{2+}$ waves.

*1995 Calcium waves, gradients and oscillations. Wiley, Chichester (Ciba Foundation Symposium 188) p 175–194*

Intercellular communication is an integral property of multicellular systems and provides a mechanism by which cells can coordinate their activities (Warner 1992). Gap junctions form a major pathway for intercellular communication (Bennett & Verselis 1992, Sáez et al 1993), but the identity or function of the intercellular messengers is not well understood. Gap junctions are permeable to molecules with molecular masses of up to about 1000 Da, which would allow a variety of molecules to act as intercellular messengers. However, the role of

*Present address: Department of Physiology, University of Massachusetts Medical Center, 55 Lake Avenue North, Worcester, MA 01655, USA.

the second messenger inositol 1,4,5-trisphosphate ($InsP_3$) in intercellular communication and $Ca^{2+}$ waves will be my focus here. Interestingly, as will be discussed, $Ca^{2+}$ itself does not appear to act as a messenger for intercellular $Ca^{2+}$ signalling (see also review by Sanderson et al 1994).

## Types of calcium waves

Calcium waves occur in a variety of cells (Jaffe 1993) but the mechanisms generating these waves are not all identical, making a classification of $Ca^{2+}$ waves as a single phenomenon inappropriate. One major distinction that I shall emphasize here is the difference between *intra*cellular and *inter*cellular $Ca^{2+}$ waves. Intracellular $Ca^{2+}$ waves are often associated with $Ca^{2+}$ oscillations and consist of a spatiotemporal increase in intracellular $Ca^{2+}$ concentration ($[Ca^{2+}]_i$) that spreads across an individual cell (Rooney & Thomas 1993). In contrast, an intercellular $Ca^{2+}$ wave is the propagation of increases in $[Ca^{2+}]_i$ through multiple cells (Sanderson et al 1994). Calcium waves can be initiated by different mechanisms. Although intracellular and intercellular $Ca^{2+}$ waves can be initiated by both mechanical and chemical stimulation, my emphasis here will be on mechanically induced intercellular calcium waves. Agonist-induced intercellular $Ca^{2+}$ waves will be briefly discussed later for comparison.

### *Mechanically induced intercellular calcium waves*

During the investigation of the mechanisms regulating the ciliary activity of airway epithelial cells, intercellular $Ca^{2+}$ waves were observed following the mechanical stimulation of an individual cell in a cultured monolayer of cells (Sanderson et al 1990). Brief distortion of the apical membrane with a glass pipette resulted in a localized increase in $[Ca^{2+}]_i$ in the stimulated cell that spread across the cell at about 25 $\mu m\, s^{-1}$. At the cell boundaries, the propagation of this $Ca^{2+}$ wave was arrested. However, after a brief delay of about 0.5–1 s, a similar $Ca^{2+}$ wave was initiated in surrounding adjacent cells from sites closest to the stimulated cell. This process was repeated several times in more distal cells, such that an increase in $[Ca^{2+}]_i$ was observed to propagate radially through many cells (Sanderson et al 1990). Similar mechanically induced intercellular $Ca^{2+}$ waves occur in cultures of mixed glial cells (Charles et al 1991), astrocytes (Enkvist & McCarthy 1993), endothelial cells (Demer et al 1993), osteoblasts (Xia & Ferrier 1992), mammary gland epithelial cells (Furuya et al 1993) and cardiac myocytes (Sigurdson et al 1992).

## The characteristics of intercellular calcium wave propagation

### *Gap junctions: the route for intercellular calcium signalling*

At the cell boundaries the propagation of a mechanically induced $Ca^{2+}$ wave is temporarily arrested, and $Ca^{2+}$ waves in adjacent cells are initiated at localized sites of cell–cell contact. Calcium wave propagation frequently follows a circuitous pathway through multiple cells, and sequential mechanically induced $Ca^{2+}$ waves follow similar cellular paths (Sanderson et al 1990, Charles et al 1991, Demer et al 1993). These characteristics of wave propagation are consistent with the idea that signals move between cells through gap junctions. This hypothesis is consistent with the presence of gap junctions in airway epithelial cells and glial cells and is supported by the inhibition of $Ca^{2+}$ wave propagation with halothane or octanol, local anaesthetics that reduce gap junction conductance (Sanderson et al 1990, Xia & Ferrier 1992). The role of gap junctions was further investigated with C6 glioma cells, which do not express significant amounts of the gap junction protein connexin 43. C6 glioma cells are not electrically coupled or dye-coupled and do not propagate intercellular $Ca^{2+}$ waves when mechanically stimulated (Charles et al 1992). Transfection of these cells with the gene for connexin 43 resulted in cell lines that were dye-coupled and expressed messenger RNA and protein for connexin 43, indicating that gap junction communication had been restored (Naus et al 1992). Significantly, these transfected cells propagated intercellular $Ca^{2+}$ waves after mechanical stimulation; the extent and rate of $Ca^{2+}$ wave propagation was proportional to the level of expression of connexin 43 (Charles et al 1992). The alteration of gap junction conductance by other means, such as through the action of protein kinase C or extracellular $K^+$, also affects the extent of $Ca^{2+}$ wave propagation in glial cells (Enkvist & McCarthy 1993). In conclusion, these results strongly indicate that intercellular $Ca^{2+}$ waves are propagated via gap junctions.

*Extracellular messengers: an alternative pathway*

Communication of $Ca^{2+}$ signals between cells may also occur through the diffusion of an extracellular messenger. It appears that basophilic leukaemia cells show this form of $Ca^{2+}$ signalling in response to mechanical stimulation by releasing ATP during exocytosis (Osipchuk & Cahalan 1992). The $[Ca^{2+}]_i$ in airway epithelial cells also increases in response to ATP, and the possibility that an extracellular messenger such as ATP could contribute to the propagation of $Ca^{2+}$ waves was investigated by mechanically stimulating cells in the presence of an extracellular fluid flow (Hansen et al 1993). Because fluid flow would bias a $Ca^{2+}$ response mediated by an extracellular messenger, the persistence of a radially propagating intercellular $Ca^{2+}$ wave in the presence of fluid flow suggests that an extracellular messenger does not mediate $Ca^{2+}$ waves in airway epithelial cells (Hansen et al 1993). Similarly, a flow of fluid had no effect on $Ca^{2+}$ waves passing between osteoblasts (Xia & Ferrier 1992), although some biasing of the $Ca^{2+}$ response was observed in glial cells (M. J. Sanderson & A. C. Charles, unpublished work).

The possibility that mechanical stimulation releases ATP, or other cytoplasmic contents, is also unlikely in view of the fact that $Ca^{2+}$ wave propagation is unaffected by suramin, a $P_2$-purinergic antagonist that attenuates the $Ca^{2+}$ response of the cells to ATP. Similarly, in young cultures that were responsive to ATP, mechanically induced $Ca^{2+}$ waves failed to propagate, again indicating that mechanical stimulation did not release ATP (Hansen et al 1993). The circuitous routes of propagation and blockage of propagation by gap junction inhibitors is also not consistent with the involvement of an extracellular messenger (Sanderson et al 1990, Charles et al 1992, Enkvist & McCarthy 1993).

A radial $Ca^{2+}$ wave could result from the release of a messenger into the intercellular space that is isolated from the influence of flowing fluid by apical tight junctions in epithelial cells. However, because ATP-induced, but not mechanically induced, $Ca^{2+}$ changes were biased by fluid flow and because ATP-sensitive young cultures (or cells near the leading edge of explants) were insensitive to heavy mechanical stimulation, the communication of intercellular $Ca^{2+}$ waves via intercellular spaces can be discounted (Hansen et al 1993). Collectively, these results strongly indicate that non-traumatic mechanical stimulation does not induce the release of an extracellular messenger that can mediate intercellular $Ca^{2+}$ wave propagation (Hansen et al 1993). However, under conditions where cell integrity is breached, an extracellular messenger may act simultaneously with gap junction communication to enhance intercellular $Ca^{2+}$ signalling.

### *Calcium: a non-essential intercellular messenger*

In the absence of extracellular $Ca^{2+}$, mechanical stimulation initiates the propagation of an intercellular $Ca^{2+}$ wave through adjacent cells. However, it is significant that under these $Ca^{2+}$-free conditions the $[Ca^{2+}]_i$ of the stimulated cell does not increase. These results indicate that a release of intracellular $Ca^{2+}$ in adjacent cells underlies the elevation in $[Ca^{2+}]_i$ associated with the communicated wave and that an increase in $[Ca^{2+}]_i$ in the stimulated cell is not required for the initiation of an intercellular $Ca^{2+}$ wave (Sanderson et al 1990, Charles et al 1991, Demer et al 1993, Boitano et al 1994a). (The absence of a rise in $[Ca^{2+}]_i$ in the stimulated cell under $Ca^{2+}$-free conditions is particularly interesting and will be considered below in greater detail.) The ability of intercellular $Ca^{2+}$ waves to propagate at similar speeds over equivalent distances when the $[Ca^{2+}]_i$ attained in each cell is markedly reduced by dantrolene (Charles et al 1993), a compound that inhibits $Ca^{2+}$-induced $Ca^{2+}$ release, also indicates that the elevation in $[Ca^{2+}]_i$ is not critical to the ability of cells to communicate a signal between each other.

If the movement of $Ca^{2+}$ between cells were solely responsible for the initiation of an intercellular $Ca^{2+}$ wave, any increase in $[Ca^{2+}]_i$ should initiate an intercellular $Ca^{2+}$ wave. However, it is commonly observed that the increase in $[Ca^{2+}]_i$ associated with $Ca^{2+}$ oscillations generally does not initiate intercellular $Ca^{2+}$ waves. Calcium oscillations occur independently, or asynchronously, of the activity of adjacent cells, even though the peak of the $[Ca^{2+}]_i$ associated with a $Ca^{2+}$ oscillation is similar to that of the $Ca^{2+}$ wave (Charles et al 1991, Boitano et al 1992). Similarly, increases in $[Ca^{2+}]_i$ in cells loaded with heparin, an agent that blocks $InsP_3$-dependent release of $Ca^{2+}$, do not initiate intercellular $Ca^{2+}$ waves (Boitano et al 1992). These results suggest that any $Ca^{2+}$ that passes through gap junctions is insufficient, by itself, to initiate $Ca^{2+}$ waves in adjacent cells.

One explanation for the failure of $Ca^{2+}$ to induce $Ca^{2+}$ waves in adjacent cells is the absence of the amplification mechanism of $Ca^{2+}$-induced $Ca^{2+}$ release via ryanodine receptors. An absence of ryanodine receptors in both airway epithelial and glial cells is indicated by the failure of both caffeine and ryanodine to initiate $Ca^{2+}$ responses in these cells (Charles et al 1993, Hansen et al 1995). Alternatively, the inability of $Ca^{2+}$ changes to initiate waves may result from a self-limiting response in which $Ca^{2+}$ inhibits communication between cells through gap junctions. However, cells displaying asynchronous $Ca^{2+}$ oscillations can propagate a mechanically induced intercellular $Ca^{2+}$ wave (Boitano et al 1992, Charles et al 1991). Another possibility is that an additional signal, induced by mechanical stimulation, is required to prime epithelial cells so that increases in $[Ca^{2+}]_i$ can initiate changes in $[Ca^{2+}]_i$ in adjacent cells. However, unless the additional signal remains closely associated with, but is not dependent on, the propagating $Ca^{2+}$ wave, it should initiate additional $Ca^{2+}$ waves from $Ca^{2+}$ oscillations in the vicinity, but this is not observed. An additional signal in the form of an electrical depolarization can be discounted because $Ca^{2+}$ waves are propagated when the extracellular $[K^+]$ is high.

Although $Ca^{2+}$ can move through gap junctions (Sáez et al 1993), the results described above indicate that the intercellular movement of $Ca^{2+}$ is *not* essential for the initiation or propagation of intercellular $Ca^{2+}$ waves. A similar conclusion has been reached on the basis of the fact that $Ca^{2+}$ diffusion, in comparison with $InsP_3$ diffusion, is greatly limited by the buffering capacity of the cell (Allbritton & Meyer 1993). Thus, $InsP_3$ is a better candidate for long-range propagation of intercellular $Ca^{2+}$ waves.

### *Inositol trisphosphate: an intra- and intercellular messenger*

Release of $Ca^{2+}$ from intracellular stores is initiated by $InsP_3$ via the $InsP_3$ receptor (Berridge 1993). Consequently, it was postulated that $Ca^{2+}$ waves were mediated by the movement of $InsP_3$ (Sanderson et al 1990) because the initiation of a $Ca^{2+}$ wave does not require an increase in $[Ca^{2+}]_i$ and the intercellular propagation of a $Ca^{2+}$ wave releases intracellular $Ca^{2+}$. The dependence of $Ca^{2+}$ waves on intracellular $Ca^{2+}$ is also demonstrated by their inability to propagate when intracellular $Ca^{2+}$ stores are depleted with thapsigargin (Charles et al 1992, Boitano et al 1994a), an inhibitor of endoplasmic reticulum $Ca^{2+}$ pumps.

The iontophoretic injection of $InsP_3$ into airway epithelial cells, under $Ca^{2+}$-free conditions, established that increases in $[InsP_3]_i$ could initiate intercellular $Ca^{2+}$ waves (Sanderson et al 1990). The role of $InsP_3$ as an intercellular messenger was further investigated by examining the effects on $Ca^{2+}$ wave propagation of intracellular heparin, an antagonist of the $InsP_3$ receptor that prevents $InsP_3$ from releasing $Ca^{2+}$ (Boitano et al 1992). Heparin was selectively loaded into a restricted zone of cells in an airway epithelial culture by pulsed high-frequency electroporation. Within the electroporation zone, where all cells had been loaded with heparin, mechanical stimulation failed to initiate a $Ca^{2+}$ wave. When a heparin-loaded cell at the electroporation border was mechanically stimulated, a $Ca^{2+}$ wave propagated through the heparin-free, adjacent cells, but the propagation of the $Ca^{2+}$ wave was severely attenuated in heparin-loaded cells (Boitano et al 1992). Similarly, a radially propagating $Ca^{2+}$ wave, initiated in heparin-free cells, was arrested when it encountered heparin-loaded cells. Gap junctional communication and signal transduction did not appear to be influenced by heparin or electroporation because the initiation and propagation of $Ca^{2+}$ waves was not impaired by loading cells with chondroitin sulphate (Boitano et al 1992), a molecule that is similar to heparin but lacks the ability to block the $InsP_3$ receptor. In view of the fact that $Ca^{2+}$-induced $Ca^{2+}$ release via the ryanodine receptor does not appear to be a viable mechanism in airway epithelial or glial cells, these results indicate that the propagation of $Ca^{2+}$ waves relies on the movement of $InsP_3$ between cells (Boitano et al 1992, Charles et al 1993, Sanderson et al 1994).

The possibility that a simultaneous movement of $Ca^{2+}$ and $InsP_3$ is required for $Ca^{2+}$ wave propagation (Berridge 1993) is, again, argued against by the asynchronous nature of $Ca^{2+}$ oscillations. Calcium oscillations are frequently induced by agonists that act by increasing $[InsP_3]_i$. Consequently, a movement of $Ca^{2+}$ through gap junctions under these conditions would be equivalent to the movement of $Ca^{2+}$ and $InsP_3$ into adjacent cells, but the fact remains that $Ca^{2+}$ oscillations do not initiate intercellular $Ca^{2+}$ waves. The situation is similar in glial cells, where the passage of an intercellular $Ca^{2+}$ wave initiates asynchronous $Ca^{2+}$ oscillations (Charles et al 1991). $InsP_3$ is proposed to spread through the cells to propagate the wave, but the additional movement of $Ca^{2+}$ from $Ca^{2+}$ oscillations does not initiate further $Ca^{2+}$ waves. Thus, an additional conclusion that can be drawn from the coexistence of $Ca^{2+}$ oscillations and intercellular $Ca^{2+}$ waves is that each process has a specific cellular mechanism. The fact that dantrolene abolishes $Ca^{2+}$ oscillations induced by either glutamate or mechanical stimulation, but does not abolish intercellular $Ca^{2+}$ waves, also supports this idea (Charles et al 1993).

**A hypothesis for the initiation and propagation of intercellular calcium waves**

From the results described above, the following hypothesis has been proposed (Fig. 1). Mechanical stimulation is believed to initiate the production of $InsP_3$ in the stimulated cell. This $InsP_3$ diffuses throughout the cell, priming the $InsP_3$ receptor and releasing intracellular $Ca^{2+}$ from the endoplasmic reticulum. Because the release of $Ca^{2+}$ via the $InsP_3$ receptor is sensitive to both $Ca^{2+}$ and $InsP_3$ (see Iino & Endo 1992), the released $Ca^{2+}$ initiates further release of $Ca^{2+}$ by $Ca^{2+}$-induced $Ca^{2+}$ release. Diffusion of $Ca^{2+}$ to neighbouring $InsP_3$ receptors initiates a $Ca^{2+}$ wave through $Ca^{2+}$-induced $Ca^{2+}$ release that spreads across the cell. It is proposed that $InsP_3$ diffuses through gap junctions to prime $InsP_3$ receptors and release internal $Ca^{2+}$ in adjacent cells. The $Ca^{2+}$ released in the adjacent cell then initiates a wave of $Ca^{2+}$-induced $Ca^{2+}$ release via primed $InsP_3$ receptors. Multicellular communication is achieved by the repetition of this process in more distal cells. Because it is postulated that

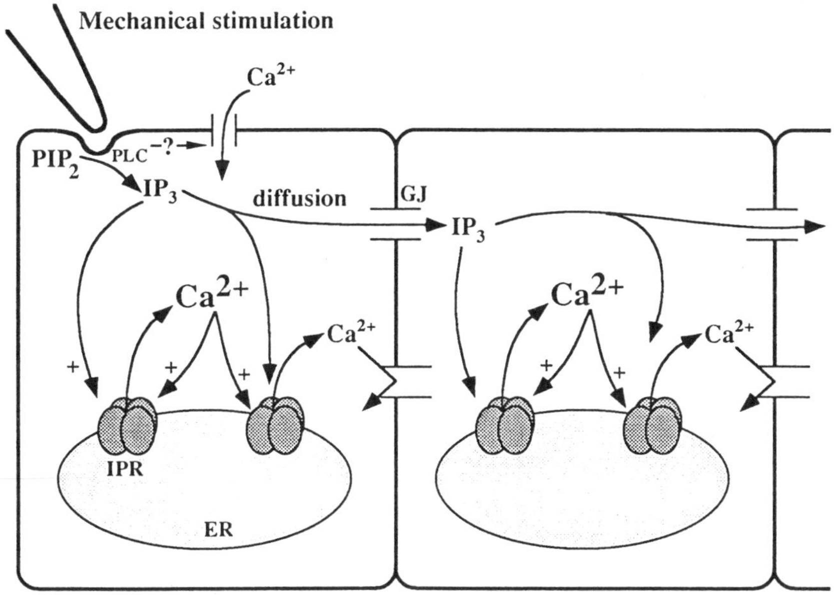


FIG. 1. A diagrammatic representation of the hypothesis for the propagation of mechanically induced intercellular $Ca^{2+}$ waves based on the diffusion of inositol 1,4,5-trisphosphate ($IP_3$) through gap junctions. Mechanical stimulation activates phospholipase C (PLC) to produce $IP_3$ from phosphatidylinositol 4,5-bisphosphate ($PIP_2$) and induces an influx of $Ca^{2+}$ through $Ca^{2+}$-conducting membrane channels. $IP_3$ releases $Ca^{2+}$ from intracellular stores in the endoplasmic reticulum (ER) via the $IP_3$ receptor (IPR). Calcium diffuses across the cell to other receptors to produce a $Ca^{2+}$ wave by $Ca^{2+}$-induced $Ca^{2+}$ release (+) via IPR. Communication of the $Ca^{2+}$ wave to the adjacent cell is achieved by the diffusion of $IP_3$ through gap junctions (GJ). Calcium does not appear to act as an intercellular messenger.

$InsP_3$ is produced only in the stimulated cell, the area over which the mechanically induced $Ca^{2+}$ wave travels is limited.

### *Mathematical modelling of intercellular calcium waves*

A valuable approach to test the validity of this hypothesis is the formulation of a mathematical model that takes into account the known biophysical parameters pertaining to the processes. A two-dimensional model of the above hypothesis, constructed by Sneyd et al (1994a,b), reproduces the experimentally observed features of the $Ca^{2+}$ waves in epithelial cells extremely well. With a slight modification to one of the kinetic parameters related to the inactivation of the $InsP_3$ receptor, the same model reproduces the $Ca^{2+}$ signalling in glial cells, including the initiation of $Ca^{2+}$ oscillations by $Ca^{2+}$ wave propagation (J. Sneyd, unpublished data). Despite this success, it should be emphasized that a single model will probably not be applicable to all cell types.

The emphasis of the mathematical model is that diffusion of $InsP_3$ from a single cell can account for a multicellular wave. The model predicts specific values for a number of important parameters, including gap junctional permeability, the ratio of the $[InsP_3]_i$ in the stimulated cell to the equilibrium dissociation constant of the $InsP_3$ receptor, and the rates and distances of wave propagation (Sneyd et al 1994b). These parameters can be measured experimentally to test the validity of this diffusional hypothesis. The possibility that a $Ca^{2+}$-independent, regenerative process underlies the propagation of $Ca^{2+}$ waves was considered, but in these regenerative models the $Ca^{2+}$ wave would not stop after a limited number of cells as experimentally observed. As a result, the diffusion of $InsP_3$ between cells via gap junctions to propagate intercellular $Ca^{2+}$ waves remains the preferred hypothesis.

### *Mechanical activation of phospholipase C*

The cellular production of $InsP_3$ generally follows exposure to agonists (Berridge 1993) but $Ca^{2+}$ waves can be initiated in the absence of agonists. Thus, mechanical stimulation may directly result in an increase in $[InsP_3]_i$. After exposure to an aminosteroid that inhibits phospholipase C (PLC), (U73122; Upjohn Laboratories, Kalamazoo, MI) mechanical stimulation of an airway epithelial cell failed to initiate an intercellular $Ca^{2+}$ wave, although $[Ca^{2+}]_i$ did increase in the stimulated cell (Hansen et al 1995). Similarly, U73122 abolished the $Ca^{2+}$ response of epithelial cells to extracellular ATP that is believed to be mediated via $P_2$-purinergic receptors, PLC and $InsP_3$ production. A similar control compound, U73343, which does not inhibit PLC, did not inhibit the initiation of $Ca^{2+}$ waves induced by mechanical stimulation. These results indicate that mechanical stimulation initiates the propagation of an intercellular $Ca^{2+}$ wave by increasing PLC activity in the stimulated cell, so bringing about an increase in $[InsP_3]_i$ (Hansen et al 1995).

The mechanism by which mechanical stimulation could activate PLC is currently unknown. However, stretching or osmotic swelling of cells can result in an elevation of $[InsP_3]_i$ (Brophy et al 1993, Dassouli et al 1993, Kulik 1991, Hoffmann et al 1993). Inducing conformational change is a common mechanism for activating enzymes and it has been proposed that stretch-activated channel proteins are gated by a conformational change induced by mechanical distortion (Sachs 1992). Thus, membrane-associated PLC may undergo a conformational change following mechanical stimulation. Phospholipase C is also activated by G proteins and these may be a site of mechanical transduction (Dassouli et al 1993). Additionally, the cytoskeleton may play a role in mechanical transduction either by focusing the forces to specific proteins via microtubules or microfilaments (Sachs 1992) or by making membrane phospholipids available for metabolism through the displacement of profilin molecules (Goldschmidt-Clermont et al 1991). However, one caveat is that the activation of PLC should be relatively independent of $[Ca^{2+}]_i$ in view of the independence of $Ca^{2+}$ wave propagation from changes in $[Ca^{2+}]_i$.

### *The calcium response of the stimulated cell*

As mentioned earlier, the response of the stimulated cell appears to differ from that of non-stimulated cells and this point requires further consideration. A hypothesis of $Ca^{2+}$ wave propagation based on the liberation of intracellular $Ca^{2+}$ through an increase in $[InsP_3]_i$ appears contradictory to the observation that $[Ca^{2+}]_i$ increases in a stimulated cell exposed to U73122 or thapsigargin or loaded with heparin. However, these observations can be explained if it is assumed that mechanical stimulation initiates an influx of extracellular $Ca^{2+}$, via $Ca^{2+}$ channels, in the stimulated cell (though this influx is not essential for the initiation of an intercellular $Ca^{2+}$ wave, as explained above). Further support for this explanation is provided by the finding that $Ni^{2+}$, a $Ca^{2+}$ channel blocker, prevents an increase in $[Ca^{2+}]_i$ of the stimulated cell in the presence of extracellular $Ca^{2+}$ when the intracellular stores of this cell have been emptied with thapsigargin (Boitano et al 1994a).

The hypothesis that there is an influx of $Ca^{2+}$ into the stimulated cell is also supported by the finding that under $Ca^{2+}$-free conditions the $[Ca^{2+}]_i$ of the stimulated cell does not increase. However, this result also appears contradictory, because any increase in $[InsP_3]_i$ produced by mechanical stimulation would be expected to release $Ca^{2+}$ from intracellular stores. This paradox could be explained if the $Ca^{2+}$ fluxes across the cell membrane were reversed under $Ca^{2+}$-free conditions. If the $Ca^{2+}$ released from internal stores by $InsP_3$ is lost from the cell via ion channels, one would predict that extracellular ion channel blockers would trap the released $Ca^{2+}$ and restore the

$Ca^{2+}$ response of the stimulated cell. In the presence of the $Ca^{2+}$ channel blockers $Gd^{3+}$, $Ni^{2+}$ or $La^{3+}$ or the voltage-gated $Ca^{2+}$ channel blockers nifedipine or nimodipine, $[Ca^{2+}]_i$ increased in the stimulated cell in the absence of extracellular $Ca^{2+}$ (Boitano et al 1994a). The effect of $Gd^{3+}$ was dose dependent: at low concentrations of $Gd^{3+}$, $[Ca^{2+}]_i$ did not increase until several seconds after stimulation, whereas at high $Gd^{3+}$ concentrations $[Ca^{2+}]_i$ increased almost immediately (Boitano et al 1994a). This result suggests that there is a more effective blockade of the channels at higher concentrations of $Gd^{3+}$. This requirement for a high concentration of $Gd^{3+}$ to restore the $Ca^{2+}$ response of the stimulated cell, together with the evidence that other types of $Ca^{2+}$ channel blockers have a similar effect, suggests that the $Ca^{2+}$ flux across the cell membrane occurs through ion channels in addition to stretch-activated channels. It seems unlikely that non-specific holes are responsible for the $Ca^{2+}$ efflux as these would not be influenced by ion channel blockers. Furthermore, fluorescent dyes do not appear to be lost from epithelial cells after mechanical stimulation.

In view of the sensitivity of the $InsP_3$ receptor to $Ca^{2+}$, it is possible that the amount of $Ca^{2+}$ efflux that is required to prevent a $Ca^{2+}$ response in the stimulated cell is small. If $[Ca^{2+}]_i$ is prevented from rising following $InsP_3$ production, $Ca^{2+}$ -induced $Ca^{2+}$ release via the $InsP_3$ receptor will not be initiated. Consequently, very little of the intracellular $Ca^{2+}$ stores of the stimulated cell will be released. The greater efficiency of blocking the channels with higher concentrations of $Gd^{3+}$ lends support to this idea. The faster $Ca^{2+}$ can accumulate in the cell, the more likely a $Ca^{2+}$ wave is to occur in the stimulated cell. The alternative possibility, that all the internal $Ca^{2+}$ stores are released and lost from the cell, seems less feasible in view of the ability to detect relatively low levels of $Ca^{2+}$ with imaging techniques.

The mechanism by which ion channels are opened in the stimulated cell is of interest. Stretch-activated channels may be involved, but it is important to note that the mechanical stimulus lasts for only 150 ms, whereas $[Ca^{2+}]_i$ in the stimulated cell, under $Ca^{2+}$-free conditions, remains low for many seconds. The gating status of most stretch-activated channels is closely correlated with the application of force. In contrast, a stretch-sensitive channel activated after a significant delay has been described in airway epithelial cells (Kim et al 1993). This channel was unresponsive to pressure in excised patches. Together, these facts suggest that this channel may in fact be a second messenger-activated channel. It is therefore tempting to speculate that a second messenger produced in response to mechanical stimulation initiates the prolonged channel activity in the stimulated cell (Boitano et al 1994a). This messenger cannot be communicated between cells (i.e., cannot be $InsP_3$ or a metabolite) otherwise $[Ca^{2+}]_i$ would not increase in adjacent cells. The production of $InsP_3$ by PLC also leads to the production of diacylglycerol (Berridge 1993). Because of its membrane domain, this molecule is likely to remain in the stimulated cell where it is ideally located to modulate ion channels, either directly or via protein kinase C. However, the increase in $[Ca^{2+}]_i$ in the stimulated cell following treatment with U73122 would argue against this activation mechanism. A persistent effect of mechanical stimulation could also be induced by activation of a G protein. These ideas must await electrophysiological investigation.

### *Movement of inositol trisphosphate through gap junctions*

A central premise for our hypothesis is that $InsP_3$ can move through gap junctions. Unfortunately, in the absence of a biological label or dye for $InsP_3$, this is difficult to prove. However, one successful approach has involved the use of caged $InsP_3$. Caged $InsP_3$ can be perfused into single endothelial cells in culture from a patch pipette. After an incubation time to allow the caged $InsP_3$ to diffuse to adjacent cells, a distal cell is illuminated with UV light capable of photolysing the caged compound. The distal cell shows an increase in $[Ca^{2+}]_i$, indicating that active $InsP_3$ was formed, releasing intracellular $Ca^{2+}$. Because only one cell received caged $InsP_3$, which has a slightly higher molecular mass than $InsP_3$, the interpretation would be that the caged compound diffused through the gap junctions (Carter et al 1994).

The ability of $Ca^{2+}$ to close gap junctions appears to pose an obstacle to the propagation of intercellular $Ca^{2+}$ waves via gap junctions. However, this paradox can be resolved by paying careful attention to the temporal aspects of these responses. Extended exposure to high concentrations of $Ca^{2+}$ ($>10\,\mu M$) results in the closure of gap junctions (Bennett & Verselis 1992, Sáez et al 1993). It also appears that physiological concentrations of $Ca^{2+}$ (in the nanomolar range) can reversibly close gap junctions but this response takes about 40 s (Lazrak et al 1994). Thus, because $Ca^{2+}$ waves propagate faster than the proposed closure times, the intercellular messenger $InsP_3$ will have diffused to adjacent cells to propagate the wave before the $Ca^{2+}$ is able to close the gap junctions. Gap junction closure would influence the backward flow of messengers and diffusion of $InsP_3$ from the stimulated cell, thereby limiting the extent of diffusion and wave propagation.

### *A comparison of mechanically induced and glutamate-induced intercellular calcium waves*

It is important to emphasize that mechanically induced $Ca^{2+}$ waves appear to be propagated by a different cellular mechanism from glutamate-induced intercellular $Ca^{2+}$ waves. Glutamate-induced intercellular $Ca^{2+}$ waves are initiated after several minutes of exposure to glutamate in purified astrocyte cultures (Finkbeiner 1992, Kim et al 1994) and propagate without degradation through many cells (up to 100 cells in a single direction); they require extracellular $Ca^{2+}$ and $Na^+$ (Kim et al 1994). In addition, glutamate-induced intercellular $Ca^{2+}$ waves propagate smoothly between cells with a constant velocity and without a delay at the cell boundary, even though their propagation requires gap junctions (Finkbeiner 1992). Similar $Ca^{2+}$ waves, in organotypically cultured brain slices, can be evoked by electrical stimulation of glutamate-releasing neurons (Dani et al 1992, see also Dani & Smith 1995, this volume). The mechanism of propagation of glutamate-induced intercellular $Ca^{2+}$ waves does not rely on $InsP_3$ but is proposed to involve a regenerative process entailing a reversal of the $Na^+/Ca^{2+}$ exchanger of the plasma membrane. However, the nature of the messenger that moves from cell to cell via gap junctions to mediate this response is unknown (Kim et al 1994).

In contrast to these late-developing, glutamate-induced, regenerative $Ca^{2+}$ waves, Kim et al (1994) have now described an intercellular $Ca^{2+}$ wave or spike that occurs immediately after addition of glutamate and is subject to delays at cell boundaries. However, the velocity of wave propagation is much faster (50–150 $\mu m\, s^{-1}$) than that of mechanically stimulated waves. Kim et al (1994) propose that these immediate spikes or waves are mediated by $InsP_3$ but is is not clear whether or not the spatiotemporal patterns arise from different cell sensitivities to glutamate. It is interesting to point out that glutamate-induced, regenerative $Ca^{2+}$ waves do not occur in mixed glial cell cultures in the continued presence of glutamate (Charles et al 1991). However, mechanically induced waves can be propagated through cells showing a sustained $[Ca^{2+}]i$ increase in response to continued exposure to glutamate. Thus, it appears that at least two distinct types of intercellular $Ca^{2+}$ wave exist in non-excitable cells.

## The physiological relevance of intercellular calcium waves

A persistent question that is, as yet, not well answered is, what is the function of these $Ca^{2+}$ waves? Several possibilities have been presented by Sanderson et al (1994). In brief, $Ca^{2+}$ waves are envisaged as carrying signals through multicellular systems. The local stimulation (mechanical, electrical or chemical) of one cell can be transmitted to surrounding cells. This form of signalling would be applicable to physiological activities such as those associated with mucociliary clearance (Sanderson et al 1990). More complex signalling may be involved in pattern formation during development. Calcium waves may also be important in the orientation and movement of cells during the early stages of wound healing (Boitano et al 1994b). An extensive $Ca^{2+}$ wave is initiated after wounding and appears to be necessary for cells to close the wound. Finally, it has been suggested that the long-distance $Ca^{2+}$ waves of glial cells may contribute to the phenomenon of spreading depression which may underlie migraine headaches (Leibowitz 1992).

**Summary**

Intercellular $Ca^{2+}$ waves exist in diverse cell types and may be a general mechanism of intercellular communication that occurs via gap junctions. One intercellular messenger that mediates intercellular $Ca^{2+}$ waves appears to be $InsP_3$, but $Ca^{2+}$ does not appear to have an important role in communication. The function of intercellular $Ca^{2+}$ waves may be to coordinate the cooperative activity of cells. The initiation of intercellular $Ca^{2+}$ waves in non-excitable cells by mechanical stimulation provides a unique experimental approach to investigate the mechanism and importance of intercellular $InsP_3$ signalling further.

*Acknowledgements*

This work was supported by the Smokeless Tobacco Research Council Inc., and the Tobacco Related Disease Research Program of the University of California and NIH grant HL 49288. I would like to thank Andrew Charles, Scott Boitano, James Sneyd and Ellen Dirksen for their extensive contributions to this research and manuscript.

**References**

Allbritton NL, Meyer T 1993 Localized calcium spikes and propagating calcium waves. Cell Calcium 14:691–697

Bennett MVL, Verselis VK 1992 Biophysics of gap junctions. Semin Cell Biol 3:29–47

Berridge MJ 1993 Inositol trisphosphate and calcium signalling. Nature 361:315–325

Boitano S, Dirksen ER, Sanderson MJ 1992 Intercellular propagation of calcium waves mediated by inositol trisphosphate. Science 258:292–295

Boitano S, Sanderson MJ, Dirksen ER 1994a A role for $Ca^{2+}$-conducting ion channels in mechanically-induced signal transduction of airway epithelial cells. J Cell Sci 107:3037–3044

Boitano S, Dirksen ER, Sanderson MJ 1994b The role of intercellular $Ca^{2+}$ signaling in epithelial cells. FASEB (Fed Am Soc Exp Biol) J 8:A664

Brophy CM, Mills I, Rosales O, Isales C, Sumpio BE 1993 Phospholipase C: a putative mechanotransducer for endothelial cell response to acute hemodynamic changes. Biochem Biophys Res Commun 190:576–581

Carter TD, Chen XY, Carlile G, Kalapothakis E, Ogden DC, Evans WH 1994 Cultured porcine aortic endothelial gap junctions: characterization and permeation by second messenger molecules. J Physiol 475:74(abstr)

Charles AC, Merrill JE, Dirksen ER, Sanderson MJ 1991 Intercellular signaling in glial cells: calcium waves and oscillations in response to mechanical stimulation and glutamate. Neuron 6:983–992

Charles AC, Naus CCG, Zhu D, Kidder GM, Dirksen ER, Sanderson MJ 1992 Intercellular calcium signaling via gap junctions in glioma cells. J Cell Biol 118:195–201

Charles AC, Dirksen ER, Merrill JE, Sanderson MJ 1993 Mechanisms of intercellular calcium signaling in glial cells studied with dantrolene and thapsigargin. Glia 7:134–145

Dani JW, Smith SJ 1995 The triggering of astrocytic calcium waves by NMDA-induced neuronal activation. In: Calcium waves, gradients and oscillations. Wiley, Chichester (Ciba Found Symp 188) p 195–209

Dani JW, Chernjavsky A, Smith SJ 1992 Neuronal activity triggers calcium waves in hippocampal astrocyte networks. Neuron 8:429–440

Dassouli A, Sulpice JC, Roux S, Crozatier B 1993 Stretch-induced inositol trisphosphate and tetrakisphosphate production in rat cardiomyocytes. J Mol Cell Cardiol 25:973–982

Demer LL, Wortham C, Dirksen ER, Sanderson MJ 1993 Mechanical stimulation induces intercellular calcium signaling in bovine aortic endothelial cells. Am J Physiol 264:H2094–H2102

Enkvist MOK, McCarthy KD 1993 Astroglial gap junction communication is increased by treatment with either glutamate or high [$K^+$]. J Neurochem 62:489–495

Finkbeiner S 1992 Calcium waves in astrocytes—filling in the gaps. Neuron 8:1101–1108

Furuya K, Enomoto K, Yamagishi S 1993 Spontaneous calcium oscillations and mechanically and chemically induced calcium responses in mammary epithelial cells. Pfluegers Arch Eur J Physiol 422:295–304

Goldschmidt-Clermont PJ, Kim JW, Machesky LM, Rhee SG, Pollard TD 1991 Regulation of phospholipase C-$\gamma$1 by profilin and tyrosine phosphorylation. Science 251:1231–1233

Hansen M, Boitano S, Dirksen ER, Sanderson MJ 1993 Intercellular calcium signaling induced by extracellular ATP and mechanical stimulation in airway epithelial cells. J Cell Sci 106:995–1004

Hansen M, Boitano S, Dirksen ER, Sanderson MJ 1995 The role of phospholipase C and ryanodine receptors in the initiation and propagation of intercellular $Ca^{2+}$ waves. Cell Calcium, in press

Hoffmann EK, Simonsen LO, Lambert IH 1993 Cell volume regulation: intracellular transmission. Adv Comp Environ Physiol 14:187–248

Iino M, Endo M 1992 Calcium-dependent immediate feedback control of inositol 1,4,5-trisphosphate-induced $Ca^{2+}$ release. Nature 360:76–78

Jaffe LF 1993 Classes and mechanisms of calcium waves. Cell Calcium 14:736–745

Kim YK, Dirksen ER, Sanderson MJ 1993 Stretch-activated channels in airway epithelial cells. Am J Physiol 265:C1306–C1318

Kim WT, Rioult MG, Cornell-Bell AH 1994 Glutamate induced signaling in astrocytes. Glia 11:173–184

Kulik TJ, Bialecki RA, Colucci WS, Rothman A, Glennon ET, Underwood RH 1991 Stretch increases inositol trisphosphate and inositol tetrakisphosphate in cultured pulmonary vascular smooth muscle cells. Biochem Biophys Res Commun 180:982–987

Lazrak A, Peres A, Giovannardi S, Peracchia C 1994 Junctional uncoupling by ATP or arachidonic acid is $Ca_i$-mediated. Biophys J 66:A149

Leibowitz DH 1992 The glial spike theory. I. On an active role of neuroglia in spreading depression and migraine. Proc R Soc Lond Ser B Biol Sci 250:287–295

Naus CCG, Zhu S, Todd S, Kidder GM 1992 Characteristics of C6 cells overexpressing a gap junction protein. Cell Mol Neurobiol 12:163–175

Osipchuk Y, Cahalan M 1992 Cell-to-cell spread of calcium signals mediated by ATP receptors in mast cells. Nature 359:241–244

Rooney TA, Thomas AP 1993 Intracellular calcium waves generated by Ins (1,4,5) P(3)-dependent mechanisms. Cell Calcium 14:674–690

Sachs F 1992 Stretch-sensitive ion channels: an update. In: Corey DP, Roper SD (eds) Sensory transduction. Rockefeller University Press, New York (Soc Gen Physiol 45th Symp) p 242–260

Sáez JC, Berthoud VM, Moreno AP, Spray DC 1993 Gap junctions. Multiplicity of controls in differentiated and undifferentiated cells and possible functional implications. In: Shenolikar S, Nairn AC (eds) Advances in second messenger and phosphoprotein research. Raven Press, New York, vol 27:163–198

Sanderson MJ, Charles AC, Dirksen ER 1990 Mechanical stimulation and intercellular communication increases intracellular $Ca^{2+}$ in epithelial cells. Cell Regul 1:585–596

Sanderson MJ, Charles AC, Boitano S, Dirksen ER 1994 Mechanisms and function of intercellular $Ca^{2+}$ waves. Mol Cell Endocrinol 98:173–187
Sigurdson W, Ruknudin A, Sachs F 1992 Calcium imaging of mechanically induced fluxes in tissue-cultured chick heart: role of stretch-activated ion channels. Am J Physiol 262:H1110–H1115
Sneyd J, Charles AC, Sanderson MJ 1994a A model for the propagation of intercellular calcium waves. Am J Physiol 266:C293–C302
Sneyd J, Wetton BTR, Sanderson MJ 1994b Intercellular calcium waves mediated by diffusion of inositol (1,4,5)-trisphosphate: a two dimensional model. Am J Physiol, in press
Warner A 1992 Gap junctions in development—a perspective. Semin Cell Biol 3:81–91
Xia SL, Ferrier J 1992 Propagation of a calcium pulse between osteoblastic cells. Biochem Biophys Res Commun 186:1212–1219

## DISCUSSION

*Irvine:* The cell in which there is no [$Ca^{2+}$] rise in the absence of extracellular $Ca^{2+}$ must be full of $InsP_3$, because it's going to send it to all the other cells. The complete absence of a $Ca^{2+}$ signal tells you that practically every $InsP_3$ receptor in that cell is very close to the plasma membrane, because $Ca^{2+}$ diffusion is very slow and whatever is happening at the plasma membrane will influence only the free $Ca^{2+}$ close to it.

For a wave still to be generated in cell number four, despite the diffusion barriers and gap junctions, there has to be quite a lot of $InsP_3$ in cell number one. So, it troubles me that the $Ca^{2+}$ wave seems to spread with a similar velocity and in a similar way right through the sequence of cells. I feel there ought to be some kind of regenerative step. Does the phospholipase C (PLC) inhibitor inhibit a wave spread by microinjection of $InsP_3$?

*Sanderson:* We haven't done that experiment. All we can determine from experiments with a PLC inhibitor is that we don't initiate a $Ca^{2+}$ wave. The involvement of regeneration of $InsP_3$ by PLC cannot be assessed because a $Ca^{2+}$ wave is not propagating in adjacent cells.

Certainly, there is thought to be a high concentration of $InsP_3$ in the stimulated cell. Our mathematical model suggests that with an $InsP_3$ receptor with an equilibrium dissociation constant of about 20 nM for $InsP_3$ and the release of $Ca^{2+}$, not necessarily at a maximal rate, an $InsP_3$ concentration in the stimulated cell of the order of 3 $\mu$M is required for wave propagation. The critical number is the ratio between the dissociation constant of the receptor and the concentration of $InsP_3$ in the stimulated cell. The ratio needs to be about 300 to produce intercellular $Ca^{2+}$ waves that propagate across four or five cells. Although there is a lot of $InsP_3$ in the stimulated cell, the release of $Ca^{2+}$ is not directly proportional to the concentration of $InsP_3$. As soon as the [$Ca^{2+}$] reaches 300–600 nM it will shut off $Ca^{2+}$ release via the receptor, so you might expect to see a $Ca^{2+}$-induced $Ca^{2+}$ wave moving across that first cell in much the same way as the $Ca^{2+}$ wave would do in more distant cells.

*Putney:* There are two variables that I don't think you have taken into account. One is how fast the $InsP_3$ is metabolized and the other is the time course of $InsP_3$ production. When you touch the cell, is it like an injection, with a big bang of $InsP_3$, or does the PLC keep making $InsP_3$ for a long time in that activated cell?

*Sanderson:* We believe that, once activated, PLC releases $InsP_3$ for several seconds and that the stimulation is not like an instantaneous injection. This possibility of prolonged PLC activity is also suggested by the failure of the stimulated cell to show an increase in $[Ca^{2+}]$ in $Ca^{2+}$-free conditions. This response appears to occur through the loss of $Ca^{2+}$ from the cell via ion channels and requires that these ion channels also remain open for some time. This opening may be controlled by a sustained level of second messenger, but this messenger must not diffuse to adjacent cells; otherwise, increases in $[Ca^{2+}]$ would not occur in adjacent cells. Consequently, a sustained production of diacylgycerol, as well as $InsP_3$, by PLC in response to mechanical stimulation could explain the response of the stimulated cell in $Ca^{2+}$-free conditions and the production of high concentrations of $InsP_3$.

*Putney:* In the stimulated cell you're probably activating the $Ca^{2+}$ pump. At that concentration, $Gd^{3+}$ blocks the plasma membrane $Ca^{2+}$ pump, as well as the $Ca^{2+}$ channels. I would guess that in that cell alone, though not in any of the others, the stimulus activates the $Ca^{2+}$ pump also.

*Sanderson:* That's quite possible. The stimulated cell does appear to behave differently to the adjacent, non-stimulated cells that are also exposed to $Gd^{3+}$. However, it is important to note that other ion channel blockers give results similar to those with $Gd^{3+}$, suggesting that the ion channels are blocked rather than the pump.

*Putney:* I don't doubt that your conclusion that $Ca^{2+}$ entry is not required is correct. The fact that when you remove external $Ca^{2+}$ you don't see a rise in the stimulated cell tells me that the pump is activated in that stimulated cell. If you then block the pump with $Gd^{3+}$, you get a larger rise in $[Ca^{2+}]$.

*Sanderson:* What would activate the pump?

*Putney:* What activates any of the signalling when you poke the cell? Do you increase diglyceride only in the stimulated cell?

*Sanderson:* Yes; one idea is that diacylglycerol is activating the channels, among other things, but doesn't move to adjacent cells.

*Michell:* We should pursue the question of the metabolism of $InsP_3$ further. There is relatively little information about this, but its half-life is probably no more than a second or two in some cells. If you're looking at a spread into four, five, six or seven cells, you're talking about 10–15 s of spreading.

*Sanderson:* In our model we set the half-life of the $InsP_3$ receptor to a few seconds and the $Ca^{2+}$ waves propagated over several cells in 10–15 s, as observed experimentally. I believe the reason for this is that the $InsP_3$ concentration in each cell, except those at the periphery of the response, is higher than that required to release $Ca^{2+}$. Consequently, although the $InsP_3$ is being broken down, an excess is still available to propagate the wave.

In addition, $InsP_3$, as a result of diffusion, suddenly appears in an adjacent cell which hasn't been stimulated. Consequently, the breakdown of $InsP_3$ may not be as fast in the unstimulated cell. One can therefore speculate that the $InsP_3$ will move across the cell before being influenced significantly by the cell. It appears to take $InsP_3$ about one second to cross a cell.

*Michell:* In that model, when you've started with a concentration of about 3 $\mu$M in your core cell, what is the level which is calculated to be driving the responses of the cells furthest out in the field?

*Sanderson:* One will get a $Ca^{2+}$ response if the [$InsP_3$] is above the level required to release $Ca^{2+}$ via the $InsP_3$ receptor, which is about 20 nM in our model. Of course, you can't use the $Ca^{2+}$ signal as an indicator of the [$InsP_3$] in all the cells, as this will look relatively similar in view of the amplification by $Ca^{2+}$-induced $Ca^{2+}$ release, via the $InsP_3$ receptor.

*Meyer:* If one uncages $InsP_3$ to give about 20 nM $InsP_3$, one does not get a big spike. If one uncages 10 $\mu$M $InsP_3$, one gets a huge spike that lasts longer. In your experiment, the response in the second cell up to the sixth cell would never look the same as in the first cell if the $InsP_3$ concentration is so different.

*Sanderson:* In thinking about how much $InsP_3$ is needed, you have to take time into account. There is not an instantaneous 10 or 20 nM signal of $InsP_3$. There's a continual movement of $InsP_3$ through the gap junction, so the concentration can actually build up. Ian Parker's work addresses that. His threshold $InsP_3$ concentration for $Ca^{2+}$ release was 60 nM; this did not give the full, maximal rate of release, but was enough to give, with time, something like 70% of the response to 600 nM $InsP_3$.

*Petersen:* The density of gap junctions varies enormously in different types of tissues. In the liver and pancreas, for example, gap junctions are very dense and it is hard to imagine there is much of a diffusion barrier.

*Sanderson:* In our preparation there are relatively few gap junctions. They represent probably less than 1% of the cell membrane area. One of the critical things about this model is that there has to be a certain degree of $InsP_3$ permeability across the membrane, otherwise the wave would simply stay in one cell. We still have to measure the exact diffusional coefficient of $InsP_3$ between cells. As our experiments with C6 cells showed, the more gap junctions there are, the greater or faster is the response, as you might expect.

*Thomas:* The spontaneous $Ca^{2+}$ oscillations in your system do not propagate. Might that be because they're not high for long enough?

*Sanderson:* I don't think so. The spontaneous oscillations usually last for 10–20 s with $Ca^{2+}$ concentrations similar to those associated with $Ca^{2+}$ waves.

*Berridge:* But the [$InsP_3$] is kept very low.

*Sanderson:* Dr Thomas's point was that if $[Ca^{2+}]_i$ is elevated for long enough, $Ca^{2+}$ might be able to diffuse through the gap junction and initiate a $Ca^{2+}$ wave. However, this is not observed, suggesting that $Ca^{2+}$ is not a major intercellular messenger. The failure of $Ca^{2+}$ to initiate $Ca^{2+}$ waves in adjacent cells is also consistent with the apparent lack of ryanodine receptors in airway epithelial cells. With respect to $InsP_3$, it is unlikely that there is a large gradient of $InsP_3$ between cells displaying spontaneous $Ca^{2+}$ oscillations. Similarly, in agonist-stimulated oscillations, [$InsP_3$] would be increased in all cells, again preventing the formation of gradients between cells and the propagation of $Ca^{2+}$ waves.

*Thomas:* Robin Irvine thought that the failure of the [$Ca^{2+}$] to increase in the source cell in the absence of extracellular $Ca^{2+}$ meant that all the $InsP_3$ receptors are peripheral and that you are emptying all the pools; but have you actually looked to see whether or not that source cell has any $Ca^{2+}$ in it, by putting in thapsigargin, for example.

*Sanderson:* The propagation of $Ca^{2+}$ waves from randomly stimulated cells in $Ca^{2+}$-free conditions indicates that all cells contain $Ca^{2+}$ before stimulation, and this has been verified by the addition of thapsigargin to cells under $Ca^{2+}$-free conditions. However, it is not clear whether or not the stimulated cell releases all of its stored $Ca^{2+}$ after mechanical stimulation in $Ca^{2+}$-free conditions. We suggest that it does not in view of the requirement for both $InsP_3$ and $Ca^{2+}$ for activation of the $InsP_3$ receptor (Boitano et al 1994). The addition of thapsigargin following mechanical stimulation might address this point, but if membrane channels remain open as suggested, a subsequent release of $Ca^{2+}$ from internal stores may not be visualized. However, we have not done this experiment. The response of the stimulated cell is intriguing and has provided a significant amount of information, but I would like to emphasize that the $Ca^{2+}$ responses of the adjacent cells are equally important. Here, there is the communication of a signal between cells that have not been directly stimulated in any way.

*Whitaker:* You suggested that your experiments with heparin introduced by electroporation showed that $InsP_3$ is diffusing from cell to cell. The regenerative $Ca^{2+}$ release response in each cell occurs via the $InsP_3$ receptor. Wouldn't you expect the heparin to block the very regenerative response which is your assay for $InsP_3$ diffusion, so that, since heparin blocks your assay for diffusion, those experiments don't actually tell you a great deal? Heparin blocks regenerative waves through the $InsP_3$ receptor, whether or not they're generated by $InsP_3$. Your heparin experiment is not a test of your hypothesis.

*Sanderson:* What the experiment shows, in precise terms, is that the process is mediated by $InsP_3$. The experiment doesn't show that $InsP_3$ diffuses from one cell to the next. Carter et al (1994) have released caged $InsP_3$ from a patch electrode into endothelial cells and allowed it to diffuse away from the electrode before flashing a cell several cells away; that cell showed an increase in intracellular [$Ca^{2+}$]. The interpretation is that the caged $InsP_3$ diffused through gap junctions to the distal cell where it could be photoreleased and release $Ca^{2+}$.

*Jaffe:* I would like to suggest a specific model to explain all this. After you poke the cell, you somehow generate $InsP_3$. The $InsP_3$ then diffuses out in what I would call a primer wave. After the primer wave, you have a conventional, fast $Ca^{2+}$ wave.

*Sanderson:* That's basically in agreement with what we have proposed. Our idea is that $InsP_3$, generated following mechanical stimulation, moves throughout the epithelial cells to release $Ca^{2+}$. In addition to the $InsP_3$-induced $Ca^{2+}$ release, there is amplification by $Ca^{2+}$-induced $Ca^{2+}$ release. The subsequent asynchronous $Ca^{2+}$ oscillations occur as the result of $Ca^{2+}$-induced $Ca^{2+}$ waves moving across the cells that have been primed with $InsP_3$. However, multiple intercellular $Ca^{2+}$ waves are not seen because further waves of $InsP_3$ are not initiated and $Ca^{2+}$ itself appears unable to initiate waves in adjacent cells. The finding that $Ca^{2+}$-induced $Ca^{2+}$ waves can actually propagate in a direction opposite to that of the initial, mechanically stimulated wave is evidence that cells have been primed with $InsP_3$ and that subsequent waves are driven by $Ca^{2+}$-induced $Ca^{2+}$ release. Experimental observations show that the frequency of the oscillations decays, suggesting there is a decrease in $[InsP_3]$ in those cells.

*Berridge:* This is like Jim Lechleiter's excitable medium (p 68). Inositol trisphosphate has created an excitable medium through which $Ca^{2+}$ can then express itself.

*Sanderson:* I agree. However, the expression of $Ca^{2+}$ in the form of repetitive waves across multiple cells is prevented, presumably by a limited diffusion of $Ca^{2+}$ through gap junctions.

*Meyer:* There are two experiments that could convince me that this is not some regenerative wave that is just incompletely primed. One would be if you could make with heparin a little wall, a couple of cells wide, of cells in which $[Ca^{2+}]$ is not raised and the wave passes through. The second experiment would be to use your PLC inhibitors and inject $InsP_3$ into one cell and demonstrate that you can get a similar kind of spread in cells further apart.

*Thomas:* A third one would be to put BAPTA (1,2-bis[*o*-aminophenoxy]ethane-*N*,*N*,*N'*,*N'*-tetraacetic acid) in some of the cells in the path of a wave.

*Sanderson:* We don't observe waves with BAPTA present, but it would be difficult to restrict the BAPTA to a few cells because of its molecular size. It is possible to restrict heparin to selected cells and I thought that the way we had done the heparin experiments was pretty good, but now you want to electroporate a zone only two cells wide. We can try!

*Lechleiter:* If, rather than tapping the cell once, you leave the mechanical stimulus in place, e.g., by breaking off the tip of the electrode, do multiple $Ca^{2+}$ waves develop? If there is a continuous stimulus, do you increase the number of cells through which the wave will travel?

*Sanderson:* We haven't tried that. You may make the wave go further if you apply on extended mechanical pulse, but I think a change in the membrane's position is needed rather than a continual pressure.

*Miyazaki:* What do you think the mechanism of signal transduction leading to activation of the $Ca^{2+}$ channel might be?

*Sanderson:* There are two things going on in the stimulated cell: the activation of PLC and the opening of ion channels. It has been suggested that stretch-activated channels can be activated by a conformational change of a protein. Similarly, mechanical stimulation might directly alter the conformation of PLC or a G protein that can activate PLC. An alternative is that mechanical stimulation actually exposes PLC's substrates. Phosphatidylinositol 4,5-bisphosphate is associated with profilin, and mechanically distorting the arrangement of the cytoskeleton might allow the enzyme to get at the substrate.

*Miyazaki:* What kind of $Ca^{2+}$ channels are involved?

*Sanderson:* The channels are probably non-specific, cation-conducting channels. Three ionic blockers, $La^{3+}$, $Gd^{3+}$ and $Ni^{2+}$, give similar results with respect to the restoration of a $Ca^{2+}$ response of the stimulated cell in $Ca^{2+}$-free conditions. Similarly, two dihydropyridine channel blockers, nifedipine and nimodipine, restored the $Ca^{2+}$ response of stimulated cells in $Ca^{2+}$-free conditions (Boitano et al 1994).

## References

Boitano S, Sanderson MJ, Dirksen ER 1994 A role for $Ca^{2+}$-conducting ion channels in mechanically-induced signal transduction of airway epithelial cells. J Cell Sci 107:3037–3044

Carter TD, Chen XY, Carlile G, Kalapothakis E, Ogden DC, Evans WH 1994 Cultured porcine aortic endothelial gap junctions: characterization and permeation by second messenger molecules. J Physiol 475:74(abstr)

# The triggering of astrocytic calcium waves by NMDA-induced neuronal activation

John W. Dani* and Stephen J. Smith

*Department of Molecular and Cellular Physiology, Beckman Center, Stanford University School of Medicine, Stanford, CA 94305, USA*

*Abstract.* It has been well established that astrocytes possess functional receptors for the excitatory neurotransmitter glutamate and respond to physiological concentrations of this substance with oscillations in cytoplasmic $Ca^{2+}$ concentrations and spatially propagating $Ca^{2+}$ waves. These findings strongly suggest that glutamate released during synaptic transmission triggers such phenomena within the perisynaptic astrocyte *in situ*. We test this hypothesis in two preparations, the organotypic hippocampal slice and hippocampal neuron–astrocyte co-cultures, using the $Ca^{2+}$ indicator fluo-3 and confocal laser microscopy. An agonist for the *N*-methyl-D-aspartate (NMDA)-preferring glutamate receptor is employed to stimulate neuronal populations specifically, leaving the astrocytic population unaffected as these cells appear to lack this glutamate receptor subtype. Such pharmacological stimulation initially elicits large $Ca^{2+}$ transients within the neuronal populations, followed by $Ca^{2+}$ spikes in surrounding astrocytes, presumably as the result of neuronal glutamate release. During continuous neuronal stimulation, the astrocyte's $Ca^{2+}$ response becomes oscillatory, with a period averaging 33 s and ranging from 15 to 50 s at 21 °C. These findings establish another form of communication within the brain, that between neurons and astrocytes, which perhaps acts to couple astrocytic regulatory responses to neuronal activity.

*1995 Calcium waves, gradients and oscillations. Wiley, Chichester (Ciba Foundation Symposiumn 188) p 195–209*

Astrocytes are responsible for many diverse aspects of regulation within the neuronal environment. The homeostasis of ions crucial for impulse conduction and synaptic transmission such as $K^+$and $Ca^{2+}$ is known to be controlled by the astrocyte (Orkand et al 1966, Bevan 1990, Barres et al 1990). In addition,

---

* Present address: Columbia University, College of Physicians and Surgeons, 630 West 168th Street, NY 10032, USA

the maintenance of extracellular pH within a narrow physiological range has been attributed to the astrocyte (Walz & Mukerji 1988a,b, Chow et al 1991, Deitmer 1992). Neurons also appear to rely on astrocytes for certain Krebs cycle intermediates, tightly tying the metabolism of the two cell types (Sagar et al 1987, Cambray-Deakin et al 1988, Sonnewald et al 1991). For synaptic transmission to occur effectively, neurotransmitters must be removed from the synaptic cleft once an impulse has passed. In many instances, astrocytes are responsible for such neurotransmitter uptake and inactivation (Vernadakis 1988). Furthermore, by ensheathing the synaptic element, astrocytes limit the activity of the released neurotransmitters (Peters et al 1991).

Interestingly, astrocytes possess receptors for many of the neurotransmitter/neuroregulatory substances that are active on neurons (Murphy & Pearce 1987, Bevan 1990, Barres 1991). Both excitatory and inhibitory receptor types have been identified, although their functional significance in the astrocyte is unknown. An example of one such receptor is that for the excitatory neurotransmitter glutamate, which initiates intracellular $Ca^{2+}$ oscillations and spatially propagating $Ca^{2+}$ waves when applied to cultured astrocyte monolayers (Pearce et al 1986, Cornell-Bell et al 1990, Jensen & Chiu 1990, Cornell-Bell & Finkbeiner 1991). These phenomena are most probably caused by rising levels of inositol 1,4,5-trisphosphate releasing intracellular $Ca^{2+}$ stores (Berridge 1990, Jacob 1990, Meyer 1991). Several glutamate receptor subtypes exist; astrocytes express both ionotropic and metabotropic quisqualate and kainate receptors, but not *N*-methyl-D-aspartate (NMDA) receptors (Usowicz et al 1989, Barres et al 1990, Bevan 1990, Cornell-Bell & Finkbeiner 1991).

Because most, if not all, synapses in the central nervous system are ensheathed by astrocytic processes and because glutamate is the major excitatory neurotransmitter in the brain, we sought to test the hypothesis that synaptically released glutamate could trigger $Ca^{2+}$ responses in surrounding astrocytes. Experiments were initially conducted in a simplified system, the neuron–astrocyte co-culture, and later extended to a more complex preparation, the organotypic hippocampal slice, where the cellular organization and intrinsic connections are maintained (Gahwiler 1981, 1984, 1988). The existence of a signalling system from neuron to astrocyte raises many exciting possibilities relating to astrocyte function and environmental regulation. We therefore felt it necessary to determine whether or not this type of communication exists between cells of the hippocampus.

## Methods

In general, hippocampal co-cultures were prepared by first growing a confluent monolayer of astrocytes, onto which neurons were seeded. One- to four-day-old Sprague–Dawley rats were used as sources for both cell types.

Both neurons and astrocytes were isolated from hippocampal areas CA1 and CA3 only. The cells were dissociated by proteolytic digestion followed by trituration. Hippocampal cells were plated at a density of $5 \times 10^5$ cells/ml on 22 mm coverslips. Because the coverslips are not coated with a protein substrate such as collagen, only glial cells can attach and grow. When the astrocytes had reached confluency (after about 7–10 days), more hippocampal cells were plated over them. This time the neurons are able to attach because a glial substrate is present. Numerous glutamatergic synapses form between pyramidal neurons during the first week of culture.

Experiments on neuron-to-astrocyte signalling were done in a continuous flow perfusion chamber with several inlets so that solutions could be changed quickly. Co-cultures were initially perfused with saline so that the resting $[Ca^{2+}]$ could be measured. After this brief baseline period, the perfusate was switched to an agonist (20 $\mu$M NMDA)-containing saline, intended to excite specific neuronal populations.

Organotypic hippocampal slices were prepared as described in Dani et al (1992) and were perfused in a continuous flow chamber like the co-cultures.

To follow $Ca^{2+}$ concentrations in the two preparations mentioned above, both neurons and astrocytes were loaded with the acetoxymethyl ester derivative of the fluorescent $Ca^{2+}$ indicator fluo-3 (Tsien 1988, Kao et al 1989, Minta et al 1989). The loading was as described by Cornell-Bell et al (1990) and Dani et al (1992). Because fluo-3 reports increases in cytosolic $[Ca^{2+}]$ as increases in fluorescence, the images represented here rely on the grey scale to encode this information. Fluorescent images were acquired with a laser confocal microscope, which proved crucial for obtaining clear images in the tissue environment of the organotypic slice (Dani et al 1992). In addition, our microscope was equipped for time-lapse video recording, so that subtle features not apparent in real time could be extracted when the sequence was played back at faster rates.

## Neuron-to-astrocyte communication within hippocampal co-cultures

The neuronal response to NMDA was characterized by a sustained $Ca^{2+}$ rise, which was especially noticeable within the fine dendritic processes (Fig 1A). Cytosolic $[Ca^{2+}]$ remained elevated over the order of minutes in the continued presence of agonist. After this initial neuronal response, numerous astrocytes within the field of view also showed $Ca^{2+}$ transients (Fig. 1B). These responses followed the neuronal responses by tens of seconds, and appeared to be initiated beneath the areas dense in neuronal processes. In contrast to the sustained $[Ca^{2+}]$ rises seen in neurons, the astrocytic $[Ca^{2+}]$ rise was brief, although oscillations were occasionally observed. Control astrocyte cultures lacking neurons never responded to NMDA, though striking responses consisting of both $Ca^{2+}$ waves and oscillations were recorded after application of glutamate.

## Neuron-to-astrocyte communication within organotypic hippocampal slices

We next wished to determine whether a similar neuron-to-astrocyte signalling system existed in the more complex environment of the organotypic hippocampal slice, a cultured preparation in which much of the hippocampal morphology is preserved and many of the intrinsic connections remain intact. In addition, astrocytes in the organotypic slice appear to differentiate normally, following the path observed *in vivo* (Del Rio et al 1991). We decided to concentrate our efforts on the stratum luxidum within hippocampal area CA3, the zone of termination for glutamatergic mossy fibres originating in the dentate gyrus (Figs 2A and 2B) (Shepherd 1979, Andersen 1975).

Baseline $[Ca^{2+}]$ measurements taken before agonist stimulation seemed to indicate higher resting concentrations in astrocytes than in neurons (Fig. 2B), although differential loading could not be completely ruled out. The neuronal response to the continuous administration of NMDA was a stepwise increase in cytosolic $[Ca^{2+}]$ (Fig. 3, curve 1), which became apparent immediately on stimulation. Following the initial neuronal response, brief $Ca^{2+}$ spikes were recorded in every astrocyte in the field (Fig. 3, curves 2–6). This astrocytic response later (hundreds of seconds) became oscillatory, continuing for the remainder of the experiment (Figs 2D and 2E). These oscillations in intracellular $[Ca^{2+}]$ were observed not only in astrocytic soma, but also in numerous astrocytic processes (Figs 2D and 2E, arrows and arrowhead respectively). The frequency of these astrocytic $[Ca^{2+}]$ oscillations was irregular, with a period averaging 33 s and ranging from 15 to 50 s. Similarly, amplitude increases varied greatly both within and between astrocytes, ranging from 50 to 200% over the baseline fluorescence. Although the pattern of fluo-3 loading appeared quite distinctive between neurons and astrocytes, definitive identification of cells was made with a correlative glial fibrillary acidic protein immunohistochemical procedure (Fig. 2C).

## Discussion

We have shown that neuronal activity elicits a prompt astrocytic response characterized by oscillations in cytosolic $[Ca^{2+}]$, both in dissociated cell culture and *in situ*. These active glial responses, which are most probably initiated by neuronal glutamate release, can be contrasted with the passive responses accompanying depolarization during neuronal $K^+$ efflux (Trachtenberg & Pollen 1970, Orkand et al 1966). We believe this to be the first documentation of an active signalling network from neuron to astrocyte within the brain. In addition, we have extended earlier findings from dissociated astrocytic cultures by demonstrating the existence of similar astrocytic $Ca^{2+}$ responses *in situ*.

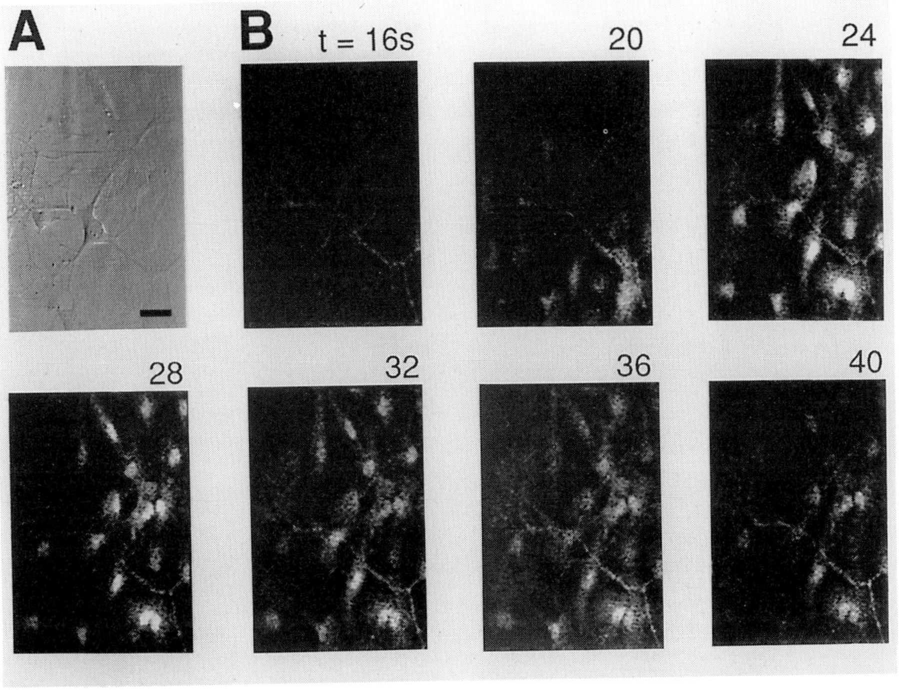


FIG. 1. Stimulation of hippocampal co-cultures with *N*-methyl-D-aspartate (NMDA) elicits a sustained $[Ca^{2+}]$ rise within the neuronal population followed by brief $Ca^{2+}$ signals in the astrocytic population. (A) A Nomarski differential interference contrast image taken from the same field as the following fluorescence images in (B). The central pyramidal-like neuron is surrounded by numerous other neurons, which have presumably made synaptic contacts onto its dendrites. Notice the extensive branching of neuronal processes and the underlying bed of astrocytes. (B) Time sequence showing the sustained rise in $[Ca^{2+}]$ observed (t = 16 s) in pyramidal neurites after application of NMDA (at t = 0 s) followed by astrocytic $Ca^{2+}$ transients (t = 20–40 s). Note how the astrocyte response appears to be initiated beneath the neuronal processes in the lower left corner of the field (t = 20 s) and spreads to adjacent astrocytes in a wave-like manner (t = 24 s). The intensity of the astrocytic response also seems to build over time (t = 28 s), but begins to die out quickly in comparison with the neuronal response (t = 40 s). Each time point shows the change in fluorescence over baseline measurements. Scale bar, 20 $\mu$m.

## Coupling astrocytic calcium responses to neuronal activity

Although we have not proven that synaptic glutamate release is the cause of the astrocytic $Ca^{2+}$ responses, several lines of evidence strongly suggest this to be the case. First, dissociated cultures enriched in astrocytes show similar repetitive $Ca^{2+}$ spiking behaviour in response to applied glutamate (100 $\mu$M)

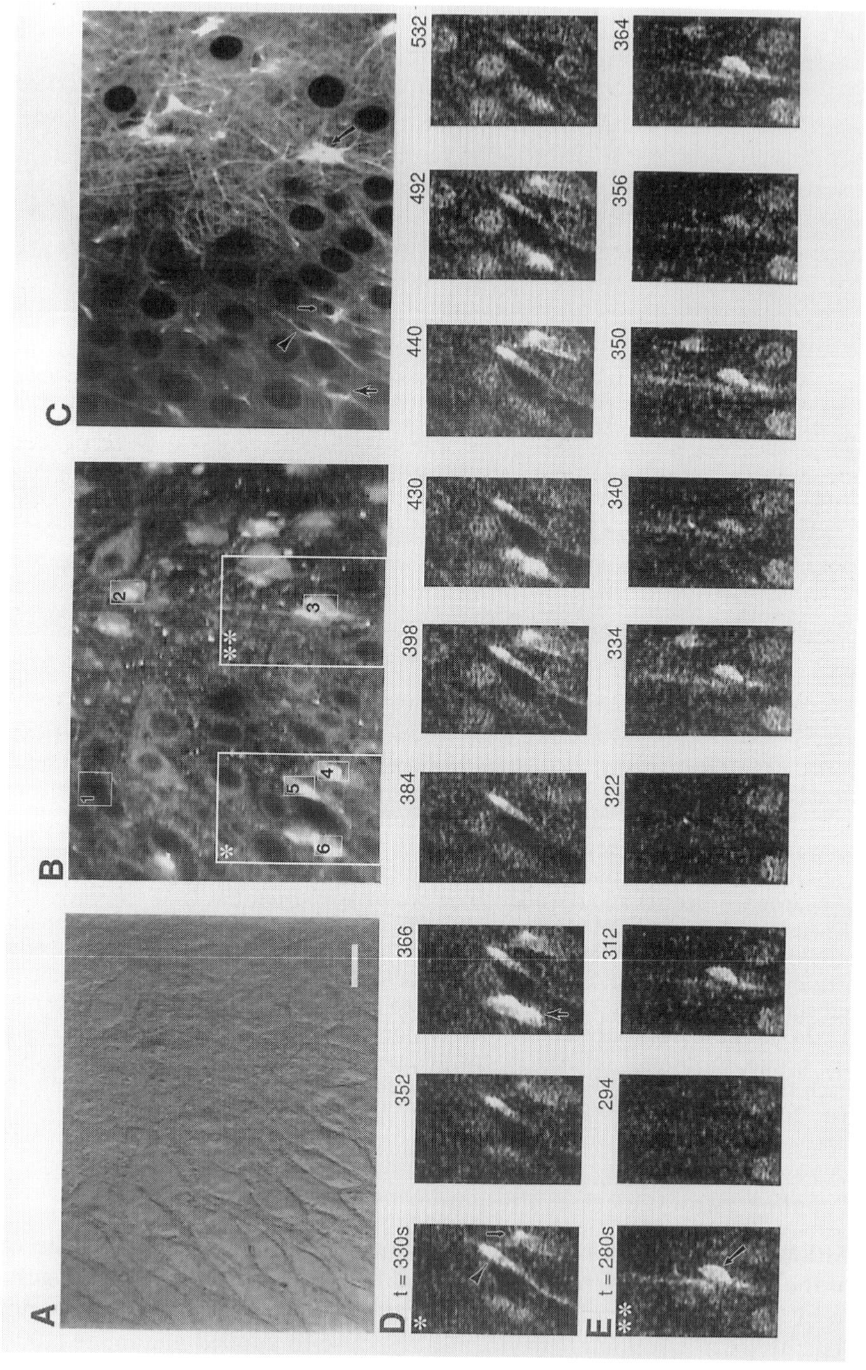
A
B
C
D
E
t = 330s
352
366
384
398
430
440
492
532
t = 280s
294
312
322
334
340
350
356
364

(Cornell-Bell & Finkbeiner 1991). The quality of astrocytic responses seen in organotypic slices differs slightly from those of cultured monolayers in that the former are less regular in frequency. This perhaps reflects the greater complexity of interactions to which the astrocyte is exposed in the tissue environment of the slice. Second, of the neurotransmitters/neuromodulators released from intrinsic connections within the hippocampus, only glutamate is able to elicit astrocytic $Ca^{2+}$ responses in culture (Straughan 1975). The only possible exception to this statement is ATP, although it has not been conclusively shown to be released within the hippocampus (Bevan 1990). Third, $K^+$ depolarization does not cause $Ca^{2+}$ oscillations in cultured astrocytes, nor does its administration (50 mM) to organotypic slices produce responses mimicking those following stimulation with NMDA. Last, we wish to emphasize again that astrocytes do not express NMDA receptors under any circumstances, as has been reported by numerous groups (Murphy & Pearce 1987, Bevan 1990, Barres 1991). For definitive proof that the agent initiating $Ca^{2+}$ responses in astrocytes is indeed glutamate we shall, unfortunately, have to wait until an appropriate glutamate receptor antagonist, one for the metabotropic subtype, is found.

## Possible biological significance

Given the vast array of neurophysiological functions attributed to astrocytes, one may begin to speculate on the purpose of neuron-to-astrocyte signalling and astrocytic $Ca^{2+}$ oscillations within this established framework. Astrocytes

---

FIG. 2. Stimulation of organotypic hippocampal slices with *N*-methyl-D-aspartate (NMDA) elicits a sustained [$Ca^{2+}$] rise in the pyramidal neurons followed by intracellular $Ca^{2+}$ oscillations in the astrocytic population. (A) A Nomarski differential interference contrast image taken from the same field as the following fluorescence images. The distinctive layer of CA3 pyramidal neurons lies to the left side of the field with the stratum lucidum and radiatum on the right side. (B) Baseline fluorescence of the slice shown in (A) loaded with fluo-3 averaged over 32 frames. The large rectangular boxes labelled with asterisks correspond to the time series shown in (D) and (E). The smaller, numbered boxes indicate areas over which fluorescence intensities shown in Fig. 3 were averaged. (C) Correlative glial fibrillary acidic protein (GFAP) immunohistochemical stain of the same field as (B). The arrows and arrowhead point to GFAP-positive cells, which showed $Ca^{2+}$ oscillations in (D) and (E); GFAP staining confirms them to be astrocytes. Note the numerous processes staining for GFAP (right side), which are confined mainly to the non-pyramidal layers. (D) Time sequence showing $Ca^{2+}$ oscillations within several cells (arrows and arrowhead), proved to be astrocytes by staining for GFAP (C). Each time point is numbered according to time after NMDA application (at $t = 0$ s) and shows the change in fluorescence over baseline measurements. (E) Time sequence, as in (D), showing $Ca^{2+}$ oscillations in a cell (arrow) later identified as an astrocyte. Scale bar, 20 $\mu$m.

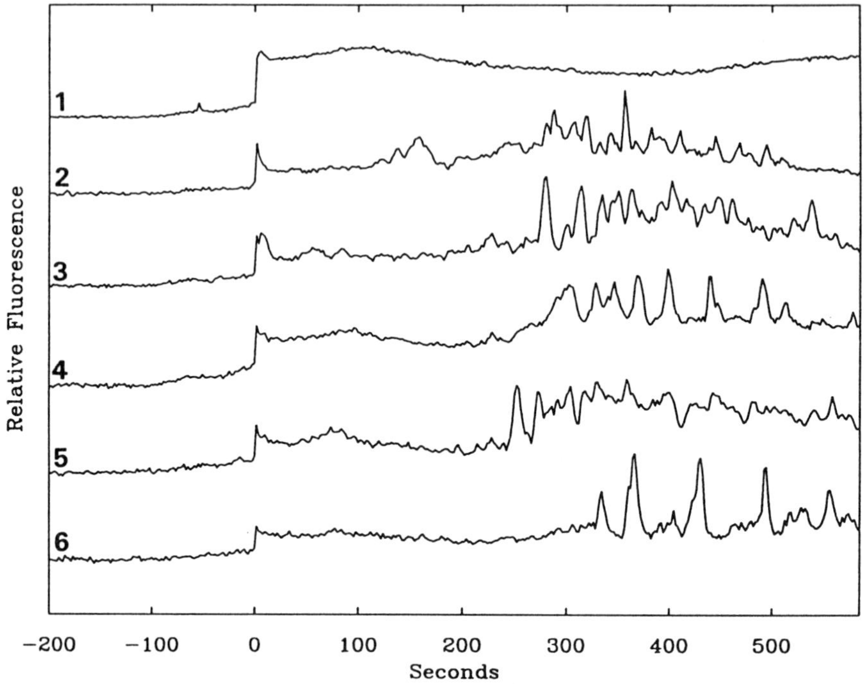


FIG. 3. Fluctuations in [$Ca^{2+}$] observed in identified cells in hippocampal area CA3 after stimulation with *N*-methyl-D-aspartate (NMDA). The changes in fluorescence signal, which were measured over image areas indicated by boxes in Fig. 2B, before and during NMDA application are depicted. For clarity, these curves are shifted arbitrarily along the ordinate axis. Curve 1 represents the typical sustained rise in intracellular [$Ca^{2+}$] experienced by neuronal populations following continuous perfusion with NMDA (beginning at t = 0 s). The other five curves (2–6) show the initial astrocytic response to neuronal activation followed by the characteristic pattern of intracellular $Ca^{2+}$ oscillations. Note the irregularity in both frequency and amplitude.

are responsible for glutamate uptake and conversion to glutamine, which when released is available for re-synthesis into glutamate by neurons (Nicklas et al 1987, Schousboe 1987, Flott & Seifert 1991). One seemingly appropriate function of a neuron-to-astrocyte signalling network would involve the regulation of this cycle. Both astrocytic glutamate uptake and glutamine release may in fact be triggered by the neuron. In addition, astrocytic $Ca^{2+}$ oscillations might serve to regulate certain metabolic steps in the conversion of glutamate to glutamine, or possibly to alter the transcription of several key genes necessary for this process. Similarly, astrocytes are thought to regulate extracellular pH and to maintain the proper extracellular ionic balances (Walz & Mukerji 1988a,b). During heavy neuronal activity, extracellular pH falls owing to build up of lactic acid and $CO_2$, whereas $[K^+]$ rises as a result of neuronal efflux (Chesler & Kraig 1987, 1989). These fluctuations within the extracellular milieu may be corrected by an astrocyte responding directly to neuronal activity, specifically synaptic transmitter release. Finally, astrocytes are the principal site of glycogen storage in the brain and are believed to provide metabolic substrates to active neurons (Sagar et al 1987, Sonnewald et al 1991). Perhaps neurotransmitter binding is the signal which initiates glycogen breakdown in the astrocyte and leads ultimately to the release of certain metabolites which then become available to the neuron.

## *Acknowledgements*

We should like to thank Mark Cooper and Alex Chernjavsky for helpful comments throughout this project. We also thank Julie Kauer for introducing us to organotypic slice culture methods. This research was supported by a National Institute of Mental Health Silvio Conte Center for Neuroscience Research Grant (MH48108), by a grant from the National Institute of Neurological Diseases and Stroke (NS28587), and by a gift from the G. Harlod and Leila Y. Mathers Charitable Foundation.

## References

Andersen P 1975 Organization of hippocampal neurons and their interconnections. In: The hippocampus, structure and development. Plenum Press, New York, vol 1:155–175

Barres BA 1991 New roles for glia. J Neurosci 11:3685–3694

Barres BA, Koroshetz WJ, Chun LLY, Corey DP 1990 Ion channel expression by white matter glia: the type-1 astrocyte. Neuron 5:527–544

Berridge MJ 1990 Calcium oscillations. J Biol Chem 265:9583–9586

Bevan S 1990 Ion channels and neurotransmitter receptors in glia. Semin Neurosci 2:467–481

Cambray-Deakin M, Pearce B, Morrow C, Murphy S 1988 Effects of extracellular potassium on glycogen stores of astrocytes *in vitro*. J Neurochem 51: 1846–1851

Chesler M, Kraig RP 1987 Intracellular pH of astrocytes increases rapidly with cortical stimulation. Am J Physiol 253:R666–R670

Chesler M, Kraig RP 1989 Intracellular pH transients of mammalian astrocytes. J Neurosci 9:2011–2019

Chow SY, Yen-Chow YC, White HS, Woodbury DM 1991 pH regulation after acid load in primary cultures of mouse astrocytes. Dev Brain Res 60:69–78

Cornell-Bell AH, Finkbeiner SM 1991 $Ca^{2+}$ waves in astrocytes. Cell Calcium 12:185–204

Cornell-Bell AH, Finkbeiner SM, Cooper MS, Smith SJ 1990 Glutamate induces calcium waves in cultured astrocytes: long-range glial signaling. Science 247: 470–473

Dani JW, Chernjavsky A, Smith SJ 1992 Neuronal activity triggers $Ca^{2+}$ waves in hippocampal astrocyte networks. Neuron 8:429–440
Deitmer JW 1992 Evidence for glial control of extracellular pH in the leech central nervous system. Glia 5:43–47
Del Rio JA, Heimrich B, Soriano E, Schwegler H, Frotscher M 1991 Proliferation and differentiation of glial fibrillary acidic protein immunoreactive glial cells in organotypic slice cultures of rat hippocampus. Neuroscience 43:335–347
Fine A, Amos WB, Durbin RM, McNaughton PA 1988 Confocal microscopy: applications in neurobiology. Trends Neurosci 11:346–351
Flott B, Seifert W 1991 Characterization of glutamate uptake systems in astrocyte primary cultures from rat brain. Glia 4:293–304
Gahwiler BH 1981 Organotypic monolayer cultures of nervous tissue. J Neurosci Methods 4:329–342
Gahwiler BH 1984 Slice cultures of cerebellar, hippocampal and hypothalamic tissue. Experientia 40:235–243
Gahwiler BH 1988 Organotypic cultures of neural tissue. Trends Neurosci 11:484–489
Jacob R 1990 Calcium oscillations in electrically nonexcitable cells. Biochim Biophys Acta 1052:427–438
Jensen AM, Chiu SY 1990 Fluorescence measurement of changes in intracellular calcium induced by excitatory amino acids in cultured cortical astrocytes. J Neurosci 10:1165–1175
Kao JPY, Harootunian AT, Tsien RY 1989 Photochemically generated cytosolic calcium pulses and their detection by fluo-3. J Biol Chem 264:8179–8184
Meyer T 1991 Cell signaling by second messenger waves. Cell 64:675–678
Minta A, Koa JPY, Tsien RY 1989 Fluorescent indicators for cytosolic calcium based on rhodamine and fluorescein chromophores. J Biol Chem 264:8171–8175
Murphy S, Pearce B 1987 Functional receptors for neurotransmitters on astroglial cells. Neuroscience 22:381–394
Nicklas WJ, Zeevalk G, Hyndman A 1987 Interactions between neurons and glia in glutamate/glutamine compartmentation. Biochem Soc Trans 15:208–210
Orkland RK, Nicholls JG, Kuffler SW 1966 Effect of nerve impulses on the membrane potential of glial cells in the central nervous system of amphibia. J Neurophysiol 29:788–806
Pearce B, Albrecht J, Morrow C, Murphy S 1986 Astrocyte glutamate receptor activation promotes inositol phospholipid turnover and calcium flux. Neurosci Lett 72: 335–340
Peters A, Palay SL, Webster H 1991 The fine structure of the nervous system. Oxford University Press, New York, p 276–295
Sagar SM, Sharp FR, Swanson RA 1987 The regional distribution of glycogen in rat brain fixed by microwave irradiation. Brain Res 417:172–174
Schousboe A 1987 Glutamine, glutamate and f-aminobutyrate in the central nervous system. Biochem Soc Trans 15:205–207
Shepherd G 1979 The synaptic organization of the brain, 2nd edn ch. 14: Hippocampus. Oxford University Press, New York, p 308–337
Sonnewald U, Westergaard N, Krane J, Unsgard G, Petersen SB, Schousboe A 1991 First direct demonstration of preferential release of citrate from astrocytes using [$^{13}$C]NMR spectroscopy of cultured neurons and astrocytes. Neurosci Lett 128: 235–239
Straughan DW 1975 Neurotransmitters and the hippocampus. In: Isaacson RL, Pribram CH (eds) The hippocampus, structure and development. Plenum Press, New York, vol 1:239–268

Trachtenberg MC, Pollen DA 1970 Neuroglia: biophysical properties and physiologic function. Science 167:1248–1251
Tsien RY 1988 Fluorescence measurement and photochemical manipulation of cytosolic free calcium. Trends Neurosci 11:419–424
Usowicz MM, Gallo V, Cull-Candy SG 1989 Multiple conductance channels in type-2 cerebellar astrocytes activated by excitatory amino acids. Nature 339: 380–383
Vernadakis A 1988 Neuron–glial interrelations. Int Rev Neurobiol 30:149–224
Walz W, Mukerji S 1988a Lactate production and release in cultured astrocytes. Neurosci Lett 86:296–300
Walz W, Mukerji S 1988b Lactate release from cultured astrocytes and neurons: a comparison. Glia 1:366–370

## DISCUSSION

*Jaffe:* The evidence for the hypothesis that these waves underly migraine is really very strong. The basic reason for thinking that is that the injury wave which underlies migraine crosses the brain at exactly the same speed as these astrocytic $Ca^{2+}$ waves (Jaffe 1991, Leibowitz 1992).

*Sanderson:* I gather that the $Ca^{2+}$ waves you see in organotypic slice preparations pass along the processes of individual cells but there do not appear to be many occasions when the $Ca^{2+}$ wave jumps or passes into adjacent cells, as was observed in your original cell cultures. Are we really dealing with intercellular waves or just intracellular waves along processes?

*Smith:* There certainly are instances of intercellular waves, but you have to look for them really hard because they don't jump out and hit you.

*Sanderson:* Waves in cultures of purified astrocytes can propagate through hundreds of cells.

*Smith:* That's correct. I've never seen that happen in the slice.

*Galione:* Could you comment on the recent report in *Science* (Nedergaard 1994) showing that $Ca^{2+}$ waves can pass into neurons from astrocytes via gap junctions.

*Smith:* I was talking about signalling from neurons to astrocytes. That's interesting, but it would be really interesting if the signal, once it has been processed through an astrocyte network and communicated over a long or short distance, could feed back onto neurons and do something. The experiments you are referring to were done by Nedergaard (1994) and by Parpura et al (1994), although I understand that at least two other groups have got the same result independently. Andy Charles was actually the first to report this in an abstract for the 1993 Society for Neuroscience meeting. Nedergaard's experiment reported in *Science* was done on a typical astrocyte–neuron co-culture, with neurons growing on a confluent sheet of astrocytes. A local, confined stimulus was given to the astrocyte to induce a $Ca^{2+}$ wave. You can do this by touching or by giving it a little electric shock. The wave then spreads through the astrocyte at 20 $\mu m/s$. When the wave reaches the position of a neuron, the neuron lights up. A large $Ca^{2+}$ transient, with a fourfold elevation in $[Ca^{2+}]$, is induced in the neuron at the precise moment that the astrocytic $Ca^{2+}$ wave reaches the site of that neuron. This kind of observation potentially closes the loop. The mechanism is not yet known. Gap junctions may be involved but Nedergaard could not demonstrate gap junctions by dye passage. She ruled out some other things, such as glutamate, but Parpura et al are saying the opposite, that glutamate and not gap junctions are involved.

*Galione:* She couldn't see transmission the other way around, from neuron to astrocyte.

*Smith:* She did not say, although other groups, including ourselves, have shown that in similar cultures (e.g., Dani & Smith 1995, this volume).

*Galione:* Is there any evidence that glial cells store and release neurotransmitters?

*Smith:* Yes. They certainly take up neurotransmitters and metabolize them. They turn $\gamma$-aminobutyrate and glutamate into glutamine and pass it back to neurons. They contain all kinds of secretory vesicles. Calcium-dependent release of both $\gamma$-aminobutyrate and glutamate from cultures of astrocytes has been demonstrated.

*Sanderson:* Andrew Charles' (personal communication) interpretation of his experiments is that rather than there being a direct coupling between the neuron and the glial cell, the $[Ca^{2+}]$ increase releases something from the glial cell which modulates the neuron.

If you culture glial cells on their own, you don't get much activity, but in co-cultures of neurons and astrocytes the spontaneous oscillations of the glial cells are increased in frequency. Something about the addition of neurons to glial cells seems to bring glial cells into a much more excitable state.

*Putney:* Dr Smith, the difference between your results and Michael Sanderson's is that you often get waves that propagate over long distances, sometimes explosively. In your system, there must be a regenerative process. This is truly a $Ca^{2+}$ wave being propagated from cell to cell, not a chemical mediator diffusing.

*Smith:* Perhaps I'm being complacent, but the difference makes perfect sense to me. We're doing something that will make $[InsP_3]$ go up everywhere.

We've dried out the tinder over the whole tundra so that one little spark can make it go . . . glutamate everywhere.

*Putney:* What about $InsP_3$? Do you think $Ca^{2+}$-activated phospholipase C is involved? You are raising [$InsP_3$] uniformly in the field, producing waves that move very quickly.

*Smith:* Glutamate definitely increases [$InsP_3$] in these cells. If [$InsP_3$] is increased everywhere, it's no longer a good messenger. Calcium is a good messenger. If I had to come down on one side or the other, I would say, why fight the idea of $Ca^{2+}$ going through? Some experiments bringing my and Michael Sanderson's two approaches together would be in order at this point. You could, for example, see whether or not a low dose of glutamate makes a mechanically induced wave go further.

*Meyer:* How do your results exclude the possibility that extracellular [$K^+$] is changed by increases in [$Ca^{2+}$] in glial cells, which, in turn, would control neurons?

*Smith:* That hypothesis still holds good. There are $Ca^{2+}$-activated $K^+$ channels in glial cells and $Ca^{2+}$-activated $Cl^-$ channels . . . .

*Smith:* There has been a hypothesis for a long time that the glial cell controls the environment of neurons by changing [$K^+$] around the neurons. That's good, but, in the context of this meeting, I say why mess around with $K^+$ when you can have $Ca^{2+}$? It's well established that the probability of releasing a neurotransmitter vesicle increases as the fourth power of the extracellular [$Ca^{2+}$]. The astrocyte has prodigious $Ca^{2+}$ transport machinery so does not have to change [$Ca^{2+}$] very much to turn transmitter release on or off. The architecture is such that it's easy to imagine the astrocyte doing this in a synapse-specific way.

*Berridge:* I think Tobias Meyer is referring to a separate phenomenon. He is asking how the astrocyte communicates with the neuron.

*Smith:* Some kind of ionic communication is a real possibility.

*Berridge:* One interesting possibility is that the intracellular $Ca^{2+}$ wave is transformed into a wave of extracellular $K^+$. That's one way of communicating information from the glia to the neurons and this might play a role in spreading depression.

*Smith:* Yes. That hasn't been ruled out.

*Fay:* Could you tell us a little more about the transition from the oscillatory behaviour to wave behaviour that you mentioned in the early part of your talk?

*Smith:* Steve Finkbeiner has some unpublished data suggesting that there is a non-$InsP_3$ second messenger that is triggered somehow by the kainate receptor to release $Ca^{2+}$ from intracellular stores. This is preliminary, but I have thought cyclic ADP-ribose might be the messenger.

*Jaffe:* On that same point, it occurred to me that during the three-minute intervals before you got waves you might be getting a synchronization of cells.

It's well known that all sorts of chaotic oscillations tend to synchronize after a while.

*Smith:* Especially because we know that even the most robust oscillators run down a little bit, getting weaker and weaker until at some point they synchronize. It's harder to synchronize really strong oscillators.

*Thomas:* I know it's early days, but do you have any evidence for what Mike Sanderson saw in terms of a delay between cells for the transit of $Ca^{2+}$ from one cell to the next?

*Smith:* That is most striking in Mike Sanderson's results in the epithelium. It is less obvious in his astrocytes, and I have never noticed it in ours. Steve Finkbeiner actually made some measurements and couldn't really see anything at the cell boundary. The geometries of these two cell types are quite a bit different. The epithelium is made up of big cuboidal cells.

*Thomas:* In our hands, there seems to be little delay in liver cells.

*Putney:* But in the liver it is a $Ca^{2+}$ wave, not an $InsP_3$-mediated wave. In the gap junctions, the $Ca^{2+}$ wave will go right through.

*Smith:* I can see that difference as being possibly accidental. How many gap junctions do these cells have? Where is the gunpowder located?

*Berridge:* You probably need a lot of gap junctions in order for a $Ca^{2+}$-induced $Ca^{2+}$ release process to spread from one cell to the next.

*Petersen:* In the liver, of course, there is a fantastic density of gap junctions.

*Sanderson:* It was originally reported that the long-range $Ca^{2+}$ waves move smoothly through cultured astrocytes with no delay at the bounday (Cornell-Bell et al 1990). Glial cells actually have quite a small number of gap junctions because only small processes go from cell to cell, whereas the liver has extensive gap junctions.

The other point about those long-range waves is that they're dependent on extracellular $Ca^{2+}$ and extracellular $Na^{+}$. Cornell-Bell et al (1990) state that there is a difference between our mechanically induced waves and glutamate-induced waves. It's interesting that we have never seen glutamate induce long-distance waves in mixed glial cultures. All this suggests that there is a significant difference between local, mechanically induced waves and long, propagating waves.

## References

Cornell-Bell AH, Finkbeiner SM, Cooper MS, Smith SJ 1990 Glutamate induces calcium waves in cultured astrocytes: long-range glial signaling. Science 247:470–473

Dani JW, Smith SJ 1995 The triggering of astrocytic calcium waves by NMDA-induced neuronal activation. In: Calcium waves, gradients and oscillations. Wiley, Chichester (Ciba Found Symp 188) p 195–209

Jaffe LF 1991 The path of calcium in cytosolic calcium oscillations: a unifying hypothesis. Proc Natl Acad Sci USA 88:9883–9887

Leibowitz D 1992 On an active role of neuroglia in spreading depression and migraine. Proc R Soc Lond B Biol Sci 250:287–295
Nedergaard M 1994 Direct signaling from astrocytes to neurons in cultures of mammalian brain cells. Science 263:1768–1771
Parpura V, Basarsky TA, Liu F, Jeftinija K, Jeftinija S, Haydon PG 1994 Glutamate-mediated astrocyte–neuron signalling. Nature 369:744–747

# Calcium oscillations in neurons

David D. Friel

*Department of Neurosciences, Case Western Reserve University, 10900 Euclid Avenue, Cleveland, OH 44106-4975 and Department of Molecular and Cellular Physiology, Stanford University School of Medicine, Stanford, CA 94305-5426, USA*

*Abstract.* Oscillations in the cytosolic free $Ca^{2+}$ concentration ($[Ca^{2+}]_i$) have been described in a variety of cells. In some cases, $[Ca^{2+}]_i$ oscillations reflect cycles of membrane depolarization and voltage-dependent $Ca^{2+}$ entry. In others, they are caused by periodic $Ca^{2+}$ uptake and release by internal stores, with little immediate requirement for external $Ca^{2+}$. A third type of $[Ca^{2+}]_i$ oscillation is typified by caffeine-induced oscillations in sympathetic neurons. Here, the oscillations depend on the interplay between $Ca^{2+}$ transport across the plasma membrane and transport by a caffeine-sensitive store. These oscillations can occur at a steady membrane potential and are blocked by ryanodine (1 $\mu$M), indicating that they do not result from voltage-dependent changes in $Ca^{2+}$ entry but do require $Ca^{2+}$-induced $Ca^{2+}$ release. Entry of $Ca^{2+}$ from the external medium is important during all phases of the oscillatory cycle except the rapid upstroke, which is dominated by $Ca^{2+}$ release from an internal store. It is proposed that caffeine-induced $[Ca^{2+}]_i$ oscillations are cyclic perturbations of $[Ca^{2+}]_i$ caused by exchange of $Ca^{2+}$ between the cytosol and the caffeine-sensitive store: net $Ca^{2+}$ loss from the store increases $[Ca^{2+}]_i$ transiently above its steady-state value ($[Ca^{2+}]_{ss}$), whereas net accumulation of $Ca^{2+}$ by the store transiently depresses $[Ca^{2+}]_i$ below $[Ca^{2+}]_{ss}$. The effects of rapid removal of $Ca^{2+}$ and caffeine on the rate of change of $[Ca^{2+}]_i$ ($d[Ca^{2+}]_i/dt$) provide estimates of the rates of net $Ca^{2+}$ entry and (caffeine-sensitive) $Ca^{2+}$ release and information on the way these rates vary during the oscillatory cycle.

*1995 Calcium waves, gradients and oscillations. Wiley, Chichester (Ciba Foundation Symposium 188) p 210–234*

Calcium is an important signalling ion, serving as a critical link between a variety of physiological stimuli and their intracellular effectors (Berridge 1993, Tsien & Tsien 1990). In neurons, stimulus-evoked changes in the cytosolic free $Ca^{2+}$ concentration ($[Ca^{2+}]_i$) regulate processes as diverse as membrane excitation, exocytosis and gene expression. Calcium transport systems that are targets for stimulation can be found either in the plasma membrane or in the membrane of internal organelles. However, it is becoming increasingly clear that the effects on $[Ca^{2+}]_i$ of stimulation depend on interactions between multiple $Ca^{2+}$ transport systems. For example, although neuronal depolarization promotes

$Ca^{2+}$ entry through voltage-sensitive $Ca^{2+}$ channels in the plasma membrane, the changes in $[Ca^{2+}]_i$ that result also reflect $Ca^{2+}$ extrusion as well as uptake and release by internal stores and buffering systems. Therefore, to understand the effects of stimulation on $[Ca^{2+}]_i$, and on the intracellular processes that are sensitive to $Ca^{2+}$, it is necessary to consider the concerted actions of multiple $Ca^{2+}$ transport systems.

Caffeine-induced $[Ca^{2+}]_i$ oscillations in sympathetic neurons provide a good example of the interplay between $Ca^{2+}$ transport across the plasma membrane and $Ca^{2+}$ uptake and release by internal stores. These oscillations were first described by Kuba & Nishi (1976), who observed periodic membrane hyperpolarizations during steady exposure to caffeine. These authors provided compelling evidence that the rhythmic hyperpolarizations result from periodic increases in $[Ca^{2+}]_i$ generated by $Ca^{2+}$-induced $Ca^{2+}$ release and the subsequent activation of a $Ca^{2+}$-sensitive $K^+$ channel in the plasma membrane. Direct measurements of periodic changes in $[Ca^{2+}]_i$ were later made by Lipscombe et al (1988).

This review describes some of the basic properties of caffeine-induced $[Ca^{2+}]_i$ oscillations in bullfrog sympathetic neurons. These cells possess a $Ca^{2+}$ store that is sensitive to both caffeine and ryanodine (Friel & Tsien 1992a). Discharging the store by exposing the neurons to caffeine transiently raises $[Ca^{2+}]_i$ above the resting level, whereas replenishment of the store, for example following caffeine removal, transiently lowers $[Ca^{2+}]_i$ below the resting concentration. It is argued that the ability of the caffeine-sensitive store to raise and lower $[Ca^{2+}]_i$ transiently through net $Ca^{2+}$ release and uptake provides the key to understanding caffeine-induced $Ca^{2+}$ oscillations. The results form the basis for a model of caffeine-induced $[Ca^{2+}]_i$ oscillations that leads to testable predictions concerning the relative rates of $Ca^{2+}$ movements between the cytosol and internal and external $Ca^{2+}$ pools during each phase of the oscillatory cycle.

## Conditions which lead to oscillations in cytosolic calcium concentration

Figure 1 illustrates $[Ca^{2+}]_i$ recordings from single fura-2-loaded bullfrog sympathetic neurons under conditions of stimulation that lead to $[Ca^{2+}]_i$ oscillations (see Friel & Tsien 1992b for methods). Exposure to a depolarizing solution of high $[K^+]$ (30 mM) produces a maintained increase in $[Ca^{2+}]_i$ (Fig. 1a); exposure to caffeine (5 mM) in the continued presence of high $[K^+]$ elicits a further transient $[Ca^{2+}]_i$ rise. However, after declining from this transient peak, $[Ca^{2+}]_i$ becomes unstable and eventually oscillates (with a cycle period of about one minute). The oscillations are quite robust, lasting in some cells for several hours. Presenting these stimuli in the reverse order produces different transient responses, but the resulting oscillations are essentially the same (Fig. 1b). The oscillations are not induced by depolarization alone and are

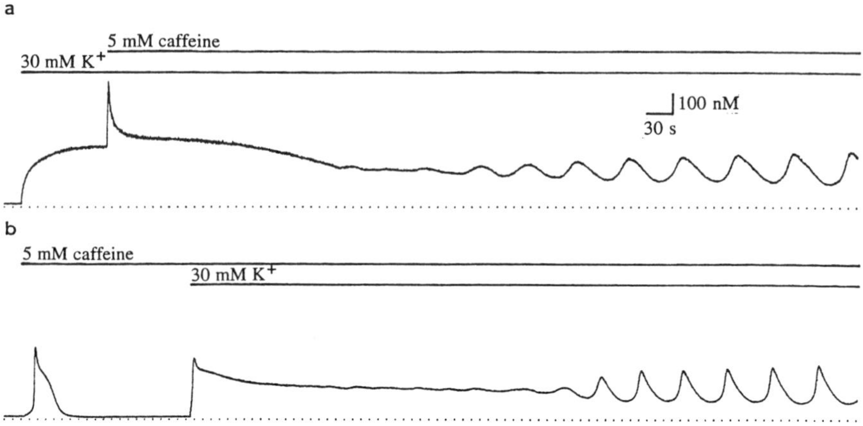


FIG. 1. Conditions which elicit $[Ca^{2+}]_i$ oscillations in single fura-2 loaded bullfrog sympathetic neurons. (a) Changes in $[Ca^{2+}]_i$ induced by high $[K^+]$ (30 mM), followed by caffeine (5 mM) (in the presence of high $[K^+]$), leading to $[Ca^{2+}]_i$ oscillations. (b) Changes in $[Ca^{2+}]_i$ in another cell stimulated in the reverse order. Dotted lines indicate zero $[Ca^{2+}]_i$. Reproduced from Friel & Tsien 1992b.

seen in only a small fraction of cells (about 5%) exposed to caffeine alone, but are almost invariably observed when caffeine treatment and depolarization are combined. This suggests that a caffeine-sensitive mechanism is critical for the oscillations and that depolarization in some way increases the reliability with which they can be elicited by caffeine.

## The role of calcium entry from the external medium

In a study of the role of $Ca^{2+}$ entry from the extracellular medium, cells in which $[Ca^{2+}]_i$ was oscillating were exposed rapidly to a $Ca^{2+}$-free external solution (nominally $Ca^{2+}$-free supplemented with 0.2 mM EGTA) (Friel & Tsien 1992b). After removal of external $Ca^{2+}$, the oscillations stop (within 0–1 cycles, depending on when $Ca^{2+}$ is removed; see below). This indicates that $Ca^{2+}$ entry from the external medium is required for continued oscillations. Entry of $Ca^{2+}$ is evidently voltage dependent, because lowering the extracellular $[K^+]$ from 30 mM to 2 mM has essentially the same effect as removal of external $Ca^{2+}$. Voltage-dependent $Ca^{2+}$ entry presumably occurs through the N- and L-type voltage-sensitive $Ca^{2+}$ channels that are present in these cells (Tsien et al 1988, Jones & Marks 1989, Boland & Bean 1993).

To determine whether the $[Ca^{2+}]_i$ oscillations reflect periodic membrane depolarization and voltage-dependent $Ca^{2+}$ entry (Friel & Tsien 1992b), $[Ca^{2+}]_i$ and membrane potential ($V_m$) were measured simultaneously using high resistance ($\approx 120\ M\Omega$) microelectrodes while $[Ca^{2+}]_i$ was oscillating (Fig. 2a).

This method of electrical recording was chosen rather than the whole-cell voltage clamp technique to minimize disruption of intracellular $Ca^{2+}$ regulation and the washing-out of voltage-dependent $Ca^{2+}$ channel activity that occurs during whole-cell dialysis (unpublished observations). Exposure to high $[K^+]$ elicits a rapid depolarization to a level between $-30$ and $-35$ mV; restoration of the control extracellular solution (2 mM $K^+$) leads to a brisk repolarization (Fig. 2a, *left*). Changes in membrane potential were complete within 100–200 ms, which is the time required to complete the solution change (Tsien & Friel 1992a). Importantly, during exposure to caffeine and high $[K^+]$, which elicited steady $[Ca^{2+}]_i$ oscillations, there was no detectable change in $V_m$ (Fig. 2a). For example, in the case illustrated in Fig. 2, $V_m$ remained at $-33 \pm 1$ mV while $[Ca^{2+}]_i$ oscillated between about 30 and 300 nM. Therefore, although membrane depolarization enhances the effectiveness with which caffeine induces $[Ca^{2+}]_i$ oscillations, changes in membrane potential *per se* are not required. This indicates that caffeine-induced $[Ca^{2+}]_i$ oscillations do not reflect voltage-dependent changes in the activity of plasma membrane ion channels.

The absence of oscillations in $V_m$ may seem puzzling, given the initial observation that caffeine induces rhythmic membrane hyperpolarizations in these cells (Kuba & Nishi 1976). One possible explanation is that with 30 mM $K^+$, in contrast to 2 mM $K^+$, $V_m$ is so close to the $K^+$ equilibrium potential ($-35$ mV at 20 °C, assuming $[K^+]_i = 120$ mM) that increases in a plasma membrane $K^+$ conductance caused by increases in $[Ca^{2+}]_i$ have little additional effect on $V_m$ (see Hodgkin & Horowicz 1959).

## The role of calcium release from internal stores

Caffeine has been used extensively in the study of $Ca^{2+}$-induced $Ca^{2+}$ release in muscle cells, where it increases the sensitivity of ryanodine receptor $Ca^{2+}$ release channels to $Ca^{2+}$ (Rousseau & Meissner 1989). These channels, which are located in the sarcoplasmic reticulum, open in response to increases in $[Ca^{2+}]_i$ and participate in $Ca^{2+}$-induced $Ca^{2+}$ release (Endo 1985). Ryanodine has been used as a specific ligand in the isolation of $Ca^{2+}$ release channels from both muscle and nerve cells (Imagawa et al 1987, Lai et al 1988, McPherson et al 1991). Ryanodine also modifies channel gating (Rousseau et al 1987), stabilizing an open subconductance state that is thought to render the sarcoplasmic reticulum leaky to $Ca^{2+}$, thereby compromising its ability to accumulate $Ca^{2+}$ and participate in $Ca^{2+}$-induced $Ca^{2+}$ release. In sympathetic neurons, caffeine releases $Ca^{2+}$ from an intracellular compartment, and this release is inhibited in a use-dependent manner by ryanodine (Thayer et al 1988, Friel & Tsien 1992a), suggesting that $Ca^{2+}$ release channels are indeed present in these cells.

To test for the involvement of $Ca^{2+}$-induced $Ca^{2+}$ release in caffeine-induced $[Ca^{2+}]_i$ oscillations, Friel and Tsien (1992b) exposed sympathetic

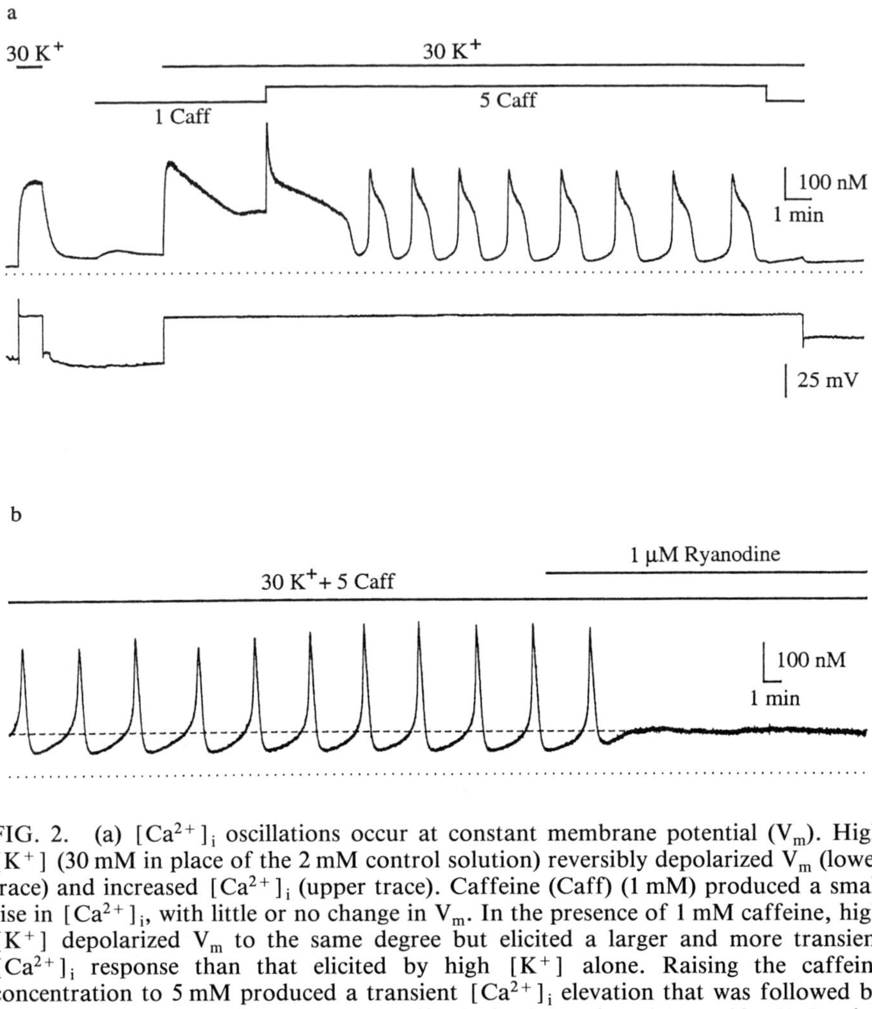


FIG. 2. (a) $[Ca^{2+}]_i$ oscillations occur at constant membrane potential ($V_m$). High $[K^+]$ (30 mM in place of the 2 mM control solution) reversibly depolarized $V_m$ (lower trace) and increased $[Ca^{2+}]_i$ (upper trace). Caffeine (Caff) (1 mM) produced a small rise in $[Ca^{2+}]_i$, with little or no change in $V_m$. In the presence of 1 mM caffeine, high $[K^+]$ depolarized $V_m$ to the same degree but elicited a larger and more transient $[Ca^{2+}]_i$ response than that elicited by high $[K^+]$ alone. Raising the caffeine concentration to 5 mM produced a transient $[Ca^{2+}]_i$ elevation that was followed by oscillations, during which $V_m$ remained effectively clamped at about −33 mV. Resting potential in this cell (B13L) was −66 mV. (b) Involvement of $Ca^{2+}$-induced $Ca^{2+}$ release in caffeine-induced $[Ca^{2+}]_i$ oscillations. Oscillations in $[Ca^{2+}]_i$ are blocked by 1 μM ryanodine; in the presence of ryanodine $[Ca^{2+}]_i$ stabilizes at a steady-state level ($[Ca^{2+}]_{ss}$, dashed line) that falls between the minimum and maximum $[Ca^{2+}]_i$ values. Dotted lines indicate zero $[Ca^{2+}]_i$. Reproduced from Friel & Tsien (1992b).

neurons to ryanodine while $[Ca^{2+}]_i$ was oscillating. Ryanodine terminated the oscillations (Fig. 2b), suggesting that the oscillations require the activity of $Ca^{2+}$ release channels. Note that during exposure to ryanodine, $[Ca^{2+}]_i$ stabilizes at a level that is between the extremes reached during the oscillatory cycle. The stability of $[Ca^{2+}]_i$ indicates that the rate at which $Ca^{2+}$ is entering the cytosol is zero. To the extent that it also indicates that the rate at which $Ca^{2+}$ is crossing the plasma membrane is zero, as would be expected for a true steady state (ss), $[Ca^{2+}]_{ss}$ provides an estimate of where within the oscillatory cycle $Ca^{2+}$ entry and extrusion across the plasma membrane balance one another. This point will be taken up below.

## Coordination of calcium entry and release during the oscillatory cycle

To examine the moment-to-moment involvement of extracellular $Ca^{2+}$, and, by inference, $Ca^{2+}$ entry across the plasma membrane, Friel & Tsien (1992b) removed external $Ca^{2+}$ rapidly at different points within the oscillatory cycle. Although $Ca^{2+}$ was required for continued oscillations, as mentioned above, the *immediate* effect of $Ca^{2+}$ removal depended critically on where in the cycle the perturbation occurred (Fig. 3a). If $Ca^{2+}$ was removed either during the slow interspike rise, or following the peak, $[Ca^{2+}]_i$ declined steadily and the oscillations stopped. However, if $Ca^{2+}$ was removed during the rapid upstroke, $[Ca^{2+}]_i$ continued to rise, although at a slightly slower rate. This demonstrates both that $Ca^{2+}$ entry from the external medium occurs during all phases of the

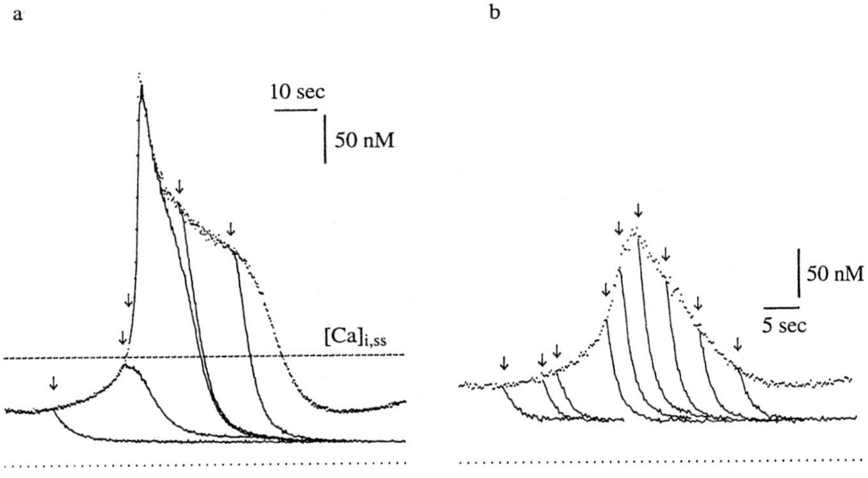


FIG. 3. (a) Effects of removing external $Ca^{2+}$ from fura-2-loaded sympathetic neurons at different points (arrows) within the oscillatory cycle. Successive perturbations were made after restoring external $Ca^{2+}$ and permitting oscillations to resume. The dashed line indicates $[Ca^{2+}]_{i,ss}$, the steady-state concentration of cytosolic free calcium ($[Ca^{2+}]_i$) determined when $[Ca^{2+}]_i$ stabilized after exposure to ryanodine (1 $\mu$M) at the end of the experiment. Oscillations were monitored in the presence of 5 mM caffeine and 30 mM $K^+$. (b) Effects of removing caffeine at different points in the oscillatory cycle (D. D. Friel, unpublished results). (a) Reproduced from Friel & Tsien 1992b.

oscillatory cycle, and that the rapid upstroke occurs in concert with $Ca^{2+}$ release from an internal store. Moreover, during the upstroke, $Ca^{2+}$ release is sufficiently powerful to raise $[Ca^{2+}]_i$ without $Ca^{2+}$ entry from the extracellular medium. Rapid membrane repolarization, such as that induced by rapid reductions in $[K^+]_o$, has essentially the same phase-dependent effects as $Ca^{2+}$ removal, further supporting the idea that $Ca^{2+}$ entry depends on the activity of voltage-dependent $Ca^{2+}$ channels in the plasma membrane.

In contrast to the phase-dependent effects of $Ca^{2+}$ removal or repolarization, sudden removal of caffeine leads to a rapid decline in $[Ca^{2+}]_i$ and abrupt termination of the oscillations no matter at what point during the oscillatory cycle it is removed (Fig. 3b). The decline in $[Ca^{2+}]_i$ appears to reflect the attenuation of an ongoing caffeine-sensitive $Ca^{2+}$ flux from an internal store, which unmasks $Ca^{2+}$ uptake by the store and causes $[Ca^{2+}]_i$ to decline (see below).

## A kinematic model of caffeine-induced calcium oscillations

Drawing on the experimental results described above, it is possible to formulate a hypothesis regarding the temporal relationship between the rates of net $Ca^{2+}$ transport across the plasma membrane, and between the cytosol and the caffeine-sensitive store. It is helpful to begin by making three assumptions. (i) That $[Ca^{2+}]_i$ changes with time under the influence of a total net $Ca^{2+}$ flux ($J_{total}$) that is the sum of the net flux across the plasma membrane ($J_{io}$) and the net flux between the cytosol and a single internal store of finite capacity ($J_{is}$). By convention, positive fluxes are directed outward from the cytosol (see inset, Fig. 4a). (ii) That there is a unique $[Ca^{2+}]_i$ encountered during the oscillatory cycle where $J_{io}$ is zero. Although there is no direct information available concerning the magnitude of this concentration, an estimate is provided by the steady $[Ca^{2+}]_i$ reached during exposure to ryanodine ($[Ca^{2+}]_{ss}$). (iii) That when $[Ca^{2+}]_i$ exceeds $[Ca^{2+}]_{ss}$, $J_{io}$ is outward relative to the cytosol (positive) and increases with $[Ca^{2+}]_i$, whereas when $[Ca^{2+}]_i$ is less than $[Ca^{2+}]_{ss}$, $J_{io}$ is inward relative to the cytosol and increases in magnitude as $[Ca^{2+}]_i$ falls. $[Ca^{2+}]_{ss}$ is analogous to a reversal potential for an ion permeation pathway, but, instead of describing a condition for the balance of net ionic fluxes, it describes a condition for the balance of net material fluxes across the plasma membrane, in particular, the rates of $Ca^{2+}$ entry and extrusion.

If these assumptions are valid, there are six critical points that separate the oscillatory cycle into phases during which the net $Ca^{2+}$ fluxes $J_{io}$ and $J_{is}$ have specific relationships (Fig. 4). At point 1, where $[Ca^{2+}]_i$ assumes its maximal value, $J_{total}$ must be zero because $[Ca^{2+}]_i$ is not changing with time. Because $[Ca^{2+}]_i$ exceeds $[Ca^{2+}]_{ss}$, $J_{io}$ is outward, so there must be an inward flux $J_{is}$ of equal magnitude arising from the store. During this phase the store is

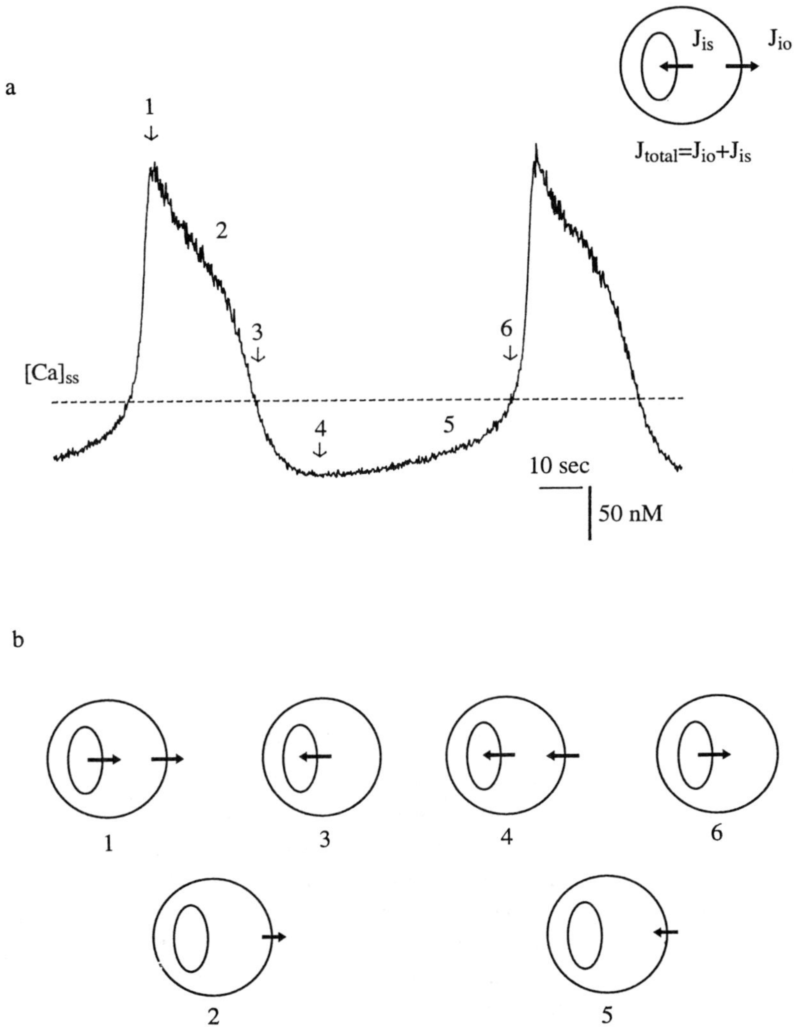


FIG. 4. A kinematic model of caffeine-induced oscillations in cytosolic free calcium concentration ($[Ca^{2+}]_i$). (a) A representative oscillatory cycle; the dashed line indicates $[Ca^{2+}]_{ss}$, the steady concentration of free cytosolic calcium reached during exposure to ryanodine. Four critical points within the cycle are indicated by arrows (1,3,4,6) along with two additional points (2,5) whose precise locations are unknown but must be located in the intervals indicated. The inset at the top right of the figure shows that the total net $Ca^{2+}$ flux, $J_{total}$, is the sum of the net flux across the plasma membrane, $J_{io}$, and the net flux between the cytosol and the internal store, $J_{is}$. Fluxes outward from the cytosol are positive. (b) Model of the direction and relative magnitudes of the net $Ca^{2+}$ fluxes across the plasma membrane ($J_{io}$) and the store membrane ($J_{is}$) at critical points (1–6). Reproduced from Friel & Tsien (1992b).

discharging. Because the store has a finite capacity, the rate at which it delivers $Ca^{2+}$ to the cytosol must eventually decline. When the inward flux $J_{is}$ becomes smaller in magnitude than the outward flux $J_{io}$, $J_{total}$ is outward and $[Ca^{2+}]_i$ declines. As $[Ca^{2+}]_i$ crosses $[Ca^{2+}]_{ss}$ from above (point 3, Fig. 4a), $J_{io}$ is zero, so $[Ca^{2+}]_i$ declines at a rate that is determined exclusively by the rate of net $Ca^{2+}$ accumulation by the store. During this phase the store is being replenished. As $[Ca^{2+}]_i$ continues to fall, $J_{io}$ becomes an inward net flux of increasing magnitude. At point 4 (Fig. 4a), where $[Ca^{2+}]_i$ reaches its minimum value, the inward flux $J_{io}$ is just balanced by the outward flux $J_{is}$. Again, because the store is finite, the rate at which it accumulates $Ca^{2+}$ must decline. When the outward flux $J_{is}$ becomes smaller in magnitude than the inward flux $J_{io}$, $J_{total}$ is inward so that $[Ca^{2+}]_i$ rises. When $[Ca^{2+}]_i$ crosses $[Ca^{2+}]_{ss}$ from below (point 6, Fig. 4a), the rate at which $[Ca^{2+}]_i$ rises is determined exclusively by the rate at which the store is losing $Ca^{2+}$. If the inward flux $J_{is}$ increases in magnitude (e.g., owing to $Ca^{2+}$-induced $Ca^{2+}$ release) faster than the opposing outward flux $J_{io}$, $[Ca^{2+}]_i$ will continue to rise to initiate a new cycle.

Because $J_{is}$ changes sign between critical points 1 and 3 and between points 4 and 6, there must also be at least two points within the oscillatory cycle where it is zero. At these points (2 and 5, Fig. 4a), which have not been specifically located, $[Ca^{2+}]_i$ is expected to change exclusively under the influence of the plasma membrane flux $J_{io}$.

## Measurements of the rates of net calcium entry and release

In terms of the model outlined above, $[Ca^{2+}]_i$ changes under the influence of a total net $Ca^{2+}$ flux that can be separated into two components, $J_{io}$ and $J_{is}$. However, each of these components is thought to reflect the properties of multiple $Ca^{2+}$ transport systems. One of the current experimental challenges is to measure the rates at which these systems individually transport $Ca^{2+}$.

Figure 5 illustrates the rationale behind one experimental approach to measuring individual components of $J_{io}$ and $J_{is}$. Under steady-state conditions, the net flux of $Ca^{2+}$ across the plasma membrane, $J_{io}$, must be zero. This presumably reflects the balance between net $Ca^{2+}$ entry and extrusion (Fig. 5a, *top*, *left*). Under these conditions, removal of external $Ca^{2+}$ is expected to nearly eliminate $Ca^{2+}$ entry and unmask an outward flux that causes $[Ca^{2+}]_i$ to decline (Fig 5a, *top*, *right*). Moreover, the initial rate at which $Ca^{2+}$ declines ($d[Ca^{2+}]_i/dt$) provides a measure of the of $Ca^{2+}$ extrusion ($J_{P1}$), and of the rate of $Ca^{2+}$ entry ($J_{L1}$) that balances extrusion in the steady state. Under stationary conditions when $[Ca^{2+}]_i$ is about 100 nM, $Ca^{2+}$ removal is followed by a rapid drop in $d[Ca^{2+}]_i/dt$ (Fig. 5a, *bottom*), which provides a measure of $J_{L1}$ and $J_{P1}$ of about 2–4 nM/s. Provided there is a linear relationship between the steady-state extrusion rate and $[Ca^{2+}]_i$, these estimates are in good agreement with the measured $Ca^{2+}$ extrusion rate in snail neurons of

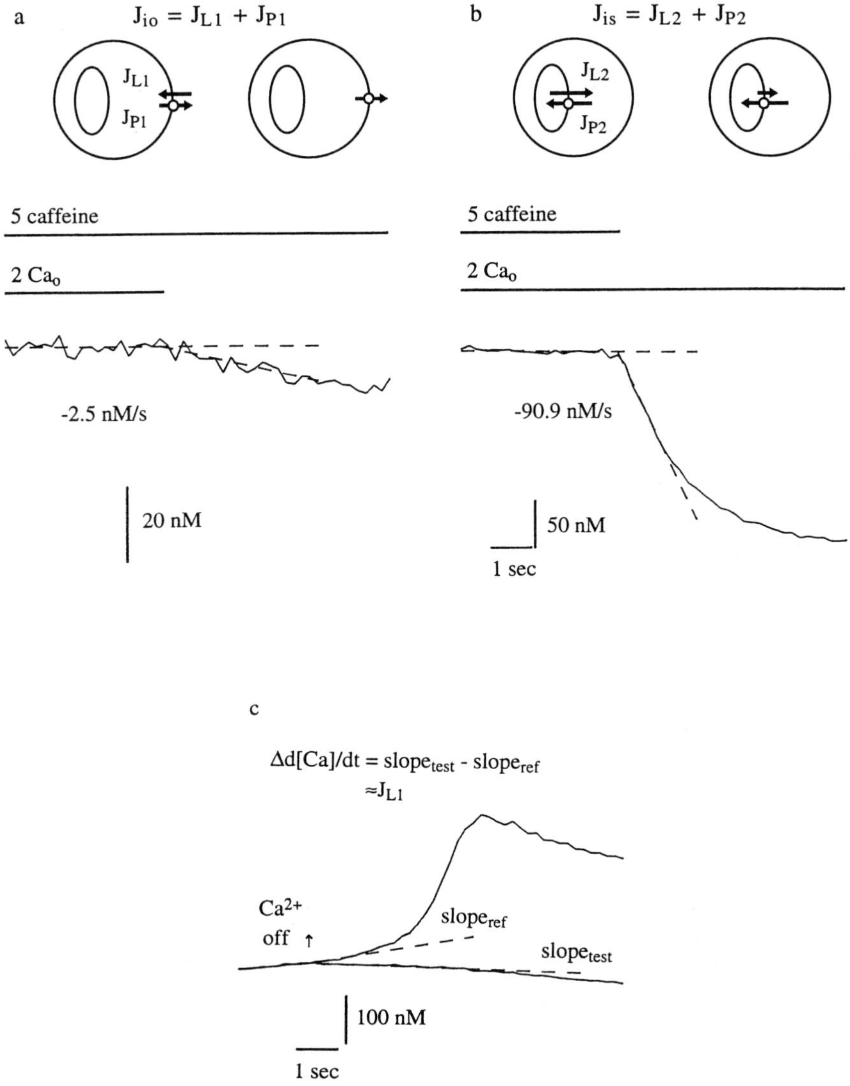


FIG. 5. Comparison between the rates of $Ca^{2+}$ entry and release. Sudden removal of external $Ca^{2+}$ (a) or caffeine (b) while the cytosolic free calcium concentration ($[Ca^{2+}]_i$) is steady produces a rapid drop in $d[Ca^{2+}]_i/dt$. Calcium was removed (a) in the presence of high $[K^+]$, and caffeine was removed (b) after oscillations had been blocked with ryanodine (1 $\mu$M). Dashed lines show the estimated initial $[Ca^{2+}]_i$ (about 115 nM) and the linear decline in $[Ca^{2+}]_i$ with time following the perturbations. (c) Method for measuring net $Ca^{2+}$ fluxes under non-steady state conditions, illustrated by the effect of $Ca^{2+}$ removal while $[Ca^{2+}]_i$ is oscillating. Reproduced from Friel & Tsien (1992b).

about 20 nM/s when $[Ca^{2+}]_i$ is steady at about 400 nM (estimated from Fig. 3 in Tepikin et al 1991). Note that these fluxes have dimensions of concentration per unit time and are therefore *concentration* fluxes.

Similarly, the effects of rapid removal of caffeine provide a way to measure the rate of $Ca^{2+}$ release from the caffeine-sensitive store (Fig. 5b). Under steady-state conditions, the net $Ca^{2+}$ flux between the cytosol and the store must be zero, reflecting a balance between net $Ca^{2+}$ uptake ($J_{P2}$) and release ($J_{L2}$) (Fig. 5b, *top*, *left*). Sudden removal of caffeine is expected to eliminate the caffeine-sensitive component of $J_{L2}$, creating an imbalance between uptake and release that leads to a net outward flux of $Ca^{2+}$ and a decline in $[Ca^{2+}]_i$ (Fig. 5b, *top*, *right*). In this case, the $d[Ca^{2+}]_i/dt$ provides a measure of the imbalance between the rates of net $Ca^{2+}$ uptake and release, and therefore of the flux that is eliminated by removing caffeine. This flux is considerably larger than $J_{L1}$, averaging about 70 nM/s. It is assumed for simplicity that removal of neither $Ca^{2+}$ nor caffeine appreciably alters the rate of $Ca^{2+}$ extrusion, although this needs to be verified experimentally.

This perturbation method also provides a way to compare the rates of $Ca^{2+}$ entry and caffeine-sensitive $Ca^{2+}$ release under *non-stationary* conditions, i.e., while $[Ca^{2+}]_i$ is oscillating. However, instead of measuring the initial rate at which $[Ca^{2+}]_i$ declines following each perturbation, it is necessary to measure the initial change in $d[Ca^{2+}]_i/dt$, $\Delta d[Ca^{2+}]_i/dt$ (Fig. 5c). This can be explained as follows. At each instant in time, $[Ca^{2+}]_i$ is expected to change under the influence of the total net $Ca^{2+}$ flux $J_{total}(t)$, as shown in **1**.

$$d[Ca^{2+}]_i(t)/dt = -J_{total}(t) = -[J_{L1} + J_{P1} + J_{L2} + J_{P2}] \qquad \mathbf{1}$$

If an experimental perturbation specifically eliminates one component of $J_{total}$, then $d[Ca^{2+}]_i/dt$ will change instantly by an amount that gives the magnitude of that component flux. For example, if $Ca^{2+}$ removal specifically eliminates $J_{L1}$, then $\Delta d[Ca^{2+}]_i/dt$ will provide a measure of $J_{L1}$. Note that removal of external $Ca^{2+}$ is not expected to eliminate the passive movement of $Ca^{2+}$ across the plasma membrane completely. However, the residual (outward) net flux would be very small compared with the (inward) net flux present when the extracellular $[Ca^{2+}] = 2$ mM, so $\Delta d[Ca^{2+}]_i/dt$ on removal of $Ca^{2+}$ should provide a reasonable estimate of $J_{L1}$.

With similar reasoning, if removal of caffeine abolishes the intracellular caffeine-sensitive leak flux, then $\Delta d[Ca^{2+}]_i/dt$ in this case provides an estimate of $J_{L2}$. On the basis of measurements taken from recordings like those illustrated in Fig. 3, the rate of caffeine-sensitive $Ca^{2+}$ release is periodic, varying nearly 10-fold during the oscillatory cycle, with magnitudes reaching 100 nM/s. In contrast, $J_{L1}$ varies with $[Ca^{2+}]_i$ over the oscillatory cycle but does not exceed 30–35 nM/s. The observation that $J_{L1}$ oscillates in parallel with

$[Ca^{2+}]_i$ raises the possibility that in intact cells voltage-dependent $Ca^{2+}$ channel activity is potentiated by increases in $[Ca^{2+}]_i$ (Friel & Tsien 1992b). Although independent measurements of $J_{P1}$ and $J_{P2}$ have not been possible so far, the sum of $J_{P1}$ and $J_{P2}$ can be determined from measurements of $d[Ca^{2+}]_i/dt$ and $J_{L1} + J_{L2}$ as shown in **2**.

$$J_{P1} + J_{P2} = -(d[Ca^{2+}]_i/dt + J_{L1} + J_{L2}) \quad \mathbf{2}$$

On the basis of measurements of this kind, it can be shown that $J_{P1} + J_{P2}$ is proportional to $[Ca^{2+}]_i$ throughout the oscillatory cycle. This is consistent with the notion that while $[Ca^{2+}]_i$ is oscillating, the rates of both uptake and extrusion of $Ca^{2+}$ are linear functions of $[Ca^{2+}]_i$ (Friel 1995).

## Summary

The results of this study provide a framework for understanding caffeine-induced $[Ca^{2+}]_i$ oscillations in sympathetic neurons in terms of the coordinated net $Ca^{2+}$ fluxes generated by different $Ca^{2+}$ transport systems. The oscillations require voltage-dependent $Ca^{2+}$ entry, like a membrane oscillator, but also require $Ca^{2+}$ release from an internal store, as do oscillations in many non-neuronal cells. Thus, caffeine-induced $[Ca^{2+}]_i$ oscillations in sympathetic neurons provide an example where $Ca^{2+}$ transport across the plasma membrane and uptake and release by an internal compartment are both required.

Do oscillations of this kind play a role in physiological $[Ca^{2+}]_i$ signalling? Nishimura et al (1991) have described spontaneous rhythmic membrane hyperpolarizations in neurons from rabbit vesical pelvic ganglia. In many respects, these oscillations resemble those described here. For example, they require external $Ca^{2+}$ and are speeded by caffeine and inhibited by ryanodine. These oscillations appear to reflect periodic increases in $[Ca^{2+}]_i$ and activation of plasma membrane ion channels selective for $K^+$ and $Cl^-$. Therefore, $[Ca^{2+}]_i$ oscillations like those described here may occur in some cells under physiological conditions. The role that such $[Ca^{2+}]_i$ oscillations play in intracellular signalling remains unclear.

Caffeine-induced $[Ca^{2+}]_i$ oscillations in sympathetic neurons may be viewed as an example of $[Ca^{2+}]_i$ regulation under conditions where $Ca^{2+}$ transport variables, such as the $[Ca^{2+}]_i$ sensitivity of $Ca^{2+}$-induced $Ca^{2+}$ release and the plasma membrane permeability to $Ca^{2+}$, are somewhat outside their normal operating ranges. In this view, $[Ca^{2+}]_i$ oscillations serve to illustrate properties of $Ca^{2+}$-induced $Ca^{2+}$ release in intact cells that otherwise manifest themselves only subtly as kinetic effects on responses to stimulation. For example, $Ca^{2+}$-induced $Ca^{2+}$ release appears to speed $[Ca^{2+}]_i$ responses elicited by membrane depolarization (Friel & Tsien 1992a). Variations in such kinetic effects may be important in modulating the temporal properties of $[Ca^{2+}]_i$ responses to physiological depolarizing stimuli.

Finally, given the observation that $InsP_3$-induced $Ca^{2+}$ release is modulated by $[Ca^{2+}]_i$ (Bezprozvanny et al 1991) and the functional similarity between $Ca^{2+}$-induced $Ca^{2+}$ release via $InsP_3$- and ryanodine-sensitive $Ca^{2+}$ channels (Berridge 1993), it is possible that the findings and experimental methods described here will contribute to an understanding of agonist-induced $[Ca^{2+}]_i$ oscillations as well.

*Acknowledgements*

This work was carried out in the laboratory of R. W. Tsien and supported by a grant from the National Institutes of Health (NS07102-11).

## References

Berridge MJ 1993 Inositol trisphosphate and calcium signalling. Nature 361:315–325

Bezprozvanny I, Watras J, Ehrlich BE 1991 Bell-shaped calcium-response curves of Ins(1,4,5)$P_3$- and calcium-gated channels from endoplasmic reticulum of cerebellum. Nature 351:751–754

Boland LM, Bean BP 1993 Modulation of N-type calcium channels in bullfrog sympathetic neurons by luteinizing hormone releasing-hormone: kinetics and voltage dependence. J Neurosci 13:516–533

Endo M 1985 Calcium release from sarcoplasmic reticulum. Curr Top Membr Transp 25:181–230

Friel DD 1995 $[Ca^{2+}]_i$ oscillations in sympathetic neurons: an experimental test of a theoretical model. Biophys J, in press

Friel DD, Tsien RW 1992a A caffeine- and ryanodine-sensitive $Ca^{2+}$ store in bullfrog sympathetic neurons modulates the effects of $Ca^{2+}$ entry on $[Ca^{2+}]_i$. J Physiol 450:217–246

Friel DD, Tsien RW 1992b Phase-dependent contributions from $Ca^{2+}$ entry and $Ca^{2+}$ release to caffeine-induced $[Ca^{2+}]_i$ oscillations in bullfrog sympathetic neurons. Neuron 8:1109–1125

Hodgkin AL, Horowicz P 1959 The influence of potassium and chloride ions on the membrane potential of single muscle fibres. J Physiol 148:127–160

Imagawa T, Smith JS, Coronado R, Campbell KP 1987 Purified ryanodine receptor from skeletal muscle sarcoplasmic reticulum is the $Ca^{2+}$-permeable pore of the calcium release channel. J Biol Chem 262:16636–16643

Jones SW, Marks TN 1989 Calcium currents in bullfrog sympathetic neurons. J Gen Physiol 94:151–167

Kuba K Nishi S 1976 Rhythmic hyperpolarization and depolarization of sympathetic ganglion cells induced by caffeine. J Neurophysiol 39:547–563

Lai FA, Erickson HP, Rousseau E, Liu QY, Meissner G 1988 Purification and reconstitution of the calcium release channel from skeletal muscle. Nature 331:315–319

Lipscombe D, Madison DV, Poenie M, Reuter H, Tsien RW, Tsien RY 1988 Imaging of cytosolic $Ca^{2+}$ transients arising from $Ca^{2+}$ stores and $Ca^{2+}$ channels in sympathetic neurons. Neuron 1:355–365

McPherson PS, Kim YK, Valdivia H et al 1991 The brain ryanodine receptor: caffeine-sensitive calcium release channel. Neuron 7:17–25

Nishimura T, Akasu T, Tokimasu T 1991 A slow calcium-dependent chloride current in rhythmic hyperpolarization in neurons of the rabbit vesical pelvic ganglia. J Physiol 437:673–690

Rousseau E, Meissner G 1989 Single cardiac sarcoplasmic reticulum $Ca^{2+}$ release channel: activation by caffeine. Am J Physiol 256:H328–H333
Rousseau E, Smith JS, Meissner G 1987 Ryanodine modifies conductance and gating behavior of single $Ca^{2+}$ release channel. Am J Physiol 253:C364–C368
Tepikin AV, Kostyuk PG, Snitsarev VA, Belan PV 1991 Extrusion of calcium from a single isolated neuron of the snail *Helix pomatia*. J Membr Biol 123:43–47
Thayer SA, Hirning LD, Miller RJ 1988 The role of caffeine sensitive calcium store in the regulation of intracellular free calcium concentration in rat sympathetic neurons *in vitro*. Mol Pharmacol 34:664–673
Tsien RW, Tsien RY 1990 Calcium channels, stores and oscillations. Annu Rev Cell Biol 6:715–760
Tsien RW, Lipscombe D, Madison DV, Bley KB, Fox AP 1988 Multiple types of neuronal calcium channels and their selective inhibition. Trends Neurosci 11:431–438

## DISCUSSION

*Jacob:* I still don't understand why removing $Ca^{2+}$, or preventing $Ca^{2+}$ influx, during the cycle just after the peak has been reached has such a profound effect on the duration of the spike, when it seems that the other $Ca^{2+}$ fluxes in your model are much larger.

*Friel:* Assume for the moment that the rate at which $Ca^{2+}$ enters the cytosol across the plasma membrane does not change very much during the oscillatory cycle, whereas the net $Ca^{2+}$ flux between the store and the cytosol varies periodically. During the rapid upstroke, the net flux from store to cytosol is large, whereas following the peak it is relatively small. As a result, the rate at which $[Ca^{2+}]_i$ changes ($d[Ca^{2+}]_i/dt$) during the upstroke is relatively insensitive to $Ca^{2+}$ removal, compared with the decline following the peak, when $Ca^{2+}$ entry constitutes a larger proportion of the total net $Ca^{2+}$ flux. Now, relax the assumption that the rate of $Ca^{2+}$ entry is constant. On the basis of the $d[Ca^{2+}]_i/dt$ induced by rapid $Ca^{2+}$ removal, it appears that the rate of $Ca^{2+}$ entry rises somewhat with $[Ca^{2+}]_i$. Thus, $Ca^{2+}$ removal has an even greater effect on $d[Ca^{2+}]_i/dt$ when $[Ca^{2+}]_i$ is high, and this, in turn, influences spike duration.

*Jacob:* How fast does the caffeine come out of the cell when you remove it?

*Friel:* I haven't measured this in sympathetic neurons, but O'Neill et al (1990) have shown that in cardiac cells caffeine application or removal leads to a rapid change in the intracellular [caffeine]. For example, under conditions where the [caffeine] in the external solution is changed over 100 ms, the intracellular [caffeine] changes within about 400–500 ms. In my experiments, solution changes are complete in about 150 ms. Therefore, if sympathetic neurons and cardiac cells have the same plasma membrane permeability to caffeine, changes in cytosolic caffeine levels would be complete in about a second.

*Jacob:* That's where you are measuring the initial rate, to establish the store movement. Might you be underestimating the time it would take?

*Friel:* Delayed washout from the cytosol after removing caffeine from the external medium would cause my estimates of the size of the caffeine-sensitive flux to be low. However, I should mention that fast caffeine removal leads to an abrupt drop in $d[Ca^{2+}]_i/dt$ (see Fig 5b, Friel & Tsien 1992), suggesting that washout is quite rapid.

*Meyer:* I have two questions: (1) In your system, does store depletion affect $Ca^{2+}$ influx? And, (2) $Ca^{2+}$ inactivation of the ryanodine receptor; do you see this in your system?

*Friel:* I don't think store depletion has an effect on $Ca^{2+}$ influx, because thapsigargin doesn't elevate the resting $[Ca^{2+}]_i$, even though it does eliminate responsiveness to caffeine.

*Berridge:* Dr Putney, are there any examples of the ryanodine receptor being used for generating capacitative $Ca^{2+}$ entry?

*Putney:* Not to my knowledge. In general, the more neuron-like a cell is, the less likely it is to have capacitative $Ca^{2+}$ entry. I would guess that Dr Friel's cells don't have capacitative $Ca^{2+}$ entry, because if they did, you wouldn't need the high $[K^+]$. The liver cell does not need to be activated independently of phospholipase C activation because the capacitative $Ca^{2+}$ entry fulfils that role, but in your system you have to have the high $[K^+]$.

*Friel:* Inactivation of the ryanodine receptor is difficult to test in intact cells. The only thing I have done so far to examine inactivation is to terminate $[Ca^{2+}]_i$ oscillations by removing external $Ca^{2+}$ and then wait various periods of time before restoring $Ca^{2+}$. The idea is that if there is inactivation and recovery is slow, the latency to the first spike following readmission of external $Ca^{2+}$ should decline as the recovery period is increased. This is not observed (Fig. 7c, Friel & Tsien 1992). However, this type of experiment would not detect inactivation if recovery is fast.

*Meyer:* Does cyclic ADP-ribose have any effect in your system?

*Friel:* I haven't tried it yet but I plan to.

*Thomas:* You described the $Ca^{2+}$ store filling as a primer to one of the spikes. Do oscillations of this type require the stores to be less than full in a basal state? In other words, is this filling stage necessary for the oscillation to affect the period, or is it possible to have an oscillation which is based solely on $Ca^{2+}$ appearing at a region in the cytosol, perhaps with $Ca^{2+}$-binding proteins in the cytosol setting the period?

*Friel:* Such a mechanism is certainly possible, but it would still be necessary to ask what causes the oscillations. With respect to caffeine-induced $[Ca^{2+}]_i$ oscillations, it appears that the amount of $Ca^{2+}$ in the caffeine-sensitive pool oscillates, with $Ca^{2+}$ levels rising during most of the interspike interval and then falling during the rapid upstroke. This is based on the magnitude of the initial change in $d[Ca^{2+}]_i/dt$ that follows rapid elevations in [caffeine] at different points in the oscillatory cycle (Figs 7a,b, Friel & Tsien 1992). However, this could reflect phase-dependent changes in the ability of caffeine to increase the stores' $Ca^{2+}$ permeability independently of changes in the stores' $Ca^{2+}$ content. Further experiments will be necessary to distinguish between these two possibilities.

*Putney:* There's nothing in your model suggesting that luminal $Ca^{2+}$ regulates sensitivity. The only thing that's regulated in your model is $[Ca^{2+}]_i$ and the $Ca^{2+}$-induced $Ca^{2+}$ release threshold.

*Friel:* The model is very simple. With one exception, all the transporters are linear in the sense that they generate $Ca^{2+}$ fluxes that are proportional to $[Ca^{2+}]$. The exception is the passive permeability of the store, which is assumed to rise with $[Ca^{2+}]_i$. Intraluminal $Ca^{2+}$ influences the driving force for $Ca^{2+}$ release, but not the store's permeability. According to the model, $[Ca^{2+}]_i$ rises during the rapid upstroke because the store releases $Ca^{2+}$ more rapidly than it can be pumped out across the plasma membrane. This phase reflects $Ca^{2+}$-induced $Ca^{2+}$ release. The $[Ca^{2+}]_i$ rise is transient simply because the store eventually becomes depleted. There is no inactivation of the $[Ca^{2+}]_i$-sensitive permeability. The $[Ca^{2+}]_i$ falls to the interspike minimum because as the store refills, $Ca^{2+}$ is accumulated more rapidly than $Ca^{2+}$ can leak into the cytosol across the plasma membrane. This decline in $[Ca^{2+}]_i$ is transient because the store eventually refills. Within each oscillatory cycle, the store is a $Ca^{2+}$ source and then a $Ca^{2+}$ sink.

Does the mechanism of $[Ca^{2+}]_i$ oscillations in sympathetic neurons have any bearing on agonist-induced oscillations in non-neuronal cells? One difference seems to be that in many non-neuronal cells, intraluminal $Ca^{2+}$ levels influence the $Ca^{2+}$ permeability of the plasma membrane. Evidently, you can't dissociate them as you can in neurons.

*Putney:* You could do the whole experiment in the absence of $Ca^{2+}$ in a system with an $InsP_3$-driven store and get several cycles. That's a substantial deviation from your model. In $InsP_3$-driven systems it doesn't matter where $Ca^{2+}$ comes from; it doesn't have to come from outside in those cells but it does in your system.

*Friel:* Are caffeine-induced $[Ca^{2+}]_i$ oscillations related to agonist-induced oscillations? The former can be accounted for reasonably well without invoking inactivation. Is it clear that inactivation is more important in $InsP_3$-driven oscillations than in those induced by caffeine in neurons?

*Berridge:* That depends on who is doing the modelling! Some people use inactivation in order to explain oscillations, but others don't.

*Stucki:* In Dr Friel's model, the spike terminates because of an emptying of the intracellular stores.

*Friel:* Not a complete emptying, though.

*Stucki:* That's true, but you always get the same monotonic decay. Earlier, we heard from Andrew Thomas about experiments in which there was a stepwise emptying of the stores, and we have done similar experiments using the $Mn^{2+}$-quench technique and following the ratio of the fura-2 signal. During the quenching of the cytosolic fluorescence, the ratio at 340:380 nm goes up, indicating that the dye now reports exclusively the luminal [$Ca^{2+}$]. Then when you add a hormone such as phenylephrine, you get a stepwise decay in the ratio which can be explained by a stepwise release of $Ca^{2+}$ from the intracellular stores. This only works with small concentrations of the hormone, i.e., in the range typical for eliciting oscillations. Higher concentrations of the hormone, which normally give rise to overstimulation, result in a monotonic decay.

*Berridge:* The critical point is, why does the spike actually begin? This brings us back to the business of initiation. Why do you get to a point where external $Ca^{2+}$ is no longer important? I would argue that it's the loading of the pool, which reduces its buffering capacity, thus allowing the cytosolic [$Ca^{2+}$] to rise to the threshold to trigger the regenerative process.

*Whitaker:* Are you assuming there is a linear buffering capacity in the cell in your model, and is that a reasonable assumption?

*Friel:* Calcium buffering is not included explicitly in the model.

*Whitaker:* The data seem to fit, so one could forget buffering!

*Friel:* It should be mentioned that formally, the model would be unchanged if multiple fast linear $Ca^{2+}$ buffers were present. The presence of such buffers would simply scale the cytosolic volume. Therefore the properties of the oscillations are consistent with the presence of fast linear buffers.

*Berridge:* Buffering is clearly important, because the store is a buffer.

*Whitaker:* I'm talking about cytoplasmic buffers.

*Berridge:* I understand, but the point is the same in terms of how you handle the constant input of $Ca^{2+}$.

*Whitaker:* No, because cytoplasmic buffers are generally modelled to be instantaneous, whereas the $Ca^{2+}$ store has buffering (uptake) kinetics that are slower than those of release.

*Friel:* Cytoplasmic buffers and membrane-delimited stores can have different effects on $[Ca^{2+}]_i$ dynamics. Fast linear cytoplasmic buffers are expected to have the same effect as an expansion in cytosolic volume, whereas a membrane-bound store that can take up and release $Ca^{2+}$ may have more complicated effects.

*Whitaker:* If they are purely linear, as you have implied, the modelling problem is easy. If they're normally non-linear, the problem becomes difficult. If your data actually fit with a linear cytoplasmic buffer, modelling is much easier.

*Cannell:* The approach Dr Friel has used is similar to that used by Negretti et al (1993) in analysing the cardiac muscle store with a pump/leak model. O'Neill & Eisner (1990) also suggested that caffeine works as a sensitizer of $Ca^{2+}$-induced $Ca^{2+}$ release, which makes a clear prediction for your class of model: that low doses of caffeine should greatly increase the frequency of spontaneous $Ca^{2+}$ spikes.

*Friel:* Increasing the [caffeine] experimentally increases the frequency of the oscillations. That's in keeping with the model.

*Berridge:* Does it decrease the amplitude?

*Friel:* Yes.

*Jaffe:* Your spikes tended to have a kind of shoulder. Can you provide a qualitative exlanation for that?

*Friel:* I could only speculate.

*Jaffe:* Could there be two different pumps?

*Friel:* That's possible. However, the shoulder that precedes the accelerated decline in $[Ca^{2+}]_i$ (Friel, this volume, Fig. 4) can be accounted for simply in terms of deactivation of $Ca^{2+}$ release channels, without postulating different types of $Ca^{2+}$ pumps. For example, as $[Ca^{2+}]_i$ declines, $Ca^{2+}$ release channels are expected to close. This would render the store less permeable to $Ca^{2+}$ and bias the pump/leak balance towards net $Ca^{2+}$ accumulation. Calcium accumulation by the store represents an outward net flux from the cytosol, and this would speed the decline in $[Ca^{2+}]_i$. The decline would be regenerative because as $[Ca^{2+}]_i$ falls, more release channels close, which further shifts the balance in favour of net accumulation.

*Cobbold:* What is the effect of increased cyclic AMP (cAMP) concentration on the experimentally observed oscillations, and can you use your model to predict what the most important target might be?

*Friel:* Exposure to 8-Br-cAMP has no detectable effect on $[Ca^{2+}]_i$ in these cells, suggesting that caffeine does not elicit oscillations simply by elevating [cAMP]. Inhibition of the oscillations by ryanodine points to a ryanodine-sensitive process as being critical.

*Williams:* In your model are you making all the rate constants simple, first-order rate constants, then modifying the first-order constants by putting the $Ca^{2+}$ effect just on the outlet system from the store?

*Friel:* Yes. The rate constant for passive $Ca^{2+}$ release from the store is assumed to be $[Ca^{2+}]_i$-dependent. Specifically, it increases sigmoidally with $[Ca^{2+}]_i$. This is a reasonable description of the $[Ca^{2+}]_i$-dependence of $Ca^{2+}$ release channel gating over the physiological range of $[Ca^{2+}]_i$ (Bezprozvanny et al 1991).

*Williams:* One would suspect the pump systems, the outer membrane one, at least, to be $Ca^{2+}$ dependent. You should therefore have had a $Ca^{2+}$-dependent rate constant there as well. You would also expect that not to be first order in $Ca^{2+}$, because the binding of $Ca^{2+}$ to calmodulin, which controls the pump, is at least second order.

*Friel:* This raises a methodological issue. One approach to understanding $[Ca^{2+}]_i$ oscillations is to construct a model that takes into consideration all known $Ca^{2+}$ transport systems, and ask whether such a model can account for the oscillations. Another approach is to select a particular type of cell and ask whether the properties of $[Ca^{2+}]_i$ regulation that have been demonstrated in that cell can account for the observed oscillations. A problem with the first approach is that it may combine $Ca^{2+}$ transporters that are not present in the cell under study, and use transport rate laws determined under non-physiological conditions. Given that the $Ca^{2+}$ transporters in sympathetic neurons have not been described in detail, should one combine results from other systems, even though they may not strictly apply, or make use of experimental observations from the cell in question to deduce the types of transporters that are present and their approximate rate laws?

*Williams:* You can say certain things. For example, as far as the buffering is concerned, if you have calsequestrin in your vesicle, in all probability you've got good buffering there. This means that the free $Ca^{2+}$ level is fairly constant, so your release has a fixed rate constant so long as you have a subsequent rate constant which is first order; your back-up would be a safe, constant store. However, the system that controls the outward rate is also assumed to be first order by you. I do not think this can be true of a $Ca^{2+}$-regulated $Ca^{2+}$ pump.

*Friel:* In the model, the rate at which $Ca^{2+}$ is released by the store is the product of a rate constant that rises sigmoidally with $[Ca^{2+}]_i$, and the difference between the luminal and cytosolic $[Ca^{2+}]$.

*Williams:* It's $Ca^{2+}$ dependent in a linear way. That's a linear binding constant, if you like.

*Friel:* But the rate constant and the rate of net $Ca^{2+}$ release both depend on $[Ca^{2+}]_i$ in a non-linear manner.

*Williams:* Yes. You then go on and you keep your first-order dependence on the pumps. That is not really what the nature of the mechanism is like, because the pump mechanism has a very different power dependence in it.

*Friel:* It seems to me that if the rate of $Ca^{2+}$ extrusion is saturable and can be described by a Hill coefficient not too different from one, the pump rate will increase nearly linearly with $[Ca^{2+}]_i$, as long as $[Ca^{2+}]_i$ is well below the $[Ca^{2+}]$ giving a half-maximal pump rate ($K_m$). Of course, as $[Ca^{2+}]_i$ rises, this approximation will break down, but the important question is: is the linear approximation valid over the range of $[Ca^{2+}]_i$ at which oscillations occur (50–400 nM)? As it turns out, using linear approximations to the rate laws for $Ca^{2+}$ uptake and extrusion, the model accounts reasonably well for the experimental observations. This argues against gross departures from linearity over this range of $[Ca^{2+}]_i$.

*Stucki:* We have exactly the same problems with our model for oscillations in hepatocytes (Somogyi & Stucki 1991). Hans-Dieter Söling has clearly shown that in hepatocytes, neither the endoplasmic reticulum ATPase nor the plasma membrane ATPase is dependent on calmodulin (Schütze & Söling 1987). Hence, you get a good approximation by choosing a linear pump model, assuming that it is operating below its $K_m$.

*Petersen:* The permeability of caffeine through membranes is enormously high (Toeschu et al 1992).

*Friel:* The solution changes are complete in about 150 ms.

*Petersen:* In that case, caffeine would go out of the cells extremely quickly. You could measure it by looking at the quench of indo-1 (Toescu et al 1992). Also, Alexey Tepikin has looked at the dependency of $Ca^{2+}$ extrusion by the $Ca^{2+}$ pump in pancreatic cells. In the range we are talking about, extrusion of $Ca^{2+}$ from the cell has a linear dependence on $[Ca^{2+}]$.

*Cannell:* Have you tried just altering the resting $[Ca^{2+}]$ and seeing whether or not the model generates the correct answers? If you double the $[Ca^{2+}]$ do you get the right effect on frequency?

*Friel:* Qualitatively, the model behaves in accord with experiment. Lowering the $[Ca^{2+}]$ in the external medium increases both the oscillatory period and the spike amplitude. However, I haven't determined whether or not the agreement is quantitative.

*Nathanson:* How sensitive is the predicted cytosolic $[Ca^{2+}]$ to changes in the rate constants?

*Friel:* That's a detailed question to which I don't yet know the answer.

*Berridge:* What's behind that question?

*Friel:* I thought the question refers to how accurate the rate constants are if they are obtained by fitting the model to the data. More work on error analysis is required before this can be answered adequately.

*Galione:* Is anything known about neurotransmitter receptors on these cells that might produce a messenger that has a similar effect to caffeine?

*Friel:* Not that I know of. Of course, potential second messengers such as cyclic ADP-ribose come to mind.

*Galione:* How does your model of $Ca^{2+}$ oscillations in sympathetic neurons compare with that of Kuba & Takeshita (1981)?

*Friel:* The oscillations that I described were first reported by Kuba & Nishi (1976) in terms of periodic membrane hyperpolarization occurring when sympathetic neurons are exposed to caffeine. Kuba & Takeshita (1981) described a model which could account for many features of the observed hyperpolarizations. The model that I have described includes the same number of compartments and the same types of transport systems, but the rate laws are somewhat simpler. For example, both $Ca^{2+}$ extrusion and uptake are accomplished by linear pumps, and the $[Ca^{2+}]_i$-sensitive permeability that is responsible for $Ca^{2+}$-induced $Ca^{2+}$ release is not sensitive to intraluminal $Ca^{2+}$ levels. The simplicity of the rate laws has facilitated the design of experiments that test the model and provide estimates of the associated rate parameters.

*Berridge:* It should be emphasized that Kuba and Nishi were modelling this before any of us were even thinking about oscillations. They deserve quite a lot of credit in that respect.

*Miyazaki:* Sympathetic neurons in physiological conditions are more or less firing, and sometimes they show bursts of spikes. I wondered how intracellular $Ca^{2+}$ behaves.

*Friel:* I'm not sure about sympathetic neurons, but there are cases in the literature of peripheral neurons undergoing periodic membrane hyperpolarizations without exposure to caffeine. These periodic hyperpolarizations are sensitive to caffeine and ryanodine, so they may operate via a similar mechanism. There are cells that under more or less physiological conditions do exhibit these kinds of oscillations, but their physiological role is not clear.

*Jacob:* When you applied caffeine in the presence of $Ca^{2+}$ you got a sharp initiation of the spike, whereas when you applied it in the absence of $Ca^{2+}$ there seemed to be a definite ramp before it took off. Can your model explain that?

*Friel:* I think this has more to do with the fact that the response to caffeine elicited in the absence of external $Ca^{2+}$ was preceded by an application of caffeine, so that the store was partially depleted. If you first expose a naive cell to caffeine in the absence of external $Ca^{2+}$, you don't see a slow rise.

*Jacob:* So if the store is partly empty, the caffeine discharges $Ca^{2+}$ more slowly and you see that.

*Friel:* Right. One interpretation is that caffeine increases the $Ca^{2+}$ permeability of the store, but the magnitude of the net flux originating from the store also depends on the content of the store when caffeine is applied. The net flux in turn determines the initial rate at which $[Ca^{2+}]_i$ rises following caffeine exposure, so that when the store is partially depleted, caffeine produces a slower $[Ca^{2+}]_i$ rise with longer spike latency.

*Berridge:* It's interesting that when you increase the [caffeine] the oscillator accelerates and the amplitude comes down. Presumably, one would explain this by taking into account the fact that the threshold for regenrative $Ca^{2+}$ release has been reduced.

*Friel:* Yes. Two things are expected to occur when the [caffeine] is elevated. First, the threshold for activating $Ca^{2+}$-induced $Ca^{2+}$ release is lowered. Second, the average level of intraluminal $Ca^{2+}$ is lowered, which reduces the driving force for net $Ca^{2+}$ release.

*Berridge:* What you imply, therefore, is that the amount that can be released is then reduced. This is interesting with regard to the hepatocyte oscillator, with which one sees a different phenomenon in that the oscillator can be accelerated, whereas the amplitude remains relatively constant. In the latter case, the agonist presumably elevates $[InsP_3]$, which should decrease the amplitude by reducing the threshold.

*Friel:* There's a difference in that elevations in $[InsP_3]$ are expected to increase the magnitude of $InsP_3$-sensitive $Ca^{2+}$-induced $Ca^{2+}$ release and also increase the plasma membrane permeability to $Ca^{2+}$.

*Jaffe:* I think that the reason for this profound difference is that, whereas in your system the rise is initiated by cytosolic $Ca^{2+}$, in the hepatocyte, as in other wave-generating systems, the rise is initiated by luminal $Ca^{2+}$ and, thus, by a completely different mechanism.

*Berridge:* I don't think many people here would agree with that.

*Putney:* It seems to me that all you need to do if you want the amplitude to be constant is to have an inactivation mechanism that is sufficiently prolonged to allow the stores to refill to the same steady-state level every time. There's no inactivation mechanism in Dr Friel's model. If, as is known for the $InsP_3$ receptor, at the top of the peak there is a $Ca^{2+}$- and an $InsP_3$-dependent inactivation that stay in place for longer than is necessary to rebuild the steady-state, you will get oscillations of constant amplitude.

*Berridge:* From what I understood, there is an argument about how quickly the reactivation occurs, that once the [$Ca^{2+}$] comes back down again and the resting level is recovered, the receptor becomes reactivated.

*Putney:* This may depend on the relative rate of reactivation and how fast you can pump the $Ca^{2+}$ back in. About 80% of the $Ca^{2+}$ that is released gets taken back up and the other 20% is made up from $Ca^{2+}$ entry.

*Berridge:* The interval between spikes can last for a minute when the oscillator is going really slowly.

*Putney:* But reactivation could occur relatively quickly, and then the interval could depend on the buffers.

*Lechleiter:* There is a potential problem. Iino & Endo's (1992) data on $Ca^{2+}$ inactivation of $InsP_3$-mediated $Ca^{2+}$ release, using caged $Ca^{2+}$, appears to be instantaneous, at least in smooth muscle. If you look at the slope changes, the moment they released caged $Ca^{2+}$, $Ca^{2+}$ release was inhibited. It wasn't slow, as was assumed in some models. The onset of the inactivation was very rapid.

*Cobbold:* Gil Weir was talking about local $Ca^{2+}$ areas around the mouth of an L-type channel. If that local build-up of [$Ca^{2+}$] is happening around an $InsP_3$ receptor as it opens, where is our positive and negative feedback then?

*Thomas:* We have found in permeabilized hepatocytes that the $InsP_3$ receptor can be inactivated by $InsP_3$ pretreatment at fixed [$Ca^{2+}$] (Hajnoczky & Thomas 1994). This was done with the $Mn^{2+}$-quench technique, because we can't measure $Ca^{2+}$ flux after we have added $InsP_3$. We can show something like a 10- to 20-fold decrease in the initial rate of quenching by pretreating with $InsP_3$. That process is accelerated in terms of the rate and onset of inactivation by $Ca^{2+}$; the higher the [$Ca^{2+}$], the more rapidly inactivation occurs. We have been able to push that to only 1–2 $\mu$M $Ca^{2+}$. If there is 100 $\mu$M $Ca^{2+}$ in the channel, and if that interacts with $InsP_3$ to inactivate the receptor, the inactivation could occur very rapidly.

*Lechleiter:* Isn't that what Finch, Turner and Goldin reported (Finch et al 1991)?

*Thomas:* But in their case $Ca^{2+}$ was moving through the channel and, therefore, changing concentration. In our experiments, at a fixed [$Ca^{2+}$], the rate of inactivation depends on how long the receptor has had $InsP_3$ bound to it. That's different.

*Putney:* It's not necessarily a different phenomenon.

*Thomas:* That's true, but one might say that the phenomenon of $Ca^{2+}$ inactivation is actually due to $InsP_3$ being bound—to a ligand-induced inactivation—but that the receptor must have $Ca^{2+}$ bound to it. You can argue either way.

*Fay:* Can you tell us anything more of the kinetics of the activation process, or about its specific $Ca^{2+}$ dependence?

*Thomas:* The $InsP_3$ receptor is not completely inactivated; it goes to a lower conductivity state, showing something like 20-fold or so lower conductivity than in the naive state. The time course depends on the free $[Ca^{2+}]$. At 300 nM, which we selected as a concentration that might be of the order of the cytosolic concentration, the half-time was about 15 s. The rate was faster at a higher $[Ca^{2+}]$, which is why it's important to know what the $Ca^{2+}$ exposure of the $InsP_3$-sensitive site is.

Also, at a submaximal dose of $InsP_3$—if one accepts that $InsP_3$ receptor activation occurs in a quantal manner such that at low $[InsP_3]$ only subpopulations of receptors are activated—the inactivation process seemed to be specific for the population of $InsP_3$ receptors that were first activated. In other words, one may add a low dose of $InsP_3$, then add $Mn^{2+}$ and look at what happens to the rate in comparison with cells to which $InsP_3$ and $Mn^{2+}$ were added together, and then look at cells to which one adds a higher $InsP_3$ dose during the $Mn^{2+}$-quench phase. Low doses of $InsP_3$ added during the pretreatment period only very slightly inactivated the $Mn^{2+}$-quench response to high $[InsP_3]$. There seems to be an incremental inactivation. That means that a wave passing through a cell may inactivate only those receptors which are opening; then we have to ask the question of how many $InsP_3$ receptors need to open for a wave to propagate.

*Fay:* What do you know about the reactivation process?

*Thomas:* It can be reversed by removing $InsP_3$ or by removing the $Ca^{2+}$, with a time course of a few seconds.

*Cobbold:* That may be important. We often see in hepatocytes with some agonists, ATP in particular, several $Ca^{2+}$ spikes a few seconds after the initial rise.

*Berridge:* I would like to end on a slightly frivolous note. The ryanodine receptor is a rather strange molecule, which has a disc shape with four openings around the rim. We all draw our channels with $Ca^{2+}$ gushing out at the top. However, Sidney Fleischer's work suggests that the $Ca^{2+}$ may pour out from the sides, suggesting that the ryanodine receptor operates much like a muckspreader.

*Williams:* The $Ca^{2+}$-ATPase is exactly like that. The $Ca^{2+}$ channel of the ATPase lies across the membrane and there is no channel in the larger, extramembrane part of it, which is the ATPase part. It is supposed that the $Ca^{2+}$ goes in and out of the channel not through the ATPase section. The $H^+$-ATP synthetase is also exactly like that. The proton comes out along the membrane, it is proposed, not through the top part of the molecule, which is an ATPase when separated. In both cases, the active part of the channel need only be short (about one nanometre long).

*Lechleiter:* With the ryanodine receptor there are four subunits. Are there four individually conducting pores or just one pore surrounding by four subunits?

*Wier:* This is an appealing idea. Alan Williams has been making conductance measurements (Tinker et al 1992). These four 'vestibules', as he calls them, are far too large to be the actual pore. It's a nice idea that $Ca^{2+}$ comes out in four different directions, thereby avoiding a large build up of $Ca^{2+}$ in the wrong place, but his work doesn't allow you to conclude that these four vestibules are pores.

*Michell:* Bob Williams said the pores are very short. What is the evidence for that?

*Williams:* In the $Ca^{2+}$-ATPase, there is in the membrane section a very short length where there are three or four COOH groups quite close to one another. All of these are essential for channel functioning. We don't feel that there need be anything more to the channel, and above and below the carboxylalin there are only rather small amino acids and water.

*Michell:* So there is no kinetic evidence based on channel behaviour. The idea is based on modelling what the bit that seems to do the work should look like in space.

*Williams:* Yes. The investigators have built a fairly precise model containing seven helices and a little channel with carboxylate groups in the middle. The ATPases that are pumping the ions must be connected to the channel part to alter the conformation, to open and shut the channel. The whole is modelled on known ATPases, known channels and sequence similarities.

## References

Bezprozvanny I, Watras J, Ehrlich BE 1991 Bell-shaped calcium-response curves of Ins(1,4,5)$P_3$- and calcium-gated channels from endoplasmic reticulum of cetrebellum. Nature 351:751–754

Finch EA, Turner TJ, Goldin SM 1991 Calcium as a coagonist of inositol 1,4,5-trisphosphate-induced calcium release. Science 252:443–446

Friel DD 1995 Calcium oscillations in neurons. In: Calcium waves, gradients and oscillations. Wiley, Chichester (Ciba Found Symp 188) p 210–234

Friel DD, Tsien RW 1992 Phase-dependent contributions from $Ca^{2+}$ entry and $Ca^{2+}$ release to caffeine-induced $[Ca^{2+}]_i$ oscillations in bullfrog sympathetic neurons. Neuron 8:1109–1125

Hajnoczky G, Thomas AP 1994 The inositol trisphosphate calcium channel is inactivated by inositol trisphophate. Nature 370:474–477

Iino M, Endo M 1992 Calcium-dependent immediate feedback control of inositol 1,4,5-trisphosphate-induced $Ca^{2+}$ release. Nature 360:76–78

Kuba K, Nishi S 1976 Rhythmic hyperpolarization and depolarization of sympathetic ganglion cells induced by caffeine. J Neurophysiol 39:547–563

Kuba K, Takeshita S 1981 Simulation of intracellular $Ca^{2+}$ oscillation in a sympathetic neurone. J Theor Biol 93:1009–1031

Negretti N, O'Neill SC, Eisner DA 1993 The relative contributions of different intracellular and sarcolemmal systems to relaxation in rat ventricular myocytes. Cardiovasc Res 27:1826–1830

O'Neill SC, Eisner DA 1990 A mechanism for the effects of caffeine on $Ca^{2+}$ release during diastole and systole in isolated rat ventricular myocytes. J Physiol 430:519–536

O'Neill SC, Donoso P, Eisner DA 1990 The role of $[Ca^{2+}]_i$ and $[Ca^{2+}]$ sensitization in the caffeine contracture of rat myocytes: measurement of $[Ca^{2+}]_i$ and $[caffeine]_i$. J Physiol 425:55–70

Somogyi R, Stucki JW 1991 Hormone-induced calcium oscillations in liver cells can be explained by a simple one pool model. J Biol Chem 266:11068–11077

Schütze S, Söling H-D 1987 Does a calmodulin-dependent $Ca^{2+}$-regulated $Mg^{2+}$-dependent ATPase contribute to hepatic microsomal calcium uptake? Biochem J 243:729–737

Tinker A, Lindsay ARG, Williams AJ 1992 A model for ionic conduction in the ryanodine receptor channel of sheep cardiac muscle sarcoplasmic reticulum. J Gen Physiol 100:495–517

Toescu EC, O'Neill SC, Petersen OH, Eisher DA 1992 Caffeine inhibits the agonist-evoked cytosolic $Ca^{2+}$ signal in mouse pancreatic acinar cells by blocking inositol trisphosphate production. J Biol Chem 267:23467–23470

# Calcium signalling during mammalian fertilization

Shunichi Miyazaki

*Department of Physiology, Tokyo Women's Medical College, 8-1 Kawada-cho, Shinjuku-ku, Tokyo 162, Japan*

*Abstract.* The fertilized mammalian egg is a nice model system for analysing spatiotemporal $Ca^{2+}$ signalling in the intact cell. Hamster eggs show repetitive $Ca^{2+}$ transients, associated in the initial response with $Ca^{2+}$ waves which begin from the site of sperm attachment and are propagated across the deep cytoplasm to the opposite pole. In unfertilized eggs, a regenerative $Ca^{2+}$ wave is induced by injection of either inositol 1,4,5-trisphosphate ($InsP_3$) or $Ca^{2+}$, and $Ca^{2+}$ oscillations are produced by continuous injection of $InsP_3$. These $Ca^{2+}$ waves and oscillations in both fertilized and unfertilized eggs are inhibited in a dose-dependent manner by a monoclonal antibody to the type 1 $InsP_3$ receptor. Ryanodine receptors (both skeletal and cardiac types) are not detected by physiological or immunoblot analyses. Positive and negative feedback between cytosolic $Ca^{2+}$ and $Ca^{2+}$ release from $InsP_3$-senstitive pools accounts for the spatiotemporal $Ca^{2+}$ signalling. In addition to intracellular $Ca^{2+}$ release, $Ca^{2+}$ entry from outside the egg is necessary to refill the $Ca^{2+}$ pools and maintain $Ca^{2+}$ oscllations. Evidence suggests that inositol 1,3,4,5-tetrakisphosphate activates the $Ca^{2+}$ influx. The signal transduction process leading to the production of $InsP_3$ and the mechanism of egg activation following the $Ca^{2+}$ response still remain to be elucidated.

*1995 Calcium waves, gradients and oscillations. Wiley, Chichester (Ciba Foundation Symposium 188) p 235–251*

A wide variety of cells show propagating and oscillatory rises in the intracellular calcium ion concentration ($[Ca^{2+}]_i$) in response to various biological activators (see Berridge 1993). These $Ca^{2+}$ waves and $Ca^{2+}$ oscillations are the general spatiotemporal pattern of $Ca^{2+}$ signals. Eggs fertilized by spermatozoa show a dramatic increase in $[Ca^{2+}]_i$. The $Ca^{2+}$ waves were first demonstrated in fish eggs (Gilkey et al 1978) and are now thought to be a universal mechanism of activation of deuterostome eggs (see Jaffe 1991). In addition to $Ca^{2+}$ waves, fertilized mammalian eggs show repetitive, transient fluctuations in $[Ca^{2+}]_i$, as demonstrated in the hamster, mouse, rabbit, pig and cow (see Miyazaki et al 1993). The relatively large size of egg cells permits microinjection of chemicals and detailed $Ca^{2+}$ image analysis; mammalian eggs are thus a nice model system by which to analyse spatiotemporal $Ca^{2+}$ signalling in the intact cell.

### Calcium signals in fertilized eggs

The rise in $[Ca^{2+}]_i$ begins at the site of sperm attachment (a point stimulation). The $Ca^{2+}$ signal is amplified at this site and is conducted throughout the egg with a velocity of 16–28 μm/s in hamster eggs (which have a diameter of 70 μm) (Miyazaki et al 1986). Confocal laser-scanning microscopy of hamster eggs with the dye calcium green revealed that the rise in $[Ca^{2+}]_i$ traverses across the deep cytoplasm, not preferentially in the cortical area (K. Shiraishi, H. Shirakawa & S. Miyazaki, unpublished work 1994). The initial $Ca^{2+}$ response is either three to four $Ca^{2+}$ transients occurring in succession at short intervals, as in hamster eggs (Figs 1A and 1B), or a prolonged $Ca^{2+}$ transient lasting for 3–4 min, as in mouse eggs (Kline & Kline 1992). Successive $Ca^{2+}$ waves pass through the hamster egg during the initial response (Fig. 1B), each starting from the same site (Miyazaki et al 1986).

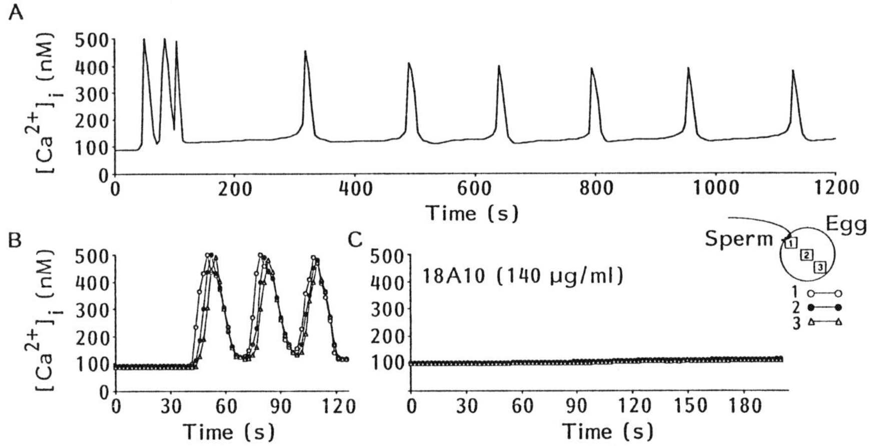


FIG. 1. Calcium waves and $Ca^{2+}$ oscillations at fertilization of hamster eggs and their blockade by the monoclonal antibody 18A10 (an antibody against the mouse inositol 1,4,5-trisphosphate receptor–$Ca^{2+}$ release channel). (A) Repetitive $Ca^{2+}$ transients recorded from the time of the sperm attachment to the egg surface (zone-free egg). $[Ca^{2+}]_i$ was averaged over the whole egg. (B) The initial response of the record in (A); $[Ca^{2+}]_i$ was measured at the sperm attachment site (○), centre (●), and opposite side (△) of the egg (see inset). There is a substantial delay in the $Ca^{2+}$ rise at the side opposite to sperm attachment but the peak $[Ca^{2+}]_i$ is identical in the three areas, indicating a propagating $Ca^{2+}$ wave. (C) Complete blockade of the sperm-induced $Ca^{2+}$ response by 18A10 (140 μg/ml in the egg). No increase in $[Ca^{2+}]_i$ occurred even at the sperm-attachment site (○). From Miyazaki 1993, with permission.

One of the immediate functions of the $Ca^{2+}$ wave is to induce exocytotic secretion from the cortical granules over the egg (Kline & Kline 1992). In mouse eggs, the secreted substance modifies sperm-binding glycoproteins ZP2 and ZP3 in the surrounding zona pellucida, preventing penetration of additional sperm (polyspermy block) (see Ward & Kopf 1993). Another critical function of the early $Ca^{2+}$ signals is to trigger the advance of the cell cycle. Mature mammalian eggs are arrested at the metaphase of the second meiosis and resume meiosis as soon as they are fertilized. This phenomenon can be prevented by loading the egg with the $Ca^{2+}$ chelator BAPTA (1,2-bis[*o*-aminophenoxyy]ethane-*N*,*N*,*N'*,*N'*-tetraacetic acid) (Kline & Kline 1992). One potential target of $Ca^{2+}$ signals is the $Ca^{2+}$-dependent protease calpain, which causes proteolysis of cytostatic factor, a factor mediating meiotic arrest in vertebrate eggs (Watanabe et al 1989). The mechanism of egg activation following the $Ca^{2+}$ response is the crucial subject that needs to be elucidated by future studies.

The basal $[Ca^{2+}]_i$ is slightly increased after the initial response (Fig. 1A). The later responses are relatively sharp $Ca^{2+}$ transients (or spikes), each of which lasts for 30–60 s, occurring at intervals of 2–4 min (Fig. 1A). The $Ca^{2+}$ spike is preceded by a slowly augmenting $[Ca^{2+}]_i$ rise, with the transition occurring at a $[Ca^{2+}]_i$ of 140–170 nM. The $[Ca^{2+}]_i$ increases almost synchronously in the whole egg without any pacemaker hot spot (Miyazaki et al 1986). Calcium transients continue for several hours; it has not yet been determined at exactly which developmental stage they cease. The majority of rabbit eggs do not show $Ca^{2+}$ transients at the pronuclear apposition stage (when male and female pronuclei are located in close apposition just before fusion) (Fissore & Robl 1993). The biological significance of such a long series of $Ca^{2+}$ transients is still unknown. In parthenogenetically activated rabbit eggs the pattern and amplitude of $Ca^{2+}$ transients induced by high-voltage electric field pulses affect later embryonic development (Ozil 1991). Pulsatile rises in $[Ca^{2+}]_i$ may be enough to activate continuously $Ca^{2+}$-dependent processes (such as enzymic activities) with slow kinetics. Because continuous elevation of $[Ca^{2+}]_i$ tends to cause irreversible cell damage, repetitive $Ca^{2+}$ spike signals may be favourable for accomplishing cellular functions without toxic side effects.

## Signal transduction leading to the calcium response

The signal transduction process following sperm–egg interaction is still unclear. A supposed signalling pathway involves the successive activation of a receptor, a GTP-binding protein, phospholipase C ($PLC\beta_1$), and cleavage of phosphatidylinositol 4,5-bisphosphate ($PtdInsP_2$) into inositol 1,4,5-trisphosphate ($InsP_3$) and diacylglycerol (Fig. 2A) (Miyazaki 1988). Recently, the sperm-binding protein was identified in sea urchin eggs (Foltz & Lennarz 1993). Mouse eggs express the cell adhesion molecule integrin $\beta_1$ (Brobel et al 1992) and

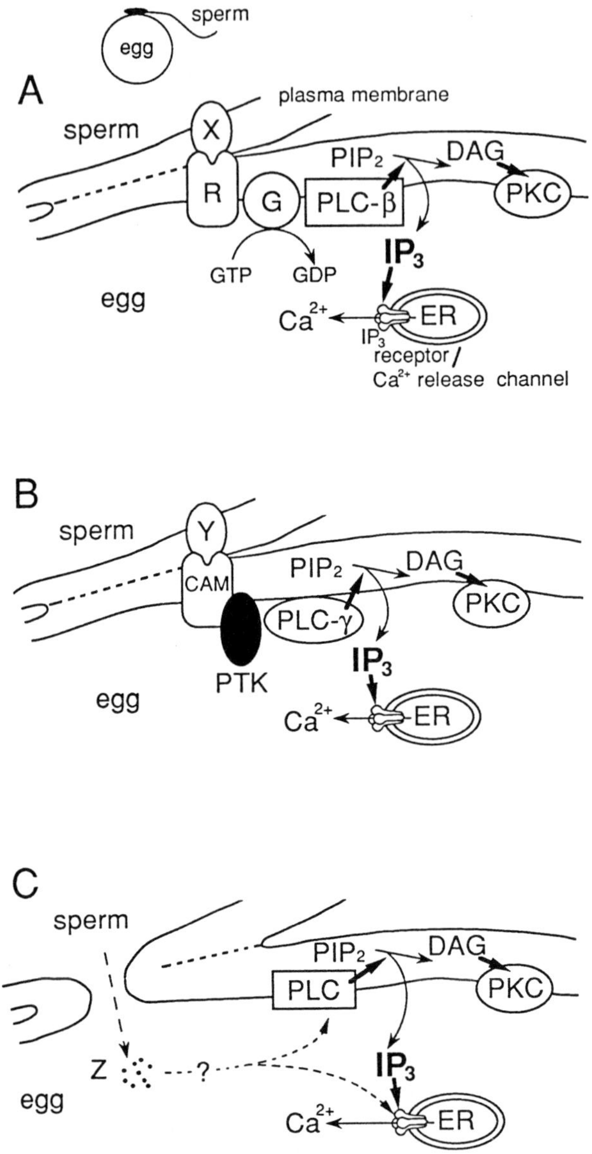


FIG. 2. Three possible mechanisms of the signal transduction pathway leading from sperm–egg interaction to $Ca^{2+}$ release. $IP_3$, inositol 1,4,5-trisphosphate; R, receptor; G, GTP-binding protein; DAG, diacylglycerol; PLC, phospholipase C; $PIP_2$, phosphatidylinositol 4,5-bisphosophate; PKC, protein kinase C; ER, endoplasmic reticulum; CAM, cell adhesion molecule; PTK, protein tyrosine kinase; X, Y, Z, putative molecules of the sperm which stimulate the signalling pathway.

guinea-pig sperm possess the surface protein PH-30, whose $\beta$ subunit has high sequence similarity to disintegrin (Tarone et al 1993). Mouse sperm express major histocompatibility complex (MHC) class II proteins, and mouse eggs possess a protein similar to CD4 (Mori et al 1991), which is an accessory molecule on T cells that binds MHC class II molecules on antigen-presenting cells. The CD4-like molecule is associated with the protein tyrosine kinase $p56^{lck}$ (Mori et al 1991), which is capable of activating $PLC\gamma_1$ (see Veillette & Davidson 1992). Thus, the cell adhesion molecule–protein tyrosine kinase interaction is a candidate signalling pathway (Fig. 2B). It is not known whether these molecules transduce signals leading to the $Ca^{2+}$ response at fertilization. It has been suggested that a cytosolic sperm factor which is transferred into the egg's cytoplasm serves as a signal (Z in Fig. 2C) (see Whitaker & Swann 1993). Injection of hamster or pig sperm extract into hamster or mouse eggs induces $Ca^{2+}$ oscillations similar to those at fertilization. The same sperm extracts cause $Ca^{2+}$ release even in *Xenopus* and sea urchin eggs (see Whitaker & Swann 1993).

## The mechanism of regenerative calcium release

The $Ca^{2+}$ wave in fertilized eggs is due to release of $Ca^{2+}$ from intracellular stores (see Jaffe 1991). The propagating $Ca^{2+}$ signal requires a regenerative process such that locally increased $[Ca^{2+}]_i$ causes $Ca^{2+}$ release in the neighbouring $Ca^{2+}$ stores in succession—that is, $Ca^{2+}$-induced $Ca^{2+}$ release. A regenerative, propagating $[Ca^{2+}]_i$ rise is evident in hamster eggs when $Ca^{2+}$ is locally injected from a micropipette with increasing current pulses while $[Ca^{2+}]_i$ is measured from fura-2 images using an image processor (Miyazaki et al 1992) (the threshold pulse is 1 nA, 2 s in Fig. 3A). Calcium-induced $Ca^{2+}$ release is known to occur in skeletal muscles through the ryanodine receptor–$Ca^{2+}$ release channel on the sarcoplasmic reticulum membrane (see Fleischer & Inui 1989). In our experiments neither caffeine nor ryanodine induced $Ca^{2+}$ release, and neither skeletal nor cardiac type ryanodine receptors were detected by immunoblot analysis (Miyazaki et al 1992). A regenerative $Ca^{2+}$ wave is also produced by injection of $InsP_3$ instead of $Ca^{2+}$, when $[Ca^{2+}]_i$ at the $InsP_3$-injection site reaches about 200 nM (Fig. 3C) (Miyazaki et al 1992). It has been demonstrated in skinned smooth muscle fibres that the rate of $InsP_3$-induced $Ca^{2+}$ release is augmented with increasing intracellular $Ca^{2+}$ concentrations up to 300 nM but inhibited by higher concentrations (Iino & Endo 1992). The $InsP_3$ receptor–$Ca^{2+}$ release channel reconstituted in planar lipid bilayers has a maximum open probability at about 200 nM $Ca^{2+}$, with sharp decreases on either side of the maximum (Bezprozvanny et al 1991). The positive feedback between $[Ca^{2+}]_i$ and $InsP_3$-induced $Ca^{2+}$ release (at relatively low $[Ca^{2+}]_i$) has been shown to operate promptly enough to cause regenerative $Ca^{2+}$ release (Iino & Endo 1992). Thus, the sensitizing effect of $Ca^{2+}$ on

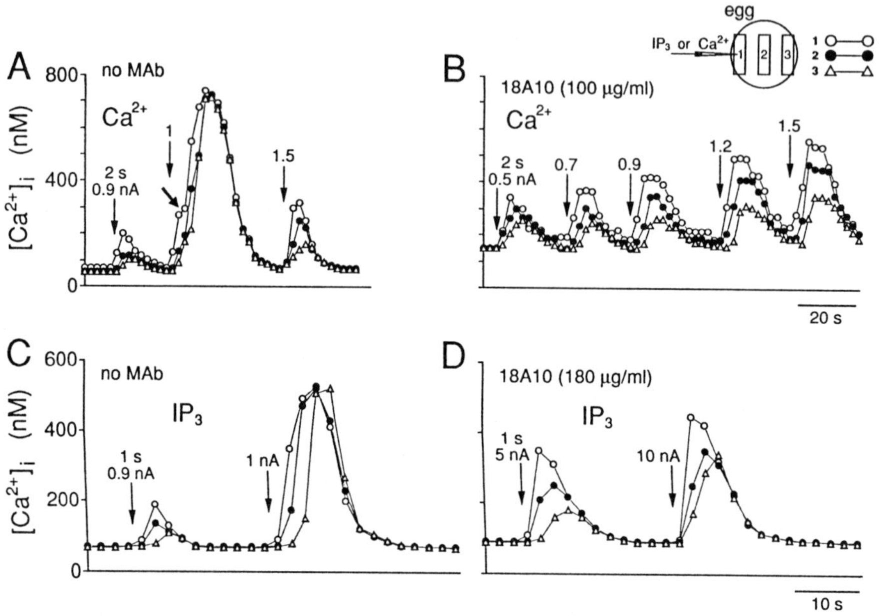


FIG. 3. Regenerative propagating $Ca^{2+}$ release induced by localized microinjection of $Ca^{2+}$ or inositol 1,4,5-trisphosphate ($IP_3$) (A and C) and its inhibition by the monoclonal antibody (MAb) 18A10 (B and D). (A) A slight change in the current pulse for injection of $Ca^{2+}$ from 0.9 to 1 nA, 2 s, changes the response from a localized $[Ca^{2+}]$ rise to a regenerative $[Ca^{2+}]_i$ rise throughout the egg. The transition from the former to the latter is suggested by an inflexion in the $[Ca^{2+}]_i$ rise at the $Ca^{2+}$ injection site (○) (thick arrow). Propagation of the latter $Ca^{2+}$ response is indicated by the delay in the $Ca^{2+}$ rise at the pole opposite to the injection site (△) (see inset). Once the regenerative $Ca^{2+}$ release has been induced, a larger pulse (1.5 nA, 2 s) fails to produce the same response, as if there is a refractory period. (B) In 18A10-treated eggs, the rise in $[Ca^{2+}]_i$ is increased approximately linearly with increasing doses of $Ca^{2+}$, and peak $[Ca^{2+}]_i$ is always the highest at the $Ca^{2+}$ injection site. These findings indicate that $Ca^{2+}$ injection-induced $Ca^{2+}$ release has been blocked. (C) $IP_3$-induced regenerative $Ca^{2+}$ release. (D) Much larger pulses of $IP_3$ injection produce smaller and graded $Ca^{2+}$ rises in the presence of 18A10. Modified from Miyazaki 1993.

$InsP_3$-induced $Ca^{2+}$ release could be the basis for $Ca^{2+}$-induced $Ca^{2+}$ release through the $InsP_3$ receptor–$Ca^{2+}$ release channel.

The sensitivity of $InsP_3$-induced $Ca^{2+}$ release is also dependent on the $Ca^{2+}$ content of the stores ('loading dependence of $InsP_3$-induced $Ca^{2+}$ release' , see Taylor 1992). The sensitivity of stores to $InsP_3$ is increased as luminal $Ca^{2+}$ increases, whereas stores depleted of $Ca^{2+}$ are less sensitive to $InsP_3$.

The $Ca^{2+}$ wave induced by locally injected $Ca^{2+}$ or $InsP_3$ can be envisaged to progress as follows (Fig. 4A). The locally increased $[Ca^{2+}]_i$ will increase

$InsP_3$-induced $Ca^{2+}$ release from the stores at the injection site by an action not only from the cytoplasmic side but also from the luminal side by overloading the stores to cause regenerative $Ca^{2+}$ release. Released $Ca^{2+}$ will stimulate $InsP_3$-induced $Ca^{2+}$ release from neighbouring stores, resulting in a $Ca^{2+}$ wave even in the deeper cytoplasm when there is a low level of $InsP_3$ or even at the resting level of $InsP_3$. A $Ca^{2+}$ wave could be produced in this manner following local stimulation by sperm or agonists (Fig. 4A).

## Inositol trisphosphate receptor-mediated spatiotemporal calcium signalling

Specific and potent inhibitors are needed to identify the $Ca^{2+}$ release mechanism which operates in functioning cells in physiological conditions. Heparin is used as a competitive inhibitor of the $InsP_3$ receptor, but it is not selective and high doses are needed (see Miyazaki et al 1993). A monoclonal antibody to the mouse $InsP_3$ receptor–$Ca^{2+}$ release channel, 18A10, which recognizes an epitope close to the region of the $Ca^{2+}$ release channel near the C-terminus of the protein, was found to inhibit $InsP_3$-induced $Ca^{2+}$ release in mouse cerebellar microsomes (Nakade et al 1991). The antibody is specific to the type 1 receptor (Yamamoto-Hino et al 1994) and reacts with the $InsP_3$ receptor–$Ca^{2+}$ release channel of hamster eggs in immunoblot analysis (Miyazaki et al 1992). Immunohistochemical staining shows that the antibody binds to the cortical area of the hamster egg and the deep cytoplasm is stained in a roughly reticular pattern, consistent with the distribution of the endoplasmic reticulum (ER) (Miyazaki et al 1992). Calcium release induced by injection of $InsP_3$ is suppressed by pre-injection of 18A10 (Fig. 3D). The mode of inhibition is non-competitive (Miyazaki et al 1992). Calcium release induced by a $Ca^{2+}$ injection is blocked by 18A10 (Fig. 3B), indicating that $Ca^{2+}$-induced $Ca^{2+}$ release occurs through the $InsP_3$ receptor–$Ca^{2+}$ release channel. Two other monoclonal antibodies, 4C11 and 10A6, which recognize epitopes distant from the $Ca^{2+}$ channel region, have no inhibitory effect. Thus, 18A10 is a useful tool with which to dissect the molecular mechanisms of $Ca^{2+}$ release.

The $Ca^{2+}$ wave induced by sperm binding is inhibited by 18A10 (Miyazaki et al 1992), with the magnitude, propagation rate and frequency of $Ca^{2+}$ transients being reduced in a dose-dependent manner. Eventually, at a certain concentration, the rise in $[Ca^{2+}]_i$ is completely blocked even at the site of the sperm attachment (Fig. 1C). Thus, $InsP_3$-induced $Ca^{2+}$ release is essential for the initiation, propagation and oscillation of the sperm-induced $Ca^{2+}$ transients. The central role of $InsP_3$-induced $Ca^{2+}$ release in the sperm-induced $Ca^{2+}$ wave has been demonstrated in *Xenopus* eggs (Nuccitelli et al 1993) and oocytes (Lechleiter & Clapham 1992). In sea urchin eggs both the $InsP_3$ receptor and the ryanodine receptor were shown to be responsible for the $Ca^{2+}$ wave (Galione et al 1993, Lee et al 1993).

## Calcium influx to maintain calcium oscillations

Repetitive $Ca^{2+}$ transients are produced in hamster eggs by various means other than sperm, including continuous injection of $InsP_3$ or the non-metabolizable analogue $InsP_3(S)_3$ (inositol 1,4,5-trisphosphorothioate), injection of GTPγS, and application of serotonin or the sulphydryl reagent thimerosal (see Miyazaki et al 1993). That these $Ca^{2+}$ oscillations are due to $Ca^{2+}$ release from $InsP_3$-sensitive $Ca^{2+}$ pools is shown by their blockage by 18A10. Interestingly, these $Ca^{2+}$ oscillations cease after perfusion with $Ca^{2+}$-free medium (Miyazaki 1991). Calcium release is never induced by injection of $InsP_3$ in a $Ca^{2+}$-free medium even with large current pulses; all that is seen is the first three or four $Ca^{2+}$ transients (Miyazaki 1991). Thus, $Ca^{2+}$ influx from outside the cell is necessary to maintain repetitive $Ca^{2+}$ release. Calcium influx is likely to provide the $Ca^{2+}$ which refills the $Ca^{2+}$ pools. If it does, greater $Ca^{2+}$ influx would be expected to refill the $Ca^{2+}$ pools more quickly. In fact, the interval between sperm-induced $Ca^{2+}$ transients is affected by the driving force for $Ca^{2+}$ influx: the interval is shortened by raising the extracellular $Ca^{2+}$ concentration, $[Ca^{2+}]_o$ (the major chemical force), or hyperpolarizing the membrane potential by passing direct current through the intracellular micropipette (the major electrical driving force) (Igusa & Miyazaki 1983). Conversely, the interval is prolonged by lowering $[Ca^{2+}]_o$ or depolarizing the membrane potential.

I interpret the process of repetitive $Ca^{2+}$ release as follows (Fig. 4). Inositol trisphosphate-sensitive $Ca^{2+}$ pools are filled at the resting state (before fertilization or other stimulations). Calcium release is initiated at the sperm-binding site by $InsP_3$ derived from $PtdInsP_2$, and the $Ca^{2+}$release is propagated in the form of a wave throughout the egg (Fig. 4A). The successive release of $Ca^{2+}$ occurring at short intervals, as in the sperm-induced initial response (see Figs 1A and 1B), probably requires a large enough amount of $InsP_3$. This phenomenon may be based on the quantal nature of $InsP_3$-induced $Ca^{2+}$ release (Meyer & Stryer 1990), or there may be reserve $Ca^{2+}$ pools with a high threshold (low sensitivity to $InsP_3$). After the initial $Ca^{2+}$ response, $InsP_3$ continues to be produced (although at a slower rate), but the majority of $InsP_3$-sensitive pools have been emptied and $InsP_3$-induced $Ca^{2+}$ release declines (Fig. 4B). Most of the released $Ca^{2+}$ ions are thought to be extruded to the outside of the egg, with few being sequestered in the pools, because $Ca^{2+}$ release cannot be induced even by high doses of $InsP_3$ in $Ca^{2+}$-free media, as described above. Meanwhile, $Ca^{2+}$ influx is enhanced and the transported $Ca^{2+}$ ions are taken up into the pools (Fig. 4C). Inositol trisphosphate-induced $Ca^{2+}$ release resumes when the luminal $Ca^{2+}$ content reaches a certain level, as reflected by the slowly augmenting rise in $[Ca^{2+}]_i$. The increased $[Ca^{2+}]_i$ sensitizes the $InsP_3$ receptor–$Ca^{2+}$ release channel in a regenerative manner, resulting in the next $Ca^{2+}$ spike (Fig. 4C). These processes are repeated, producing $Ca^{2+}$

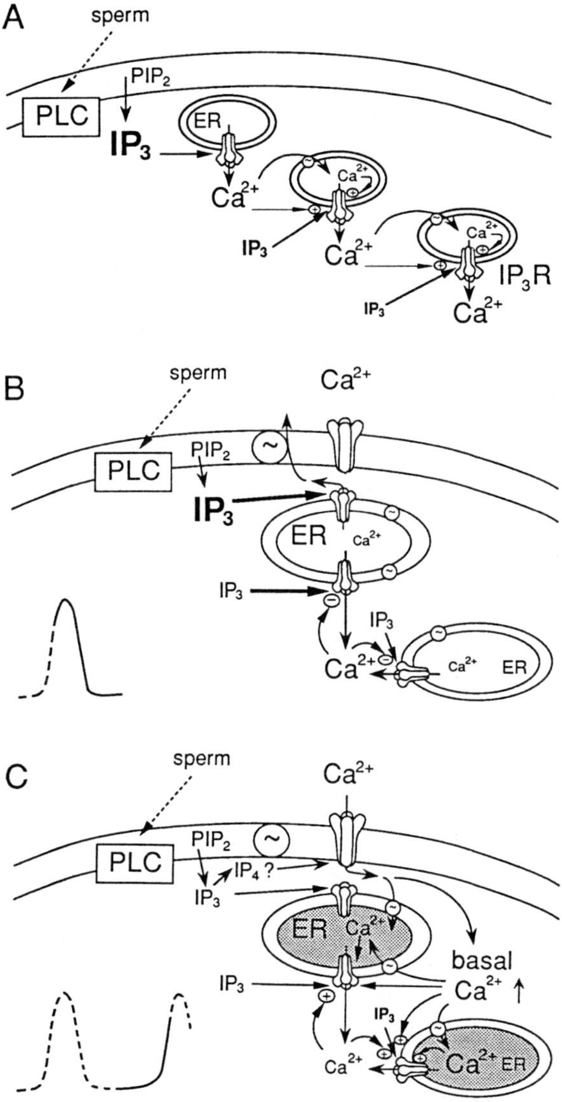


FIG. 4. Calcium waves (A) and $Ca^{2+}$ oscillations (B and C) mediated by $Ca^{2+}$ release from inositol 1,4,5-trisphosphate ($IP_3$)-sensitive pools. In B and C, the corresponding phase of the $Ca^{2+}$ oscillations is shown by a solid curve. In B, after the initial response, less $IP_3$ is produced and most of the $IP_3$-sensitive pools have been emptied; $IP_3$-induced $Ca^{2+}$ release therefore declines. The $Ca^{2+}$ released passes to the outside of the cell. In C, $IP_3$-induced $Ca^{2+}$ release has been resumed after influx of $Ca^{2+}$ and refilling of the stores. Calcium oscillations ensue. ER, endoplasmic reticulum; $IP_3R$, $IP_3$ receptor; $IP_4$, inositol 1,3,4,5-tetrakisphosphate; $PIP_2$, phosphatidylinositol 4,5-bisphosphate; PLC, phospholipase C.

oscillations if the $InsP_3$ level is only slightly increased. In this model, the extracellular medium is a kind of $Ca^{2+}$ compartment and the cortical $Ca^{2+}$ pools serve as a leading pacemaker.

**The calcium influx pathway**

At least two $Ca^{2+}$ influx pathways have been proposed to be coupled to $Ca^{2+}$ release or depletion of the pools. One is the inositol phosphate-activated $Ca^{2+}$ influx pathway. Irvine (1991, 1992) has proposed the involvement of an inositol 1,3,4,5-tetrakisphophate ($InsP_4$) receptor–$Ca^{2+}$ channel which is coupled with the $InsP_3$ receptor–$Ca^{2+}$ release channel on the endoplasmic reticulum in close apposition to the plasma membrane and is open when the two receptors are dissociated from each other, i.e., when both $InsP_3$ and $InsP_4$ bind their receptors or when the two receptors are both free from $InsP_3$ and $InsP_4$. The other proposed pathway is the inositol phosphate-independent $Ca^{2+}$ influx pathway which is operative when $InsP_3$-sensitive pools are emptied ('capacitative $Ca^{2+}$ influx', see Putney 1992). A diffusible factor which is released from $Ca^{2+}$ pools and activates $Ca^{2+}$ influx has been identified (Randriamampita & Tsien 1993).

To identify the $InsP_4$-activated $Ca^{2+}$ influx pathway in the hamster egg, we injected $InsP_4$ with direct current while the membrane potential was clamped at the resting level with the single-electrode voltage clamp method (Fig. 5) (Shirakawa et al 1993). Continuous injection of $InsP_4$ caused repetitive $Ca^{2+}$ release, although $InsP_4$ is much less effective than $InsP_3$. The $InsP_4$ injection-induced $Ca^{2+}$ release was blocked by the monoclonal antibody 18A10, indicating that it is mediated by the $InsP_3$ receptor–$Ca^{2+}$ release channel. However, it is not known whether $InsP_4$ acts directly on the $InsP_3$ receptor or whether some fraction of the $InsP_3$ derived from $InsP_4$ induces $Ca^{2+}$ release. In the presence of 18A10 the basal $[Ca^{2+}]_i$ did not increase in response to $InsP_4$ injection, but an increase was detected after $Ca^{2+}$ uptake into $InsP_3$-sensitive pools was prevented by applying an inhibitor of the $Ca^{2+}$-ATPase, thapsigargin (Fig. 5) (Shirakawa et al 1993). This $[Ca^{2+}]_i$ rise is due to $Ca^{2+}$ influx, because it is abolished in $Ca^{2+}$-free media and the magnitude of the $Ca^{2+}$ rise is increased by raising $[Ca^{2+}]_o$ or by hyperpolarizing the membrane (Shirakawa et al 1993). Thus, we have obtained evidence for the involvement of an $InsP_4$-activated $Ca^{2+}$ influx pathway in hamster eggs. The expected $Ca^{2+}$ inward current (Fig. 5) was not clearly recorded, because leak current changes during long-term voltage clamp and also because of the involvement of a $Ca^{2+}$-activated $K^+$ current.

Activation of $Ca^{2+}$ influx by co-existing $InsP_4$ and $InsP_3$ has been demonstrated in lacrimal cells (see Irvine 1991, 1992). However, it is hard to show evidence that the $Ca^{2+}$ release component is completely eliminated. Our

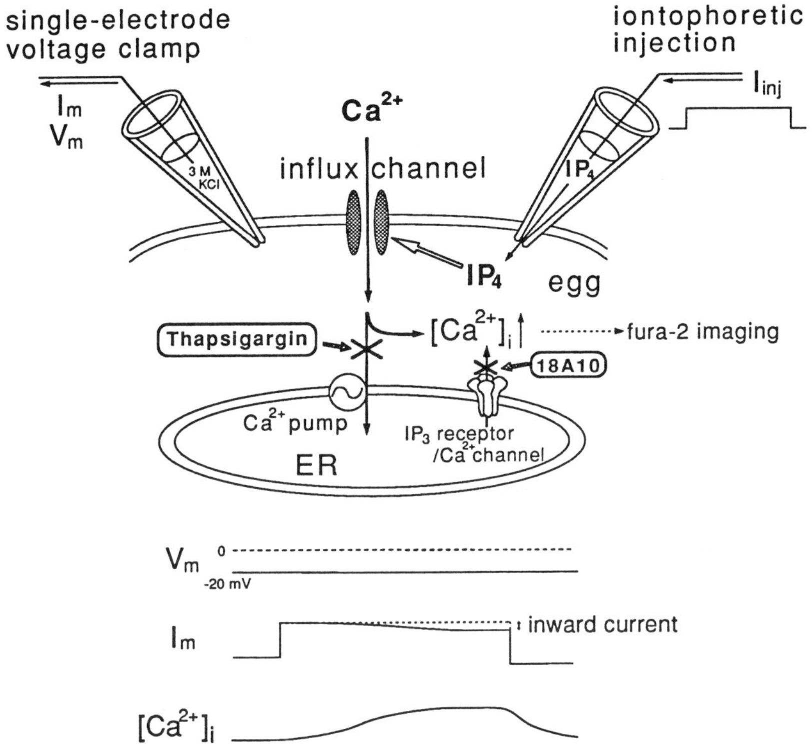


FIG. 5. The identification of the inositol 1,3,4,5-tetrakisphosphate ($IP_4$)-activated $Ca^{2+}$ influx pathway causing an increase in $[Ca^{2+}]_i$. $V_m$, holding membrane potential; $I_m$, membrane current. $IP_4$-injection-induced $Ca^{2+}$ release was blocked by the monoclonal antibody 18A10. Basal $[Ca^{2+}]_i$ increases during injection of $IP_4$, when $Ca^{2+}$ uptake into inositol 1,4,5-trisphosphate ($IP_3$)-sensitive pools is blocked by thapsigargin, an inhibitor of the $Ca^{2+}$-ATPase.

tentative model for $Ca^{2+}$ influx which leads to the refilling of $InsP_3$-sensitive pools is that $InsP_3$ is continuously produced during fertilization and that some fraction is converted to $InsP_4$, which activates the $Ca^{2+}$ influx pathway in a manner not always synergistic with the activation of the $InsP_3$ receptor–$Ca^{2+}$ release channel.

## *Acknowledgements*

I thank H. Shirakawa and K. Shiraishi for collaboration in the experiments on $InsP_4$-activated $Ca^{2+}$ influx and in confocal microscopy. This work was supported by Grants-in-Aid of General Scientific Research from the Japan Ministry of Education, Science and Culture.

## References

Berridge MJ 1993 Inositol trisphosphate and calcium signalling. Nature 361:315–325

Bezprozvanny I, Watras J, Ehrlich BE 1991 Bell-shaped calcium-response curves of Ins(1,4,5)$P_3$- and calcium-gated channels from endoplasmic reticulum of cerebellum. Nature 351:751–754

Brobel CP, Wolfsberg TG, Turck CW, Myles DG, Primikoff P, White JM 1992 A potential fusion peptide and an integrin ligand domain in a protein active in sperm–egg fusion. Nature 356:248–252

Fissore RA, Robl JM 1993 Sperm, inositol trisphosphate, and thimerosal-induced intracellular $Ca^{2+}$ elevation in rabbit eggs. Dev Biol 159:122–130

Fleischer S, Inui M 1989 Biochemistry and biophysics of excitation–contraction coupling. Annu Rev Biophys Biophys Chem 18:333–364

Foltz KR, Lennarz WJ 1993 The molecular basis of sea urchin gamete interactions at the egg plasma membrane. Dev Biol 158:46–61

Galione A, McDougall A, Busa WB, Willmott N, Gillot I, Whitaker M 1993 Redundant mechanisms of calcium-induced calcium release underlying calcium waves during fertilization of sea urchin eggs. Science 261:348–352

Gilkey JC, Jaffe LF, Ridgeway DB, Reynolds GT 1978 A free calcium wave traverses the activating egg of the medaka, *Oryzias latipes*. J Cell Biol 76:448–466

Igusa Y, Miyazaki D 1983 Effects of altered extracellular and intracellular calcium concentration on hyperpolarizing responses of the hamster egg. J Physiol 340:611–632

Iino M, Endo M 1992 Calcium-dependent immediate feedback control of inositol 1,4,5-trisphosphate-induced $Ca^{2+}$ release. Nature 360:76–78

Irvine RF 1991 Inositol tetrakisphosphate as a second messenger: confusions, contradictions, and a potential resolution. Bioessays 13:419–429

Irvine RF 1992 Is inositol tetrakisphosphate the second messenger that controls $Ca^{2+}$ entry into cells? Adv Second Messenger Phosphoprotein Res 26:161–185

Jaffe LF 1991 The path of calcium in cytosolic calcium oscillations: a unifying hypothesis. Proc Natl Acad Sci USA 88:9883–9887

Kline D, Kline JT 1992 Repetitive calcium transients and the role of calcium in exocytosis and cell cycle activation in the mouse egg. Dev Biol 149:80–89

Lechleiter JD, Clapham DE 1992 Molecular mechanisms of intracellular calcium excitability in X. laevis oocytes. Cell 69:283–294

Lee HC, Aarhus R, Walseth TF 1993 Calcium mobilization by dual receptors during fertilization of sea urchin eggs. Science 261:352–355

Meyer T, Stryer L 1990 Transient calcium release induced by successive increments of inositol 1,4,5-trisphosphate. Proc Natl Acad Sci USA 87:3841–3845

Miyazaki S 1988 Inositol 1,4,5-trisphosphate-induced calcium release and guanine nucleotide-binding protein-mediated periodic calcium rises in golden hamster eggs. J Cell Biol 106:345–353

Miyazaki S 1991 Repetitive calcium transients in hamster oocytes. Cell Calcium 12:205–216

Miyazaki S 1993 IP3 receptor-mediated spatial and temporal $Ca^{2+}$ signalling of the cell. Jpn J Physiol 43:409–434

Miyazaki S, Hashimoto N, Yoshimoto Y, Kishimoto T, Igusa Y, Hiramoto Y 1986 Temporal and spatial dynamics of the periodic increase in intracellular free calcium at fertilization of golden hamster eggs. Dev Biol 118:259–267

Miyazaki S, Yuzaki M, Nakada K et al 1992 Block of $Ca^{2+}$ wave and $Ca^{2+}$ oscillation by antibody to the inositol 1,4,5-trisphosphate receptor in fertilized hamster eggs. Science 257:251–255

Miyazaki S, Shirakawa H, Nakada K, Honda Y 1993 Essential role of the inositol 1,4,5-trisphosphate receptor/$Ca^{2+}$ release channel in $Ca^{2+}$ waves and $Ca^{2+}$ oscillations at fertilization of mammalian eggs. Dev Biol 158:62–78

Mori T, Guo MW, Mori E 1991 Expression of CD4-like structure on mouse egg vitelline membrane and its signal transductive roles through $p56^{lck}$ in fertilization. Am J Reprod Immunol 24:97–103

Nakade S, Maeda N, Mikoshiba K 1991 Involvement of the C-terminus of the inositol 1,4,5-trisphosphate receptor in $Ca^{2+}$ release analysed using region-specific monoclonal antibodies. Biochem J 277:125–131

Nuccitelli R, Yim DL, Smart T 1993 The sperm-induced $Ca^{2+}$ wave following fertilization of the *Xenopus* egg requires the production of Ins(1,4,5)$P_3$. Dev Biol 158:200–212

Ozil J-P 1991 The parthenogenetic development of rabbit oocytes after repetitive pulsatile electrical stimulation. Development 109:117–127

Putney JW Jr 1992 Inositol phosphate and calcium signalling. Adv Second Messenger Phosphoprotein Res 26:143–160

Randriamampita C, Tsien RY 1993 Emptying of intracellular $Ca^{2+}$ stores releases a novel small messenger that stimulates $Ca^{2+}$ influx. Nature 364:809–814

Shirakawa H, Nakada K, Honda Y, Miyazaki S 1993 Regulation of calcium influx pathway by inositol 1,3,4,5-tetrakisphosphate in hamster eggs. In: Abstracts of the 32nd International Congress of Physiological Sciences, p 196

Tarone G, Russo MA, Hirsch E et al 1993 Expression of $\beta 1$ integrin complexes on the surface of unfertilized mouse oocyte. Development 117:1369–1375

Taylor CW 1993 Kinetics of inositol 1,4,5-trisphosphate-stimulated $Ca^{2+}$ mobilization. Adv Second Messenger Phosphoprotein Res 26:109–142

Veillette A, Davidson D 1992 *Src*-related protein kinases and T-cell receptor signalling. Trends Genet 8:61–66

Ward CR, Kopf GS 1993 Molecular events mediating sperm activation. Dev Biol 158:9–34

Watanabe N, Vande Woude GF, Ikawa Y, Sagata N 1989 Specific proteolysis of the c-*mos* proto-oncogene product by calpain on fertilization of *Xenopus* eggs. Nature 342:505–511

Whitaker M, Swann K 1993 Lighting the fuse at fertilization. Development 117:1–12

Yamamoto-Hino M, Sugiyama T, Hikichi K et al 1994 Cloning and characterization of human type 2 and 3 inositol 1,4,5-trisphosphate receptors. Recept Channels 2:9–22

## DISCUSSION

*Lu:* Did you block endogenous production of inositol 1,4,5-trisphosphate ($InsP_3$) when you injected $InsP_3$ into eggs? Does such blockade affect the results?

*Miyazaki:* No, we didn't. It is very difficult to block the endogenous phospholipase C activity.

*Thomas:* One prediction from the theory that inositol 1,3,4,5-tetrakisphosphate ($InsP_4$) enhances the re-entry of $Ca^{2+}$ into the egg is that in the presence of $InsP_3(S)_3$, where you get sustained oscillations, $InsP_4$ should increase the frequency of the oscillations because it should speed the refilling process.

*Miyazaki:* Unfortunately, we haven't done combined injections of $InsP_4$ and the non-metabolizable $InsP_3$.

*Putney:* You said that in the absence of external $Ca^{2+}$ you don't get many oscillations. I would guess that you don't get many oscillations with $InsP_3(S)_3$.

*Miyazaki:* I think the stores are full before stimulation, so even in $Ca^{2+}$-free medium we get a few oscillations when we first apply $InsP_3(S)_3$. The oscillations, however, eventually disappear in $Ca^{2+}$-free medium.

*Putney:* They should disappear just as quickly in the presence of $Ca^{2+}$ as in $Ca^{2+}$-free medium if $InsP_3$ is responsible for the entry. Your initial premise was that $Ca^{2+}$ entry is needed to sustain oscillations and the conclusion was that $InsP_4$ is the signal for entry; taking away the signal for entry should be equivalent to taking away $Ca^{2+}$.

*Miyazaki:* $InsP_3(S)_3$ itself probably cannot cause $Ca^{2+}$ influx.

*Putney:* So you are saying that the oscillations are not maintained even in the presence of $Ca^{2+}$.

*Miyazaki:* $InsP_3(S)_3$ cannot produce continuous $Ca^{2+}$ oscillations.

*Berridge:* During the spike and the recovery, some of the $Ca^{2+}$ following each spike gets pumped out of the cell. If no $Ca^{2+}$ is coming back in, these oscillations should run down as if the cell were in $Ca^{2+}$-free medium.

*Sanderson:* But the endogenous phospholipase C (PLC) is still active.

*Miyazaki:* The initial $Ca^{2+}$ responses and later $Ca^{2+}$ responses have different properties. When we inject $InsP_3(S)_3$ we get an initial response with high frequency. There may be some reserve $Ca^{2+}$ stores which have a relatively low sensitivity to $InsP_3$. However, we can't induce $Ca^{2+}$ release from such stores later. The endogenous PLC is not activated enough without the sperm, so that later $Ca^{2+}$ responses depend on $Ca^{2+}$ influx under continuous stimulation by the sperm.

*Friel:* If you remove external $Ca^{2+}$ during the steady-state oscillations, after either fertilization or continuous $InsP_3$ injection, do the oscillations terminate fairly quickly?

*Miyazaki:* No; it takes one or two cycles.

*Friel:* If you take away external $Ca^{2+}$ during the initial phase, do you see more cycles before the oscillations stop?

*Miyazaki:* Yes. The initial responses don't necessarily need $Ca^{2+}$ influx. They probably involve $Ca^{2+}$ release from some further reserves.

*Berridge:* A similar situation exists during stimulation of the liver cell. When a liver cell is activated, its total $Ca^{2+}$ content declines such that it actually has less $Ca^{2+}$ at equilibrium than when it started. The mammalian egg may be similar. It may have a lot of $Ca^{2+}$ in its stores, which is mobilized during the initial long transient. Once the store has been depleted, subsequent transients become dependent on external $Ca^{2+}$.

*Cobbold:* Is it true that a $Ca^{2+}$-free medium will activate mouse eggs parthogenetically?

*Jaffe:* In many other kinds of eggs, it is possible to induce parthenogenesis with what is called $Ca^{2+}$-free medium in the old literature, but only after the egg is returned to a $Ca^{2+}$-containing medium (Jaffe 1985). I infer that the $Ca^{2+}$-free medium somehow permeabilizes the egg so that when it is returned into media containing $Ca^{2+}$ there is an abnormal $Ca^{2+}$ entry which triggers a fertilization wave.

*Miyazaki:* The initial response can occur in $Ca^{2+}$-free medium because it depends mainly on $Ca^{2+}$ release. The intial response is enough to cause cortical granule exocytosis and progression of the cell cycle.

*Cobbold:* That's when the sperm is there?

*Miyazaki:* That's right. Without sperm, mouse eggs can be activated by keeping them in $Ca^{2+}$-free and $Mg^{2+}$-free medium or $Ca^{2+}$-free medium, although extrusion of the second polar body is not seen in most cases (Surani & Kaufman 1977). In our experiments in hamster eggs, $[Ca^{2+}]_i$ was increased little when $Ca^{2+}$-free medium was introduced or $Ca^{2+}$-free medium was replaced by normal medium. Efficient parthenogenetic activation of mammalian eggs can be achieved by applying high-voltage pulses to induce repetitive $Ca^{2+}$ transients which mimick the $Ca^{2+}$ spikes occurring at fertilization. In rabbit eggs, the magnitude and frequency of electrical pulses affect later development at stages such as the rate of compaction or blastocyst formation, or the rate of explantation to a host mother (Ozil 1991). You can make the embryo develop up to 10 days of pregnancy.

*Irvine:* You said that injection of $InsP_4$ caused repetitive $Ca^{2+}$ release but that $InsP_4$ is less effective than $InsP_3$. One might expect the two compounds to have a similar biological potency if $InsP_4$ is biologically relevant. In some other systems the $InsP_4$ does seem to get some help from $InsP_3$, and its effects can even depend on $InsP_3$. Have you tried using lower doses of $InsP_4$ in conjunction with a non-metabolizable $InsP_3$ analogue?

*Miyazaki:* We had to inject more $InsP_4$ (3 nA current through a micropipette containing 500 $\mu$M $InsP_4$) than $InsP_3$ (0.2 nA and 200 $\mu$M); $InsP_3$ can easily access its target, the endoplasmic reticulum. In contrast, injection of a much larger amount of $InsP_4$ should be necessary to cause a measurable increase in the basal $[Ca^{2+}]_i$ through activation of $Ca^{2+}$ influx over the large area of the plasma membrane. We haven't yet injected $InsP_4$ together with $InsP_3(S)_3$. We have injected $InsP_4$ together with a small amount $InsP_3$; this combined injection didn't enhance the increase in the basal $[Ca^{2+}]_i$ remarkably.

*Putney:* Does it matter which isomer of $InsP_4$ you use?

*Miyazaki:* I haven't looked at the effects of different isomers yet.

*Jaffe:* You think that the later spikes occur synchronously within the whole egg. Would your method of observation have enabled you to have seen a wave? Did you look at short enough intervals?

*Miyazaki:* The minimum time resolution we tried was 250 ms with fura-2 and 30 ms with aequorin. Within that interval we have never seen spreading $Ca^{2+}$ rises in later $Ca^{2+}$ spikes.

*Jaffe:* In other eggs of that size, a postfertilization wave takes two to three seconds to cross from pole to pole.

*Berridge:* That's what Dr Miyazaki saw during the initial transients.

*Jaffe:* But he didn't see it later.

*Berridge:* It is possible that during the first few transients the sperm imposes a directionality on the signalling system to give clearly defined waves. However, as the influence of the sperm begins to disperse throughout the whole egg, the signals become much more synchronous. A similar phenomenon occurs during the action of thimerosal. Thimerosal induces this kind of synchronous spiking activity because it acts by sensitizing all the stores to create a uniform excitable medium.

*Putney:* That's true with $InsP_3$, which diffuses so quickly that during normal spiking and wave generation there are no $InsP_3$ gradients.

*Berridge:* But when Dr Miyazaki injects $InsP_3$ at one point he can generate a wave. I suspect that the $InsP_3$ can trigger a wave before it actually diffuses.

*Putney:* I'm just saying that in classical models of waves in hepatocytes and anywhere else, the sensitizing stimulus, presumably $InsP_3$, has to be uniformly distributed, yet the wave propagates because it's $Ca^{2+}$ which carries the wave.

*Berridge:* What would happen in the hepatocyte if you took all the vasopressin receptors and put them in one little locus?

*Thomas:* We don't know that's not the case, actually.

*Berridge:* If that is the case, you have to explain why stimulation with fluoride also reveals a distinct initiation point.

*Thomas:* But in the case of $Ca^{2+}$ waves elicited by aluminium fluoride you need only postulate that the G protein-linked phospholipase C is co-localized with the hormone receptors in the $Ca^{2+}$ wave initiation domain of the cell.

*Miyazaki:* I should comment on an important point. Injection of sperm extract causes continuous $Ca^{2+}$ oscillations just like the $Ca^{2+}$ responses seen at fertilization (Swann 1990). The remarkable effect of sperm extract is to sensitize $Ca^{2+}$-induced $Ca^{2+}$ release (Swann 1990). If you inject $Ca^{2+}$ into these eggs by any means, you get a very quick, large explosion of $Ca^{2+}$. The sensitivity to $Ca^{2+}$ also increases about 10-fold in fertilized hamster eggs (Igusa & Miyazaki 1983). The later response may be affected by an excitatory sperm factor to produce a synchronous $Ca^{2+}$ rise.

*Jaffe:* Do you think it's possible that with the later pulses, which seem to be synchronous, that you might have a wave which comes in from the whole surface and moves to the centre?

*Miyazaki:* It's hard to detect a gradient of cytosolic $[Ca^{2+}]$ in the later response.

*Berridge:* Even with confocal microscopy?

*Miyazaki:* Yes. We used the dye calcium green, not ratio imaging, so that I can't be exact about the $[Ca^{2+}]_i$ rise. The fluorescence intensity tended to be higher in the cortical area than in the deep cytoplasm but, at present, I hesitate to conclude that the $[Ca^{2+}]_i$ rise in the cortex precedes that in the centre.

*Whitaker:* We've made confocal measurements of $Ca^{2+}$ in immature and mature mouse oocytes. We see absolutely no spatial inhomogeneity in the later oscillations after fertilization or those in immature oocytes. Obviously, you can never rule out completely the possibility that the transient is a wave because of finite time resolution; you can estimate only a minimum velocity. We think that in mouse oocytes the minimum velocity is more than 500 $\mu$m/s, which means it's not really a wave, because the oocyte is only 100 $\mu$m in diameter.

*Jaffe:* Would you have seen a radial wave?

*Whitaker:* Yes.

*Nathanson:* Dr Miyazaki, when you inject a non-metabolizable $InsP_3$ analogue, do you also see these synchronized oscillations?

*Miyazaki:* With our point injection method we get propagating $Ca^{2+}$ rises two or three times, then we get synchronous $Ca^{2+}$ rises.

## References

Igusa Y, Miyazaki S 1983 Effects of altered extracellular and intracellular calcium concentration on hyperpolarizing responses of the hamster egg. J Physiol 340:611–632

Jaffe LF 1985 The role of calcium explosions, waves and pulses in activating eggs. In: Metz CB, Monroy A (eds) Biology of fertilization, vol 3: The fertilization response of the egg. Academic Press, Orlando, FL, p 144

Ozil J-P 1991 The parthenogenetic development of rabbit oocytes after repetitive pulsatile electrical stimulation. Development 109:117–127

Surani MA, Kaufman MH 1977 Influence of extracellular $Ca^{2+}$ and $Mg^{2+}$ on the second meiotic division of mouse oocytes: relevance to obtaining haploid and diploid parthenogenetic embryos. Dev Biol 59:86–90

Swann K 1990 A cytosolic sperm factor stimulates repetitive calcium increases and mimics fertilized hamster eggs. Development 110:1295–1303

# Regulation of nuclear calcium concentration

Tobias Meyer, Nancy L. Allbritton* and Elena Oancea

*Department of Cell Biology, Duke University Medical Center, Durham, NC 27710 and *Department of Neurobiology, Stanford University Medical School, Stanford, CA 94305, USA*

*Abstract.* Transient increases in nuclear calcium concentration have been shown to activate gene expression and other nuclear processes. It has been suggested that nuclear calcium signals are controlled by a mechanism that is independent of calcium signalling in the cytosol. This would be possible if calcium diffusion is slow and a separate calcium release mechanism is localized to the nuclear region. Alternatively, the nuclear envelope could act as a diffusion barrier for calcium ions released either inside or outside the nucleus. It has also been proposed that inositol 1,4,5-trisphosphate ($InsP_3$) can be generated inside the nucleus and that there are calcium release channels in the inner membrane of the nuclear envelope. Most of the experimental evidence supporting these hypotheses is based on the calibration of nuclear and cytosolic calcium concentrations. However, recent studies suggest that the local calibration of calcium indicators may not be accurate. We propose that nuclear calcium signals can be investigated by a different approach that does not rely on accurate calibration of indicators. We have developed calcium indicators that minimize facilitated calcium diffusion and are localized to either the nucleus or the cytosol. Using the diffusion coefficient of calcium ions, and measuring the delay between cytosolic and nuclear calcium increases, we show that the nuclear envelope is not a substantial barrier for calcium ions in PC12 (phaeochromocytoma) cells. This suggests that nuclear and cytosolic calcium signals equilibrate rapidly in these cells.

*1995 Calcium waves, gradients and oscillations. Wiley, Chichester (Ciba Foundation Symposium 188) p 252–266*

Previous studies have suggested that nuclear calcium signals mediate gene transcription, DNA synthesis and DNA repair as well as other nuclear functions (Bachs et al 1992, Karin 1992). In one pathway, $Ca^{2+}$–calmodulin mediates the phosphorylation of the transcription factors cyclic AMP responsive element-binding protein (CREB) and C/EBP$\beta$, probably through activation of nuclear calmodulin kinase II (Sheng et al 1990, Wegner et al 1992, Peunova & Enikolopov 1993). Such studies of nuclear calcium functions have generated a wide interest in the mechanism by which nuclear calcium concentration is regulated. Three approaches have been used to investigate nuclear calcium signalling: (1) nuclear and cytosolic calcium concentrations have been measured with fluorescent $Ca^{2+}$ indicators (Williams et al 1987, Hernandez-Cruz et al 1990, Przywara et al 1991, Waybill et al 1991, Yamada et al 1991, Birch et al 1992, Nakato et al 1992, Furuno et al 1993, Al-Mohanna et al 1994); (2) the uptake and release of $^{45}Ca^{2+}$ by isolated nuclei has been characterized (Malviya et al 1990, Matter et al 1993); and (3) enzymes of the phosphoinositide and calcium signalling pathways have been localized to the nucleus in biochemical and immunological studies (Payrastre et al 1992, Martelli et al 1992).

Calcium imaging studies have suggested that calcium concentrations in the nucleus and cytosol can differ for extended periods. In hepatocytes, some neurons, tumour mast cells and other cells, peak nuclear calcium concentrations during calcium transients have been reported to be higher than peak concentrations in the cytosol (Hernandez-Cruz et al 1990, Przywara et al 1991, Waybill et al 1991, Yamada et al 1991, Birch et al 1992, Nakato et al 1992, Furuno et al 1993). This led to the suggestion that there are calcium channels in the inner membrane of the nuclear envelope. In smooth muscle cells and some neurons, maximal nuclear calcium concentrations were reported to be lower than those in the cytosol (Williams et al 1987, Al-Mohanna et al 1994). A main conclusion of all these studies was that the nuclear envelope can act as a diffusional barrier for $Ca^{2+}$, allowing calcium gradients across the nuclear envelope to be sustained for prolonged periods. However, the dependence of the properties of calcium indicators on the local environment complicates their calibration and interferes with the interpretation of these studies (Hove-Madsen & Bers 1992, Blatter & Wier 1990, Connor 1993). A similar barrier function for the nuclear envelope was suggested in studies of $^{45}Ca^{2+}$ flux measured in nuclei (Malviya et al 1990, Matter et al 1993). However, $Ca^{2+}$ release sites could exist at the inner or outer membrane of the nuclear envelope, or calcium could be released from calcium stores near the nucleus. The unknown location of the $Ca^{2+}$ release sites complicates the interpretation of these experiments.

The identification of phosphatidylinositol-phosphate kinase and phospholipase C$\beta$ inside the nucleus (Martelli et al 1992, Payrastre et al 1992) led to the suggestion that nuclear phosphatidylinositol 4,5-bisphosphate is used as a substrate by phospholipase C for the production of nuclear inositol 1,4,5-trisphosphate ($InsP_3$). It is conceivable that the $InsP_3$ receptor localized in the inner membrane of the nuclear envelope could mediate nuclear functions of $InsP_3$.

We shall first discuss our recent measurements of calcium mobility, indicating that the relatively slow diffusion of calcium is still too fast to explain the existence of localized nuclear calcium signals, which means that differences between nuclear and cytosolic calcium concentrations can be explained only if the nuclear membrane is a substantial barrier for calcium ions. We shall then concentrate on the various hypotheses that have been proposed for the mechanism of nuclear calcium signalling and discuss a new approach to test them. A cytosolic and a nuclear calcium indicator are introduced as tools to elucidate the mechanisms of nuclear calcium signalling (Allbritton et al 1994). We have also developed an inhibitor of the $InsP_3$ receptor that is largely excluded from the nucleus (Allbritton et al 1994). Measurements with these new probes in PC12 (rat adrenal phaeochromocytoma) cells and other cell types suggest that the nuclear envelope is not a substantial barrier for calcium ions and that $InsP_3$ does not release calcium from inside the nucleus.

## Calcium diffusion and localized calcium signals

To improve understanding of localized calcium signals in the absence of membrane barriers, we determined the buffering and diffusion of calcium in the cytosol (Allbritton et al 1992). The diffusion of $^{45}Ca^{2+}$ was determined by measuring the movement of a point source of $^{45}Ca^{2+}$ within a tube filled with cytosolic extract of *Xenopus* oocytes. We found that the diffusion coefficient of $Ca^{2+}$ is much lower in the cytosol (around 10–30 $\mu m^2/s$) than in an aqueous solution (400 $\mu m^2/s$). The viscosity of cytosolic extracts from *Xenopus* oocytes is about twice that of water but the greater viscosity of the cytosol is insufficient to explain this large reduction in mobility (Allbritton et al 1992). The reduction in mobility could be explained by the buffering of calcium by calcium-binding proteins and by other negatively charged molecules such as phospholipids. Sequestration into calcium stores occurs on a much slower time scale and is not included in this definition of calcium buffering. Buffering of calcium appears to vary in different cell types, ranging from $B \approx 30$ (*Xenopus*; Connor 1993) to $B \approx 1000$ (neutrophils; von Tscharner et al 1986). This buffering ratio changes with the free calcium concentration as shown in **1**.

$$B([Ca^{2+}]_{free}) = \delta[Ca^{2+}]_{total} / \delta[Ca^{2+}]_{free} \qquad \mathbf{1}$$

We estimated that the period during which unbuffered calcium diffusion occurs is as short as 30 $\mu s$ in *Xenopus* oocyte extracts, defining a 0.1 $\mu m$ range for unbuffered calcium movement (Allbritton et al 1992). Over longer periods, calcium diffusion is much slower owing to calcium buffering. In 200 ms, buffered calcium travels an average of 2 $\mu m$. Thus, if there is a similar calcium diffusion coefficient in nucleoplasm and cytoplasm and if calcium diffuses freely out of the nucleus, calcium signals generated in the nucleus could not be localized to the nucleus for more than a few hundred milliseconds. If nuclear calcium signals do indeed differ from cytosolic ones for about one second or longer, the nuclear envelope must impose a substantial diffusional barrier for calcium ions.

The diffusion studies in *Xenopus* oocyte extracts also showed that the results of localized calcium imaging studies can be affected by the calcium indicators

(Allbritton et al 1992). Calcium diffusion can be accelerated by introduction of fast-diffusing calcium-binding molecules. The calcium indicator fura-2 (50 $\mu$M) increased calcium diffusion to 75 $\mu m^2/s$. This phenomenon of acceleration of calcium diffusion by a mobile buffer has been termed facilitated calcium diffusion. These experiments suggest that a large fraction of the native calcium-buffering molecules either are not diffusing or are diffusing at a slow rate. It should be mentioned that the experimental use of calcium indicators also alters calcium responses by increasing the cell's buffering capacity. Increased calcium buffering slows rapid calcium rises and may reduce peak calcium concentrations.

Facilitated calcium diffusion can be minimized by using calcium indicators of large molecular size. Dextran-coupled calcium indicators are available in a range of molecular sizes and hence are useful in studies of localized calcium signals. In addition, the dextrans are not sequestered by intracellular organelles, but their disadvantage is that they have to be microinjected.

## Models for the regulation of nuclear calcium signals

Several different models have been proposed for the control of nuclear calcium concentration. They give different answers to two fundamental questions: (1) how permeable are nuclear pores to calcium ions and can the permeability to calcium be regulated; and (2) are calcium channels or pumps localized to the inner membrane of the nuclear envelope (Fig. 1)? It is possible that the answers to these questions depend on cell type.

Studies of nuclear pores have established that nuclear import of molecules larger than about 60 kDa requires ATP and is temperature dependent (Paine et al 1975, Silver 1991). In contrast, cytosolic molecules of 20–50 kDa molecular mass can diffuse into the nucleus in an energy-independent manner (Paine et al 1975, Silver 1991). The equilibration of these small molecular mass proteins and dextrans was found to be on a time scale of a minute and in some cases much longer. Thus, although the nuclear envelope is not impermeant to calcium ions, it is conceivable that it is a substantial barrier to calcium ions. This might allow calcium gradients between the nucleus and the cytosol to be established for short periods.

What are the parameters that describe the rate of calcium equilibration across the nuclear envelope? The movement of calcium into the nucleus can be described by the permeability, P ($\mu$m/s), of the nuclear membrane to calcium ions, as shown in **2**, where V is the volume of the nucleus, A is the surface

$$d[Ca_N^{2+}]/dt = A \times P \times ([Ca_C^{2+}] - [Ca_N^{2+}])/V \qquad \mathbf{2}$$

area of the nucleus, and $Ca_C$ and $Ca_N$ are the concentration of calcium in the cytosol and nucleus, respectively. Alternatively, the movement of calcium

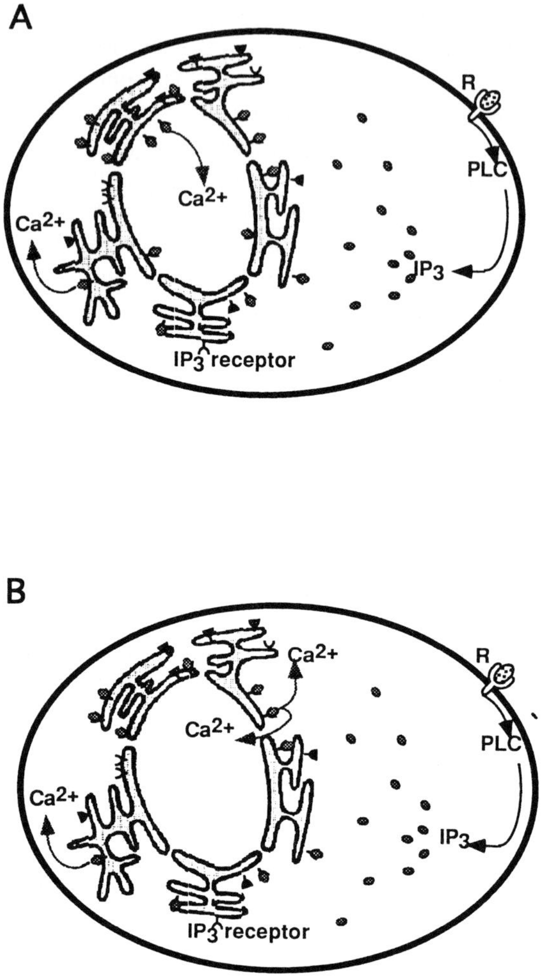


FIG. 1. Schematic representations of two models of nuclear calcium regulation. (A) Nuclear calcium ions originate from calcium channels at the inner nuclear membrane. (B) Nuclear calcium ions are released in the cytosol and diffuse into the nucleus through nuclear pores. R, agonist receptor; PLC, phospholipase C; $IP_3$, inositol 1,4,5-trisphosphate.

into the nucleus can be described by a permeability coefficient for a nuclear pore, $p_{np}$ ($\mu m^3/s$) as shown in **3**, where N is the number of nuclear pores

$$d[Ca^{2+}_N]/dt = A \times p_{np} \times ([Ca^{2+}_C] - [Ca^{2+}_N])/N \qquad \mathbf{3}$$

and $p_{np} = V \times P/N$. If the cytosol is substantially larger than the nucleus and active calcium transport mechanisms are not present in the nucleus, equation **2** can be solved as shown in **4**, where $\tau$ is the relaxation time constant for the nuclear calcium concentration following an increase in the cytosolic calcium concentration

$$[Ca_N^{2+}] = [Ca_N^{2+}]_o + ([Ca_C^{2+}]_o - [Ca_N^{2+}]_o)\exp(-t/\tau) \qquad \mathbf{4}$$

and $\tau = V/(P \times A)$; $[Ca_C]_o$ is the calcium concentration in the cytosol after a step increased and $[Ca_N]_o$ is the initial nuclear calcium concentration.

For a spherical nucleus of 6 $\mu$m diameter, $V/A \approx 1\ \mu m$. If the measured equilibration time ($\tau$) between the nucleus and cytosol is about one second for a particular molecule, the permeability value becomes $P = 1\ \mu m/s$. Earlier studies have measured the permeability values for molecules larger than calcium. One determined that a 3 kDa dextran has a $P \approx 0.2\ \mu m/s$ and a 20 kDa dextran a $P \approx 0.03\ \mu m/s$ (Paine et al 1975; for review see Peters 1986). For our model nucleus, the corresponding equilibration times ($\tau$) for these dextrans would be 5 s and 35 s, respectively. If 10 000 nuclear pores were present per nucleus, the permeability per pore would be: $p_{np} = 7 \times 10^{-4}\ \mu m^3/s$ (3 kDa dextran) and $p_{np} = 4 \times 10^{-5}\ \mu m^3/s$ (20 kDa dextran).

If the nuclear pores limit access by size alone, these quantitative considerations indicate that the permeability value for calcium should be higher than about 0.2 $\mu$m/s because of the calcium ion's small molecular size. However, these estimates do not include the buffering of calcium ions in the nucleus, which could substantially reduce the permeability value, as shown in **5**, where B is the nuclear calcium

$$\tau = (P \times A)/(B \times V) \qquad \mathbf{5}$$

buffering ratio as discussed above (equation **1**). These considerations leave open the possibility that nuclear calcium signals equilibrate with the cytosol on a time scale of many seconds. Active calcium pump mechanisms and a potential down-regulation of the permeability of nuclear pores could further help to sustain calcium gradients.

If calcium channels and calcium pumps at the inner membrane of the nuclear envelope are to have functional significance, the nuclear envelope must impose a substantial diffusion barrier. If this is the case, the direct release of calcium into the nucleus could explain the reported higher nuclear than cytosolic calcium signals. Furthermore, the observation of lower nuclear than cytosolic signals could be explained as a dampening of cytosolic calcium signals by the nuclear envelope. To investigate the mechanism of nuclear calcium signalling, we have made new calcium indicators and a cytosolic inhibitor of the $InsP_3$ receptor (Allbritton et al 1994).

## Design of cytosolic and nuclear calcium indicators

Dextran-based calcium indicators can be used to reduce the spreading of localized calcium transients by facilitated diffusion. In addition, large molecular

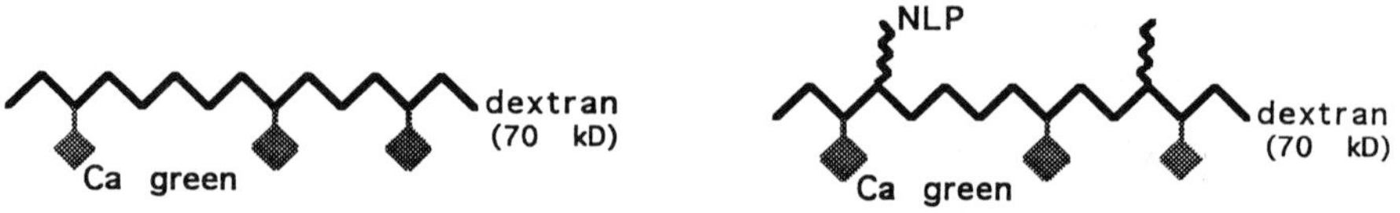


FIG. 2. Schematic representation of a cytosolic (*left*) and a nuclear (*right*) calcium indicator. Dextran–calcium green of molecular mass greater than about 70 kDa is excluded from the nucleus. Dexrans coupled to calcium green and a nuclear localization peptide (NLP) (bifunctional dextrans) are rapidly accumulated in the nucleus after microinjection into the cytosol.

mass dextrans are excluded from the nucleus and can be used as cytosolic calcium indicators (Fig. 2, *left*). When we removed the small molecular mass components from 70 kDa and 500 kDa dextran–calcium green dyes (provided by M. Kuhn, Molecular Probes), these dyes were excluded from the nucleus. A Centricon-100 (Amicon) appears to be suitable for removing the small molecular mass components that can freely diffuse into and out of the nucleus. In the studies described below, these dextran–calcium greens of large molecular mass were used as cytosolic calcium indicators.

Nuclear calcium indicators were made by synthesizing a bifunctional dextran in which calcium green and a nuclear localization peptide are coupled to the same dextran matrix (Fig. 2, *right*) (Allbritton et al 1994). On average, more than three nuclear localization peptides were attached to each 70 kDa dextran molecule. When this indicator was injected into developing zebrafish embryos (E. Reinhard, H. Yokoe, N. L. Allbritton, M. Kuhn & T. Meyer, unpublished work 1994), rat basophilic leukaemia (RBL) cells (Allbritton et al 1994), PC12 cells or myocytes (E. Oancea & T. Meyer, unpublished work 1994), the indicator localized to the respective nuclei within minutes.

## Measurements of nuclear calcium concentration

We analysed the delays between cytosolic and nuclear calcium transients using the cytosolic and nuclear calcium indicators described above. PC12 cells were microinjected with a mixture of the nuclear and cytosolic calcium green–dextran indicators to give a few-fold higher nuclear than cytosolic fluorescence signal, so that the nucleus could be located during the calcium recordings. We assembled a computer-controlled flow chamber to generate two-second depolarizing pulses of KCl, producing rapid increases in the cytosolic free calcium concentration. Nuclear calcium responses followed the cytosolic ones within a few hundred milliseconds (Fig. 3). Thus, nuclear pores in PC12 cells appear to be highly permeable to calcium ions. The apparent permeability value is greater than 2 $\mu$m/s.

A

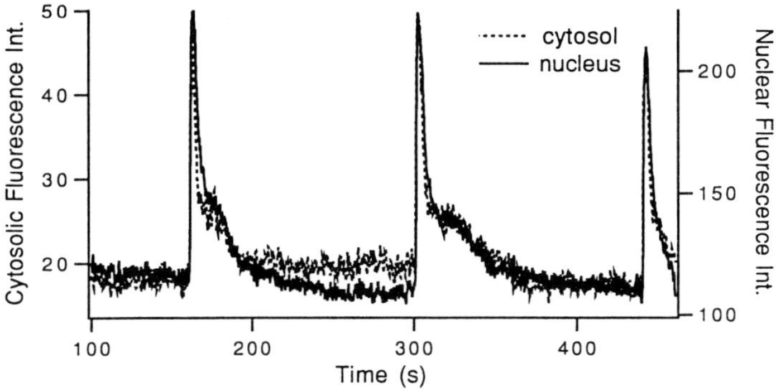


B

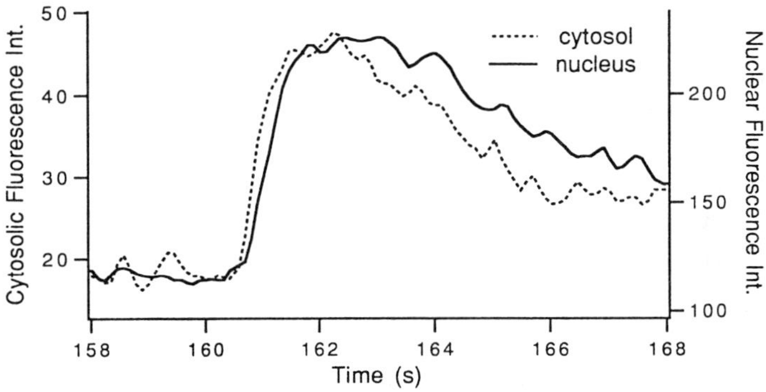


FIG. 3. Nuclear and cytosolic changes in [$Ca^{2+}$] in a PC12 cell following brief depolarizing pulses. PC12 cells were microinjected with a mixture of the nuclear and cytosolic forms of dextran–calcium green (see Fig. 2). The mixture was chosen to produce a higher nuclear than cytosolic indicator concentration. Fluorescence traces were recorded with a Noran Odyssey imaging system. A computer-controlled flow chamber was used to switch the extracellular buffer from 135 mM NaCl/5 mM KCl to 5 mM NaCl/135 mM KCl. (A) The depolarizing pulses led to rapid transient increases in the cytosolic calcium concentration. (B) A time expansion of (A). The increase in nuclear calcium concentration lagged behind that in the cytosol by a few hundred milliseconds.

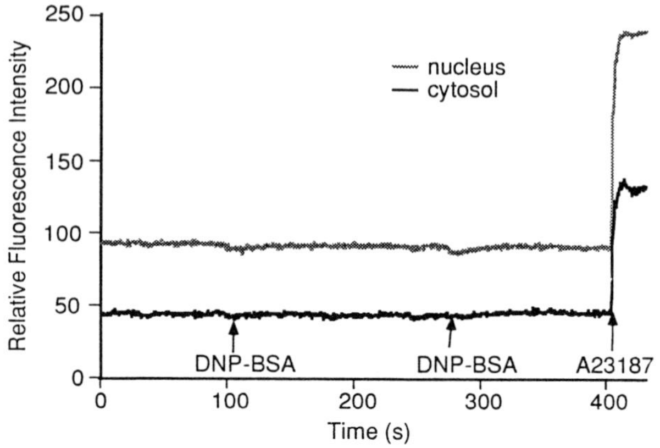


FIG. 4. Effects on calcium concentration of antigen stimulation of rat basophilic leukaemia cells microinjected with dextran–heparin. Without this cytosolic inhibitor of the inositol 1,4,5-trisphosphate receptor, this antigen stimulus would lead to repetitive calcium spikes. In the presence of dextran–heparin, both cytosolic and nuclear calcium responses to antigen stimulation are blocked. DNP-BSA, dinitrophenol coupled to bovine serum albumin; A23187, a calcium ionophore.

We used the same strategy of nuclear exclusion to make an inhibitor of the $InsP_3$ receptor that is excluded from the nucleus (Allbritton et al 1994). Instead of calcium green, we coupled heparin to the dextran matrix. Figure 4 shows that the microinjection of this inhibitor blocks not only cytosolic, but also nuclear calcium responses to receptor-mediated $InsP_3$ production. In a subsequent study in RBL cells, we showed that this cytosolic inhibitor of the $InsP_3$ receptor can block nuclear calcium responses following the uncaging of $InsP_3$ (Allbritton et al 1994).

## Summary

These studies of nuclear calcium signalling suggest that the nuclear envelopes of PC12 cells and RBL cells have a calcium permeability of greater than 2 $\mu$m/s. Thus, the nuclear envelope in these cells is not a substantial barrier to calcium ions. This makes it unlikely that nuclear and cytosolic calcium concentration gradients exist for time scales of one second or longer. In addition, our measurements do not support the hypothesis that functional $InsP_3$ receptors exist inside the nucleus.

The results of this study are not in agreement with several previous studies which suggested that there are independent calcium signalling mechanisms inside the nucleus. It is conceivable that the previously observed prolonged differences between nuclear and cytosolic calcium concentrations can be explained by difficulties in the local calibration of calcium indicators.

*Acknowledgements*

Supported by the David and Lucile Packard Foundation, an NIH grant (GM48113 to T. Meyer) and a National Institute of Mental Health grant (MH45324 to L. Stryer).

## References

Allbritton NL, Meyer T, Stryer L 1992 Range of messenger action of calcium ion and inositol 1,4,5-trisphosphate. Science 258:1812–1815

Allbritton NL, Oancea E, Kuhn M, Meyer T 1994 Source of nuclear calcium signals. Proc Natl Acad Sci USA, in press

Al-Mohanna FA, Caddy KWT, Bolsover SR 1994 The nucleus is insulated from large cytosolic calcium ion changes. Nature 367:745–750

Bachs O, Agell N, Carafoli E 1992 Calcium and calmodulin function in the cell nucleus. Biochim Biophys Acta 1113:259–270

Birch BD, Eng DL, Kocsis JD 1992 Intranuclear $Ca^{2+}$-transients during neurite regeneration of an adult mammalian neuron. Proc Natl Acad Sci USA 89:7978–7982

Blatter LA, Wier WG 1990 Intracellular diffusion, binding and compartmentalization of the fluorescent calcium indicators indo-1 and fura-2. Biophys J 58:1491–1499

Connor JA 1993 Intracellular calcium mobilization by inositol 1,4,5-trisphosphate: intracellular movement and compartmentalization. Cell Calcium 14:185–200

Furuno T, Hamano T, Nakanishi M 1993 Receptor-mediated calcium signal playing a nuclear third messenger in the activation of antigen-specific B-cells. Biophys J 64: 665–669

Hernandez-Cruz A, Sala F, Adams PR 1990 Subcellular calcium transients visualized by confocal microscopy in a voltage-clamped vertebrate neuron. Science 247: 858–862

Hove-Madsen L, Bers DM 1992 Indo-1 binding to protein in permeabilized ventricular myocytes alters its spectral and calcium binding properties. Biophys J 63:89–97

Karin M 1992 Cell signal transduction from cell surface to nucleus in development and disease. FASEB (Fed Am Soc Exp Biol) J 6:2581–2590

Malviya AN, Rogue P, Vincendon G 1990 Stereospecific inositol 1,4,5-[$^{32}P$]-trisphosphate binding to isolated rat liver nuclei: evidence for inositol trisphosphate receptor-mediated calcium release from the nucleus. Proc Natl Acad Sci USA 87: 9270–9274

Martelli AM, Gilmour RS, Bertagnolo V, Neri LM, Manzoli L, Cocco L 1992 Nuclear localization and signalling activity of phosphoinositidase $C_{\beta}$ in Swiss 3T3 cells. Nature 358:242–245

Matter N, Ritz M, Freyermuth S, Rogue P, Malviya AN 1993 Stimulation of nuclear protein kinase C leads to phosphorylation of nuclear inositol 1,4,5 trisphosphate receptor and accelerated calcium release by inositol 1,4,5 trisphosphate from isolated rat liver nuclei. J Biol Chem 268:732–736

Nakato K, Furuno T, Inagaki K, Teshima R, Terao T, Nakanishi M 1992 Cytosolic and intranuclear calcium signals in rat basophilic leukaemia cells as revealed by a confocal fluorescence micropscope. Eur J Biochem 209:745–749

Paine PL, Moore LC, Horowitz SB 1975 Nuclear envelope permeability. Nature 254: 109–114

Payrastre B, Nievers M, Boonstra J et al 1992 A differential location of phosphoinositide kinases, diacylglycerol kinase, and phospholipase C in the nuclear matrix. J Biol Chem 267:5078–5084
Peters R 1986 Fluorescence microphotolysis to measure nucleocytoplasmic transport and intracellular mobility. Biochim Biophys Acta 864:305–359
Peunova N, Enikolopov G 1993 Amplification of calcium-induced gene transcription by nitric oxide in neuronal cells. Nature 364:450–453
Przywara DA, Bhave SV, Bhave A, Wakade TD, Wakade AR 1991 Stimulated rise in neuronal calcium is faster and greater in the nucleus than in the cytosol. FASEB (Fed Am Soc Exp Biol) J 5:217–222
Sheng M, McFadden G, Greenberg ME 1990 Membrane polarization and calcium induce c-*fos* transcription via phosphorylation of transcription factor CREB. Neuron 4:571–582
Silver PA 1991 How proteins enter the nucleus. Cell 64:489–497
von Tscharner V, Deranleau DA, Baggiolini M 1986 Calcium fluxes and calcium buffering in human neutrophils. J Biol Chem 261:10163–10168
Waybill MM, Yelamarty RV, Zhang YL et al 1991 Nuclear calcium gradients in cultured rat hepatocytes. Am J Physiol 261:E49–E57
Wegner M, Cao Z, Rosenfeld MG 1992 Calcium-regulated phosphorylation within the leucine zipper of C/EBP$\beta$. Science 256:370–373
Williams DA, Becker PL, Fay FS 1987 Regional changes in calcium underlying contraction of smooth muscle cells. Science 235:1644–1648
Yamada H, Mizuguchi J, Nakanishi M 1991 Antigen receptor-mediated calcium signals in B cells as revealed by confocal fluorescence microscopy. FEBS (Fed Eur Biochem Soc) Lett 284:249–251

## DISCUSSION

*Nathanson:* If there's no inositol 1,4,5-trisphosphate ($InsP_3$)-sensitive $Ca^{2+}$ release mechanism in the nucleus, I guess you would predict that $Ca^{2+}$ will travel across the nucleus purely by diffusion at a rate that is dependent on the size of the nucleus and on local uptake mechanisms. Calcium might even be able to diffuse all the way to the centre of the nucleus.

*Meyer:* In rat basophilic leukaemia (RBL) cells the nuclei aren't particularly large, but they have a U-shaped appearance. There is probably no point in the nucleus that is more than 2 $\mu$m away from the nuclear envelope.

*Thomas:* We've looked at propagation of waves through the hepatocyte nucleus (Lin et al 1994), which is a nice, spherical nucleus that occupies more than 80% of the nuclear region in the *z* direction of cells in primary culture. The $Ca^{2+}$ waves propagate through the nucleus at a rate that is not slower than that in the cytoplasm and may even be a little faster. This suggests either that there's less $Ca^{2+}$ buffering in the nucleus than in the cytosol, allowing a full response to a relatively small quantity of $Ca^{2+}$ entering around the nuclear envelope, or that there is some form of amplification going on that we're not seeing. One possibility is that the nuclear envelope is a $Ca^{2+}$ store (we have data showing that) which is $InsP_3$ sensitive. Your results clearly show that $InsP_3$

does not release $Ca^{2+}$ directly into the nucleus, but the nuclear envelope may, within its local environment, be a large $Ca^{2+}$ store that provides extra $Ca^{2+}$ for transfer into the nucleoplasm, which has no stores itself.

*Meyer:* We don't know enough about perinuclear membrane structures in RBL cells to decide whether this could be the case.

*Thomas:* The hepatocyte nuclear envelope itself is an $InsP_3$-sensitive $Ca^{2+}$ pool.

*Fay:* If the nuclear envelope is an $InsP_3$-sensitive $Ca^{2+}$ store, would you not be short-circuiting the ability of that store to regulate [$Ca^{2+}$] locally when you open $InsP_3$-sensitive $Ca^{2+}$ channels? To look at the issue of the way in which $Ca^{2+}$ moves across that complex barrier involving pumps and channels, you should investigate transit time when you release caged $Ca^{2+}$.

*Meyer:* The closest to that was our experiment in PC12 cells in which we gave a depolarizing stimulus (Fig. 3). We have a computer-controlled flow chamber into which we can flow KCl for a brief time and then take it off. Again, with this protocol the cytosolic and nuclear responses appear to be identical.

*Fay:* Was this done with non-confocal microscopy?

*Meyer:* Yes; to record nuclear signals, we use a lot more nuclear dye than cytosolic dye, which minimizes the cytosolic contribution. We have also done confocal measurements. Unfortunately, this reduces the signal:noise ratio for fast kinetics measurements. From what we can see, the results with the confocal microscope are similar.

*Fay:* If, instead of doing a delay analysis, you look at the relationship of the amplitude of the signals in the two compartments, comparing on a population basis the dextran–heparin cells with the non-heparin-treated cells, do the $Ca^{2+}$ signals move up and down together in the same way?

*Meyer:* We haven't done an amplitude analysis because we do not trust the accuracy of the calibrations in the cytosol and the nucleus. However, we do not believe that the nuclear envelope in the cells we studied can dampen nuclear $Ca^{2+}$ responses. An additional argument against this is that any type of dampening would lead to delays in the nuclear versus cytosolic responses. In a large number of experiments, we have never found a delay significantly slower than 200 ms. Al-Mohanna et al (1994) have proposed that the nuclear envelope closes above 300 nM. However, these measurements did not show a delay in the nuclear response, relying instead on a calibration protocol for the amplitude of nuclear and cytosolic signals.

*Fay:* What [$Ca^{2+}$] have you achieved in your experiments?

*Meyer:* We went up to 700–800 $\mu$M in PC12 cells and couldn't see a delay.

*Cannell:* We have shown that the $Ca^{2+}$ transients in myocytes are essentially spatially uniform at the peak of the transient (Cannell et al 1987). Within 80 ms, the $Ca^{2+}$ becomes distributed everywhere. In our confocal imaging, we've also seen essentially the same thing. The nuclear compartment simply lags behind the cytoplasmic compartment and is ultimately indistinguishable in terms of free [$Ca^{2+}$].

*Meyer:* So you also don't find that the nuclear envelope is cutting out $Ca^{2+}$ ions at high $[Ca^{2+}]$?

*Fay:* Dr Wier's $Ca^{2+}$ waves in cardiac cells pass around and don't seem to penetrate the nucleus.

*Wier:* Our images are non-confocal, but we agree with what Dr Cannell said. There is a delay between cytoplasmic and nuclear $Ca^{2+}$ during a normal $Ca^{2+}$ transient but eventually nuclear and cytosolic $[Ca^{2+}]$ seem to reach the same level.

*Lechleiter:* Hasn't there been a report that $Ca^{2+}$ waves slow down at the nucleus?

*Wier:* Those results are consistent with a slight delay of less than 10 ms.

*Nathanson:* We've also looked at $Ca^{2+}$ signalling in the hepatocyte nucleus using confocal line-scanning microscopy (unpublished work). We too find that as the wave propagates across the hepatocyte there is a delay of tens of milliseconds at the most before we see signalling in the nucleus; then we see an increase in the nucleus that otherwise matches the increase in the cytosol.

*Thomas:* Did you look at two points within the nucleus to see whether there was a difference between them? In other words, did you look at the time course once the $Ca^{2+}$ wave was in the nucleus as it propagated through?

*Nathanson:* We haven't analysed that in any detail. We see patterns ranging from what looks like a simultaneous increase to a slight gradient.

*Petersen:* As Dr Kasai's work shows, in pancreatic acinar cells wave propagation in the basal part of the cell is considerably slower than in the other parts. It is worth remembering that a major part of the basal part of the cell seems to be occupied by the nucleus.

*Kasai:* I don't see any slowing or hastening of $Ca^{2+}$ waves that could be attributed to the nucleus.

*Sanderson:* In airway epithelial cells the $Ca^{2+}$ waves propagating across the cell don't seem to bypass the nucleus. The cells are fairly thin, so if there were a delay in wave propagation into the nucleus, we would expect to see a $Ca^{2+}$ wave go around the nucleus. We haven't observed this.

*Jaffe:* As I understand, it takes about 200 ms between the release of $InsP_3$ and the appearance of $Ca^{2+}$.

*Meyer:* We measure nuclear and cytosolic signals about 2 $\mu$m apart. The delay of less than 200 ms is consistent with simple spatial diffusion and the actual delay caused by the nuclear envelope could be much less. If we uncage $InsP_3$ locally in the cell, the responses at different cellular locations show no appreciable delay in RBL or PC12 cells.

*Jaffe:* You uncage $InsP_3$ here and then see $Ca^{2+}$ there, with a delay of about 200 ms.

*Meyer:* No. We measured the delay between cytosolic and nuclear responses. We choose points which are equally far away from the point where we uncage $InsP_3$.

*Whitaker:* Have you calibrated the $Ca^{2+}$ indicator dye independently in the nucleus and cytoplasm? Al-Mohanna et al (1994) did this in dorsal root ganglion neurons and came to the conclusion, as Dr Fay did, that there is a slight, time-dependent buffering of $Ca^{2+}$ in the nucleus, but that $Ca^{2+}$ just diffuses from the cytoplasm to the nucleus through the nuclear pores. You don't believe there's a regenerative $Ca^{2+}$ release mechanism in the nucleus, yet you measure $Ca^{2+}$ concentrations that are higher in the nucleus than in the cytoplasm.

*Meyer:* There's ample evidence that fluorescent dyes bind to proteins in cells. The nuclear composition and the charges on nuclear components are different from those in the cytosol. We think, therefore, that the calibration is inaccurate. A separate calibration in the nucleus is probably necessary.

*Whitaker:* If one looked at your results naively, one would think there must be a regenerative mechanism, whereas your point is in fact more subtle. It would be easy to be misled.

*Petersen:* Your results would not have been entirely predicted from the previous literature. One piece of evidence against the idea that there are large nuclear pores comes from the studies of isolated liver nuclei by Nicotera et al (1989,1990), who showed ATP-dependent $Ca^{2+}$ uptake in isolated nuclei and release in response to $InsP_3$. O. Gerasimenko in my laboratory has tried to look at these experiments again (unpublished work), loading the nucleus with the acetoxymethyl ester of fura. He has found that the fura is actually in a ring just on the outside of the nucleus. In fact, the distribution of fura corresponds exactly to that of an endoplasmic reticulum marker. It is now clear to us that the ATP-dependent $Ca^{2+}$ uptake and $InsP_3$-evoked $Ca^{2+}$ release are due to transport of $Ca^{2+}$ into and from the nuclear envelope and not to transport to and from the nucleoplasm.

*Meyer:* I want to emphasize that I don't believe that $Ca^{2+}$ doesn't have important functions in the nucleus. Calmodulin is highly accumulated there and some of the isoforms of protein kinase C are translocated from the nucleus to the cytosol or vice versa. There are significant changes in some of the phosphatidylinositol signalling molecules in the nucleus and $Ca^{2+}$–calmodulin-dependent protein kinase can phosphorylate CREB inside the nucleus.

*Williams:* Have you ever shown the $Ca^{2+}$ transients to have any direct effect on any kind of expression within the nucleus?

*Meyer:* I don't know whether or not anyone has studied this. Some $Ca^{2+}$-mediated gene expression can occur in a minute or less.

*Williams:* I know that, but I would like to know whether the spiking has anything to do with it or whether spiking is just some other way of changing [$Ca^{2+}$] permanently.

*Meyer:* Everybody would like to find some function, but I don't think there are any good experiments yet that conclusively demonstrate an advantage of $Ca^{2+}$ spiking.

*Berridge:* In this case you would need a marker for gene expression in a cell that's spiking. That's actually asking quite a lot at this stage, but it probably will come.

*Meyer:* We have a model in which $Ca^{2+}$–calmodulin-dependent protein kinase can use a trapping mechanism to respond to different frequencies of $Ca^{2+}$ spikes. We found that calmodulin binding tightens more than 1000-fold on phosphorylation of Thr-286 of the kinase. This effect is due to a transphosphorylation reaction of one subunit to another, generating a kind of positive feedback that makes the enzyme more responsive to spikes with high frequency.

## References

Al-Mohanna FA, Caddy KWT, Bolsover SR 1994 The nucleus is insulated from large cytosolic calcium ion changes. Nature 367:745–760

Cannell MB, Berlin JR, Lederer WJ 1987 Intracellular calcium in cardiac myocytes: calcium transients measured using fluorescence imaging. In: Mandel LJ, Eaton DC (eds) Cell calcium and control of membrane transport. Rockefeller University Press, New York, p 201–214

Lin C, Hajnoczky G, Thomas AP 1994 Propagation of cytosolic calcium waves into the nuclei of hepatocytes. Cell Calcium 16:247–258

Nicotera P, McConkey DJ, Jones DP, Orrenius S 1989 ATP stimulates $Ca^{2+}$ uptake and increases the free $Ca^{2+}$ concentration in isolated rat liver nuclei. Proc Natl Acad Sci USA 86:453–457

Nicotera P, Orrenius S, Nilsson T, Berggren P-O 1990 An inositol 1,4,5-trisphosphate-sensitive $Ca^{2+}$ pool in liver nuclei. Proc Natl Acad Sci USA 87:6858–6862

# Final discussion

*Stucki:* What is the advantage of spiking? There are three ideas. The first is that a frequency-modulated signal is superior to an amplitude-modulated signal. You can test this on your radio; choose FM or AM and you will notice the difference. The second idea was that perhaps oscillatory $Ca^{2+}$ might use less energy than elevated steady-state $Ca^{2+}$. The third possibility is that many systems become desensitized by high $[Ca^{2+}]$. Therefore the $[Ca^{2+}]$ has periodically to return to low levels in order to allow recovery of these systems.

As a mitochondriologist, as I used to be, I was of course interested in the energetic aspect. There is a complete lack of experimental data which would support the suggestion that frequency modulation has energetic advantages. Frustrated by this lack of experimental data, I sat down at my computer and explored where any possible advantages might lie, in an attempt to follow Einstein's dictum.

To look at energetic advantages, you first have to define reference states. I compared an elevated, non-oscillatory steady-state of $[Ca^{2+}]$ with an oscillatory one. The whole trick consists now in choosing the steady-state $[Ca^{2+}]$ and the amplitude of the oscillations such that the mean $[Ca^{2+}]$ is the same for both cases. Obviously, under such conditions, the same amount of ATP is used for pumping $Ca^{2+}$ through membranes, provided that the stoichiometry of the pump is independent of the $[Ca^{2+}]$, which is a fair assumption. I call these states isoenergetic.

In order to see now whether there are any energetic advantages of an oscillatory $[Ca^{2+}]$, we have to look at the function representing the response exerted by $Ca^{2+}$. This matter is too technical to be presented here, but since I am now writing a paper about this, let me just summarize the final result. The response resulting from an oscillatory $[Ca^{2+}]$ is indeed bigger than that resulting from an isoenergetic, steady-state $[Ca^{2+}]$, provided that the response function is convex. For a Hill-type response function this is the case, if the mean $[Ca^{2+}]$ is below $K_a$. To illustrate this, take the case of calmodulin. Let us assume an affinity of calmodulin for $Ca^{2+}$ of $0.6\,\mu M$, a mean $[Ca^{2+}]$ of $0.3\,\mu M$, and a Hill coefficient of four. Then the regulatory effect mediated via calmodulin goes up by a factor of 2.7 in the case of an oscillatory $[Ca^{2+}]$ compared with a non-oscillatory, steady-state $[Ca^{2+}]$. Note that this depends on the amplitude of the oscillations in a non-trivial way.

Perhaps I could summarize this through a banking analogy. If you tell a banker that you have a mechanism which gives you an interest rate of 250%, he would immediately deal with you. The other side of the coin, however, is cash flow. It may be that the total turnover of ATP for this regulation is neglible, such that an interest rate of about 200% of almost nothing would not be an interesting thing. The point is, how large is the energy cost of regulation? Many years ago, Benno Hess (personal communication) made a kind of theoretical extrapolation, and came up with a figure of about 15% of total ATP turnover. Thus it may be that the energetic advantages of $Ca^{2+}$ oscillations are important.

*Petersen:* You asked what advantage oscillations have over a steady-state increase. One simple point, which is seldom made, is that if local signalling is important, which I believe it is, oscillations are obviously advantageous because you can't really have local sustained signals. Local signals are clearly energetically favourable because only a small amount of $Ca^{2+}$ is released (Petersen et al 1994).

*Jaffe:* You *can*, in theory, have local sustained $Ca^{2+}$ signals. We can see steady $Ca^{2+}$ grades.

*Petersen:* You can make systems like that but they're certainly not common and it's not easy to do it.

*Jaffe:* I disagree; I think they are very common, but the methods for reavealing them are not widely used.

*Meyer:* You have to separate the source from the sink to get a standing gradient. In some cells you could just have more pumps and channels at one end or the other.

*Jaffe:* That's just one example. I showed a large series of examples—in fish eggs, fly eggs, in root hairs in plants, pollen grains and probably nerve cells. Wherever there is local activity you tend to get fairly steady $Ca^{2+}$ gradients.

*Williams:* But the $Ca^{2+}$ gradient question is a quite separate one from this pulse question. The pulses are of interest because of the time element, whereas you are talking about the steady-state gradient. However, I totally agree with Jaffe that it's bound to be the case that the pumps and the channels are separated, and when they're separated, there's always a gradient.

*Putney:* If one treats cells overnight with very low hormone concentrations, which never give a macroscopic or population $Ca^{2+}$ response in a laboratory experiment, genes may be induced, but only if there is $Ca^{2+}$ in the medium. I would guess that there may be some very low $Ca^{2+}$ signal that needs to be on for a very long time for that to happen. If there were a graded system, you could extrapolate from dose–response curves to find that the $[Ca^{2+}]$ would increase proportionately very little. Obviously, I don't think a biological system could read that. In other words, there's a background, a noise, in a graded system. In a digital system, in contrast, the background is zero. Cells might be spiking at a very low frequency, but that could be read. Any frequency of spikes could be read, provided you designed the right molecules with the right phosphorylation and dephosphorylation kinetics. I believe that that is the main advantage of spiking. It gives a zero noise level and a much larger dynamic range over which $Ca^{2+}$-mobilizing hormones may act.

*Berridge:* A similar situation occurs in fertilized mammalian eggs that remain quiescent for a long time, then quite spontaneously they generate a single spike at the $G_2$/M boundary.

*Irvine:* In zebrafish that doesn't happen, otherwise Toby Meyer would have seen it (as a rise in nuclear $Ca^{2+}$) in his experiments.

*Berridge:* That's a different situation because in the zebrafish embryo certain cells begin to display spontaneous activity, perhaps connected with the developmental programme. The point I want to stress, however, is that single spikes can have strong information content.

*Thomas:* Low doses of bombesin can induce rhythmic spikes in insulinoma cells with a period of 30 min. It's hard to understand how that could work through the refractory periods we've been talking about.

*Putney:* If you translated that into an average [$Ca^{2+}$] change, it would be impossible to read.

*Whitaker:* I quite agree that there's a difference between repetitive spikes, which we could think of as being frequency encoded, and the sort of spikes that we can see during the cell cycle. There is a problem. If these are both signals, and of course one could argue that the mitotic spikes aren't, why don't they interfere with one another? How can you have more than one sort of $Ca^{2+}$ signalling system in the cell?

*Berridge:* With reference to the cell cycle, we know that different proteins are being expressed at different times. The signal may thus have a different meaning at different points in the cycle.

*Whitaker:* That has always been our argument (Whitaker & Patel 1990), but in the zebrafish, the repetitive spikes don't seem to be associated with mitotic events. It may well be that in zebrafish $Ca^{2+}$ is not important for mitosis, but if it is, you would have to postulate two sets of signals: one set of local $Ca^{2+}$ signals that are controlling mitosis, and a second set of $Ca^{2+}$ signals, repetitive spikes that are doing something else and controlling cell fate during development.

*Berridge:* I would like to pursue what Tobias Meyer was saying earlier about calmodulin kinase. This is an extraordinarily interesting protein, which might have a special role to play in decoding spiking patterns. Could you tell us a little more about how it might work as a molecular counter?

*Meyer:* We recently published two articles postulating that calmodulin kinase has a role in measuring the frequency of $Ca^{2+}$ spikes (Meyer et al 1992).

Several laboratories have been working on calmodulin kinase, which has now turned out to be an octamer or decamer, depending on which subunits are present. These subunits are arranged in a rosette fashion. Each of the subunits can bind a calmodolin molecule. Because this protein is highly abundant in neurons, and because it becomes highly autophosphorylated upon stimulation, it was proposed that it is a key player in memory and in storing information. In many neurons there are repetitive [$Ca^{2+}$] rises triggered by bursts of action potentials similar to $Ca^{2+}$ spikes in non-neuronal cells. Calmodulin kinase is also present in non-neuronal cells and could have roles in responding to repetitive $Ca^{2+}$ stimuli.

We looked at one autophosphorylation site in particular, Thr-286. Phosphorylation makes the enzyme partially autonomous, so that it can phosphorylate substrates even in the absence of calmodulin. Following the binding of calmodulin a *trans*-phosphorylation from one subunit to another occurs. Calmodulin needs to be bound to the donor and the acceptor subunit. You could get a chain reaction in which each of the subunits phosphorylates the next one. The result of this autophosphorylation is a calmodulin kinase with an affinity for calmodulin of 100 pM, rather than the 100 nM affinity it began with. There is a 1000-fold strengthening of the affinity of calmodulin, and, as long as $[Ca^{2+}]$ doesn't go down to 100 $\mu$M, calmodulin stays almost irreversibly bound.

*Williams:* Is that $Ca^{2+}$-bound calmodulin?

*Meyer:* At least some of the $Ca^{2+}$ sites have to be bound. If you lower $[Ca^{2+}]$ to 100 nM, the $Ca^{2+}$ comes off relatively rapidly and calmodulin comes off with a delay of about 20 s. When calmodulin comes off, secondary autophosphorylation sites become autophosphorylated and make it impossible for calmodulin to re-bind.

We have no *in vivo* data so far, only *in vitro* data. The *in vitro* kinetic data predict that if you have a low frequency of spikes, Thr-286 autophosphorylation is competed by phosphatases and the chain reaction would never really start. There will be a particular frequency, however, where subunits become efficiently phosphorylated; a kind of all-or-none phenomenon. Only if $Ca^{2+}$ spikes occur at high frequency will the enzyme become fully activated. This is a simple model, not taking into account that if the spikes are too far apart, you can get inhibitory autophosphorylation of sites that prevent calmodulin binding. Our next step is to look actually in hippocampal neurons for frequency-dependent phosphorylation.

*Berridge:* Is the calmodulin always bound to the enzyme or is it recruited during each spike? My feeling is that if you don't have calmodulins attached to all the subunits, then the autophosphorylation will essentially stop.

*Meyer:* It will stop. You have to recruit more calmodulin.

*Berridge:* Therefore, the significance of spiking might be as a mechanism to enable the enzyme to accumulate calmodulin.

*Meyer:* So by keeping $[Ca^{2+}]$ high, you couldn't trap more calmodulin.

*Berridge:* The molecule may thus function as a spike counter, by virtue of attracting a limited number of calmodulin molecules during each $Ca^{2+}$ transient.

*Meyer:* So far, there is no proof that the concentration of free calmodulin is that low in the cells. If this were the case, then calmodulin kinase would indeed act in that fashion as a frequency detector that responds only to a particular range of frequencies.

*Friel:* Would it be reasonable to say that a requirement for an enzyme to read [$Ca^{2+}$] oscillations would simply be that it is activated with elevations in $Ca^{2+}$ at a rate that's fast compared with the time that [$Ca^{2+}$] is elevated, and is deactivated slowly while [$Ca^{2+}$] is low between spikes. In fact, on the basis of kinetic arguments, an enzyme could read [$Ca^{2+}$] oscillation frequency simply by having the appropriate kinetics of activation and deactivation.

*Jacob:* The only problem with that is that this combination of kientics means that the enzyme has a very high affinity for $Ca^{2+}$, so that very small increases over basal [$Ca^{2+}$] would activate it. The problem is that you also want to be able to detect low-frequency spiking. It's a question of finding something that is not activated by a very small rise in [$Ca^{2+}$], requiring more complicated kinetics.

*Berridge:* My feeling is that in the liver the activation of phosphorylase is probably very quick and can occur in a few seconds. I would imagine that the off-rate is just as fast, thus enabling the enzyme to detect a single $Ca^{2+}$ transient.

*Thomas:* Population measurements of $Ca^{2+}$ and of phosphorylase activation suggest there is a delay but it's pretty small, of the order of a fraction of a second. I don't have data on inactivation, but it is probably slower.

*Berridge:* I would imagine that during the course of a two-second spike, phosphorylase may be activated to give some glycogen breakdown. If another spike doesn't arrive for a while, then the whole system will inactivate and only a small bolus of glucose will be released. In order to release more glucose, the frequency of $Ca^{2+}$ spiking must be increased so as to increase the time the enzyme is in its active state.

*Thomas:* In the context of the liver, we are talking about pumping glucose out into the blood where it's mixed and diluted, and so on. With little pulses of glucose in the second time scale from different cells, it wouldn't matter whether there were pulses or continuous release.

*Nathanson:* If oscillation frequency differentially regulates cell function, surely different enzymes must have different sensitivities and different on- and off-rates.

*Berridge:* That would depend on how many enzymes you are trying to regulate with $Ca^{2+}$ in one cell. A secretory cell is not doing much more than secreting.

*Nathanson:* In the hepatocyte, though, there's clear evidence that $Ca^{2+}$ is important for glucose metabolism as well as for regulation of secretion.

*Thomas:* But there's another level of control. In a phosphorylation reaction, with a calmodulin-activated kinase, the key factor that determines the relaxation rate has nothing to do with the dissociation of $Ca^{2+}$ from calmodulin, or any $Ca^{2+}$-dependent process of activation. It will depend on the rate at which the phosphatase acts. Even without any regulation of the dephosphorylation pathways, if two enzymes are regulated identically in the kinase domain by $Ca^{2+}$, but the phosphatase rates are 10-fold different, the time average of the activities of these two enzymes can be at different levels. They will exhibit differences in sensitivity to the $Ca^{2+}$ oscillation frequency.

*Williams:* When we talk about binding constants, it's *effective* binding constants in a system we are talking about. The bound protein is called 'calmodulin', as if it is the same thing in all places and all the time *in vivo*. Once the $Ca^{2+}$-bound protein binds, whatever it binds to, it's a different protein and the binding constants for $Ca^{2+}$ change. If you say that the bound form of calmodulin is 100-fold stronger in its binding than in its unbound form, the effective $Ca^{2+}$ binding constant is 100-fold bigger. In other words, bound calmodulin here has a binding constant for $Ca^{2+}$ of $10^8$, not $10^6$. If $Ca^{2+}$ binds when calmodulin is free, it has a fast off-rate of $10^3$/s, but if the $Ca^{2+}$ calmodulin binds such that $K = 10^8$, then $Ca^{2+}$ comes off with great difficulty. The off-rate for that $Ca^{2+}$ is at least $10^{-1}$ s.

*Meyer:* It's more complicated because there are four $Ca^{2+}$ binding sites per calmodulin molecule.

*Williams:* I'm talking about the cooperative unit, not about just one $Ca^{2+}$. The cooperative constant is what we actually talk about most of the time. If it is a four-$Ca^{2+}$ system, which is observed in calmodulin or troponin, this observation is extremely interesting, because it does mean that a $Ca^{2+}$ effect can be sustained for any length of time according to the difference in binding constant between free calmodulin and bound calmodulin, due to the changes in protein conformation with the number of $Ca^{2+}$ ions bound. Some proteins will bind $Ca^{2+}$ in the cell for a long, long time because they've got high binding constants (in the order of $K = 10^{10}$), and others will bind $Ca^{2+}$ for quite a long time because they've got, deliberately, slow protein kinetics with a low binding constant (such as $K = 10^6$). Slow kinetics could be achieved by making $k_{on}$ slow, as in phospholipase, for example. A phospholipase could be kept active, with its $Ca^{2+}$ bound, after the pulse had gone, even though $[Ca^{2+}]$ has dropped. I'm not saying that that is the real situation, but it could easily be achieved. Now at this time all the things that aren't wanted any more could have been switched off, provided they release $Ca^{2+}$ quickly because the $[Ca^{2+}]$ has gone down. This would be a great idea for a pulse system; you match your pulse rates to the protein kinetics. Such a system is based on the kinetics of $Ca^{2+}$ binding, not on the thermodynamic $Ca^{2+}$-binding constants.

*Parker:* As a physiologist, what's interesting to me is the read-out that ion channels, rather than enzymes, give of the $Ca^{2+}$ spikes. The $Ca^{2+}$-activated $Cl^-$ channel in the *Xenopus* oocyte approximately signals $d[Ca^{2+}]/dt$ rather than the absolute $Ca^{2+}$ level. We see spikes of current during the rising phase of the $Ca^{2+}$ transients. The current read out is acting as a high-pass filter, preferentially transmitting spike information, yet blocking steady elevations of $[Ca^{2+}]$. For example, at high $[InsP_3]$, $[Ca^{2+}]$ becomes high uniformly in the oocyte and stays high for a long time but this is associated with little or no current. As $[InsP_3]$ drops, you get $[Ca^{2+}]$ spikes and $Ca^{2+}$ waves, which are accompanied by spikes of current. What the current is *doing* is another question, but it is selectively transmitting spike information.

*Berridge:* But it's the rate of change that is important?

*Parker:* Yes, it's the rate of change that determines the current. For a given spike amplitude you get a different current depending on the speed of the rising phase.

*Fay:* Is that because of the local nature of $Ca^{2+}$ signals and the presence of diffusion gradients, and that in reality the signal is $[Ca^{2+}]$ at the effector site, not the rate of change?

*Parker:* We think not. We have made confocal measurements of $[Ca^{2+}]$ within 1–2 $\mu$m of the inner membrane surface, and the current still corresponds to the rate of rise of $[Ca^{2+}]$ not the $[Ca^{2+}]$. We think the differentiation of the $Ca^{2+}$ signal by the CT current is an intrinsic property of the channels, but we need to do patch-clamp measurements to confirm that.

*Meyer:* If this is a general feature of $Cl^-$ channels, doesn't that pose a question for many of the experiments using $Cl^-$ channel measurements to study $[Ca^{2+}]$ increases in cells? Is the $Ca^{2+}$ dependence of $Cl^-$ channels different in different systems?

*Berridge:* Yes; but it still functions as a detector.

*Meyer:* Do $Cl^-$ channel measurements catch only the rising phase?

*Petersen:* There are differences between different types of $Cl^-$ channels. Marty (Evans & Marty 1986) has done long-term infusions of well-buffered $Ca^{2+}$ solutions in lacrimal cells and has measured sustained $Cl^-$ currents; the higher the $[Ca^{2+}]$ is, the more $Cl^-$ current there is. There is certainly not complete inactivation in that particular $Cl^-$ channel; I suspect there are a lot of different $Cl^-$ channels, which will react in different ways. It can't be a general rule that $Cl^-$ channels can detect only changes in $[Ca^{2+}]$.

*Fay:* Using near-membrane $Ca^{2+}$ indicators in single smooth muscle cells, where we are looking below the resolution of optical methods, we see that the $[Ca^{2+}]$ directly beneath the membrane seems to reflect the magnitude, in this case, of $Ca^{2+}$ currents and their kinetics rather than the integral of the $Ca^{2+}$current (Etter et al 1994). This suggests brief gradients result from the influx of $Ca^{2+}$ directly beneath the membrane, presumably where the channels are being affected by $Ca^{2+}$.

*Parker:* In our case, the $Ca^{2+}$ is released from internal stores and we think there is some separation of a few $\mu$m between these stores and the plasma membrane. It is a possibility, although we feel it unlikely, that such a gradient could exist over a time scale of hundreds of milliseconds.

*Berridge:* Perhaps this is an appropriate point for Dr Jaffe to mention his luminal hypothesis.

*Jaffe:* I want to briefly present the arguments which, in my mind, support the idea that $Ca^{2+}$ waves are immediately initiated by a rise in luminal $[Ca^{2+}]$. The idea, essentially, is that there are two different modes of $Ca^{2+}$-induced

$Ca^{2+}$ release. One is a cytosolic mode, which we're all familiar with. The postulated second mode takes place inside the lumen and is very different from the cytosolic mode. The average [$Ca^{2+}$] in the lumen is thought to be in the millimolar, not the micromolar, range. Also, the mechanism does not require as rapid a change as the cytosolic one does. A very slow rise in luminal [$Ca^{2+}$] to a trigger level would suffice.

There's a good deal of evidence, particularly from Fabiato's work on heart cells, that for cytosolic [$Ca^{2+}$] changes to trigger $Ca^{2+}$ release, [$Ca^{2+}$] has to change very fast. Unless [$Ca^{2+}$] rises very quickly, a secondary mechanism whereby $Ca^{2+}$ inhibits its own release soon takes over. The rates of rise needed to trigger release are much higher than the rates of rise of cytosolic [$Ca^{2+}$] that you see in typical records. In typical records just before a spike one often sees a small rise in cytosolic [$Ca^{2+}$], but these rise times are pretty slow, of the order of several seconds. Spikes are typically separated by substantial periods of the order of a minute, and these small rises are a substantial fraction of a minute. The rise times needed in Fabiato's experiments are the order of 100- to 1000-fold faster than one sees in such cells. That argues against a cytosolic trigger, and suggests a luminal one.

The main positive argument for a luminal trigger lies in experiments by Ikemoto et al (1991). These experiments are not well enough known. Ikemoto et al took classical vesicles from the sarcoplasmic reticulum, made in the manner that all muscle physiologists know, and they measured the [$Ca^{2+}$] inside the vesicles with an adsorption dye. They then added caffeine, among other agents, and saw the expected fall in luminal [$Ca^{2+}$] as it was released. Prior to the fall, they saw a clear, large *rise* in luminal [$Ca^{2+}$]. The rise occurred over a few tenths of a second. In another experiment, instead of vesicles they used flat pieces of membrane. You may say that membrane always forms vesicles, but that's not true. Muscle physiologists know well that if you treat these little vesicles with detergent in the right way you can get flat pieces of membrane called junctional face membranes. They added caffeine to these and saw a very large rise in free [$Ca^{2+}$]. That [$Ca^{2+}$] rise cannot occur through opening of a channel because this membrane is flat. The mechanism seems to involve a conformational change in the calsequestrin membrane complex. When Ikemoto et al removed the calsequestrin, there was no response to caffeine, and the response returned when the calsequestrin was put back on. The calsequestrin itself does not respond to caffeine, and the membrane does not, but the complex does. This is a complex protein machine, which releases large amounts of $Ca^{2+}$ in response to caffeine by a conformational mechanism. A little bit after that, the channels do open, but it's not clear whether the channel opening is due to the rise in [$Ca^{2+}$] or whether the two are dual consequences of the same change in the protein machine.

The imaginative step that I took, which may not be right, is to argue that this conformational change is brought about not just by caffeine, which is an artificial stimulus, but by $Ca^{2+}$ itself. When [$Ca^{2+}$] inside the lumen rises to some critical level, this highly cooperative conformational change occurs. Calsequestrin is known to bind 20–50 $Ca^{2+}$ per molecule and can undergo a gross conformational shift which releases $Ca^{2+}$.

*Petersen:* The Fabiato argument that you use does not necessarily apply to all types of $Ca^{2+}$ release channels. Fabiato's work was on a particular type of ryanodine receptor, and we now know from studies by Györke & Fill (1993) that there is rapid adaptation with this receptor. In the case of the $InsP_3$ receptor–$Ca^{2+}$ release channels, the steady-state measurements published by Bezprozvanny et al in 1991 show that a rise in [$Ca^{2+}$] within a certain range gives rise to an increased open-state probability. That's what is needed for a feed-forward mechanism. There is no reason to believe that a time course like the one you have suggested is incompatible with those results. Your argument could apply to certain channels that adapt very quickly, but with other channels there's no problem with the cytosolic oscillator theory.

*Jaffe:* Those experiments by Bezprozvanny et al (1991) were done in reconstituted lipid membranes. Such a method does not give a reliable indication of what happens in a whole cell.

*Berridge:* It's no less reliable than having a sheet and a bit of calsequestrin!

*Jaffe:* No. A junctional face membrane is a far more complete piece of a real cell than is a reconstituted lipid membrane.

*Jacob:* Has this increase in $Ca^{2+}$ been induced by $Ca^{2+}$ release rather than by caffeine?

*Jaffe:* That wasn't attempted.

*Thomas:* When Ludwig Missiaen, Colin Taylor and Mike Berridge were talking about the $Ca^{2+}$ sensitization of the $InsP_3$ receptor from the luminal phase (Missiaen et al 1991), there was an idea that a wave might be propagated by $Ca^{2+}$ being pumped into the endoplasmic reticulum, sensitizing $InsP_3$ receptors from the luminal phase to allow the receptors to open (Taylor & Richardson 1991). We did an experiment in hepatocytes, waiting until we had $Ca^{2+}$ waves propagating through the liver cell then bathing the whole preparation in thapsigargin, which rapidly prevented $Ca^{2+}$ from being taken up into the stores. Obviously, we expected a release of $Ca^{2+}$, due to the leak rate from the intracellular stores, but, in fact, almost all of the cells showed a rapid release of $Ca^{2+}$, which cannot be accounted for by the normal thapsigargin-induced leak and, thus, implies an activated $Ca^{2+}$ channel mechanism. The $Ca^{2+}$ increase still swept through the cell as a wave, but we know it was triggered by thapsigargin. Our interpretation is that $Ca^{2+}$ sequestration into the $Ca^{2+}$ stores is not necessary for the mechanism of wave propagation in the hepatocyte.

*Jaffe:* Are you saying that thapsigargin induces an immediate wave?

*Thomas:* It triggers $Ca^{2+}$ waves in all the cells, but I should stress that this happens only in cells that have been pretreated with hormone. If the cells have not seen hormone, we just get a slow upward drift of [$Ca^{2+}$], which reflects the leak rate. I hesitate to say it induced the wave, but it certainly did not prevent it from spreading.

*Jaffe:* Thapsigargin allows one wave to spread.

*Thomas:* Only one wave, of course.

*Berridge:* The point is that since the pump has been poisoned, there's no opportunity to put further $Ca^{2+}$ into the lumen to do the triggering.

*Putney:* Dr Jaffe's model doesn't need the pump.

*Berridge:* But it does need the $[Ca^{2+}]$ to get to that critical threshold necessary to initiate the release process.

*Putney:* There are many arguments to eliminate a luminal trigger that involves pumping. It is possible, for example, to get three or four spikes in a zero-$Ca^{2+}$ medium. If luminal $Ca^{2+}$ is to increase in this condition, you would have to arrange for it to be released at one site and taken up at another. What Dr Jaffe is saying, extrapolated to an $InsP_3$ system, is that all $InsP_3$ has to do is to somehow cause a conformational change so the $Ca^{2+}$ falls off the calreticulin.

*Berridge:* Dr Jaffe is talking about $Ca^{2+}$-induced $Ca^{2+}$ release.

*Putney:* He was talking about caffeine-induced $Ca^{2+}$-induced $Ca^{2+}$ release in skeletal muscle, but if you wanted to carry that over to $InsP_3$ initiating a spike in the luminal mode with the $InsP_3$ receptor, then you could say that $InsP_3$ causes the $Ca^{2+}$ to dissociate from calreticulin. If it did, you could get a luminal activation without having to put any $Ca^{2+}$ in through the pump.

*Thomas:* I thought the argument was that $Ca^{2+}$ itself going into the stores caused the conformational change.

*Jaffe:* I am proposing that there are two kinds of $Ca^{2+}$-induced $Ca^{2+}$ release. I'm proposing that essentially the same mechanism can be set off either from the outside by a rapid rise in cytosolic $[Ca^{2+}]$, or from the inside by a slow rise in luminal $[Ca^{2+}]$. Dr Putney is thinking more about hormonal stimuli which raise $[InsP_3]$, and is thinking of $InsP_3$ acting directly on the endoplasmic reticulum; he's suggesting that yet another stimulus besides $Ca^{2+}$ outside or inside could trigger the same explosive mechanism. Why not?

*Fay:* There seems to be some discomfort with this sort of luminal mechanism. However, we're in an uncomfortable position because of what Dr Wier told us about coupling $Ca^{2+}$ currents to sarcoplasmic reticulum $Ca^{2+}$ release by $Ca^{2+}$-induced $Ca^{2+}$ release in cardiac muscle. We're missing something in our understanding at present. I understood that $Ca^{2+}$-induced $Ca^{2+}$ release is not an all-or-none process in cardiac muscles but is graded, which doesn't fit with a simple notion of $Ca^{2+}$ at the cytosolic face of the ryanodine receptor controlling $Ca^{2+}$ flux through that channel.

*Wier:* We excluded the cells in which $Ca^{2+}$-induced $Ca^{2+}$ release was not graded, as shown by the existence of waves. In a simple-minded sense, one could say that the conditions we used corresponded to the one mode that Dr Jaffe mentioned, the rapid triggering by cytoplasmic $Ca^{2+}$, and that the mode that we didn't study could be the result of the other, luminal mechanism. I don't really know anything about propagation of $Ca^{2+}$ waves in cardiac cells that would allow me to exclude his hypothesis. However, you can't get definitive information by looking at the little rise or fall in [$Ca^{2+}$] that might occur before the wave.

*Cannell:* We don't see any evidence for $Ca^{2+}$ in activation of $Ca^{2+}$ release during normal excitation–contraction coupling. Spark frequency goes up when you raise resting [$Ca^{2+}$], which is in complete contradiction to the predictions of the model of $Ca^{2+}$ inactivation of $Ca^{2+}$-induced $Ca^{2+}$ release, since Fabiato's inactivation model would predict a decrease in spark frequency. Also, the time course of propagating $Ca^{2+}$ waves is identical to that of electrically evoked $Ca^{2+}$ transients. So, we're seeing the same mechanism, $Ca^{2+}$-induced $Ca^{2+}$ release, in both waves and transients.

*Jaffe:* In one of Fabiato's experiments, done under just the right circumstances, you can actually induce $Ca^{2+}$ waves by lowering the [$Ca^{2+}$]; just the phenomenon that you thought doesn't occur. Fabiato reported on this carefully.

*Williams:* Calsequestrin, as a protein, is exactly the opposite of calmodulin. It is an anticooperative $Ca^{2+}$-binding system. As the level of $Ca^{2+}$ in the stores is lowered, $Ca^{2+}$ release from the protein will get more and more difficult. Calsequestrin is a strange buffer. It binds 40 $Ca^{2+}$ ions, but the effective binding constants start at $10^3$ mM and go all the way down to $10^5$ mM for the last $Ca^{2+}$ ions. As you are emptying the stores you are completely changing the nature of the lumen buffer. Using an anticooperative system inside the lumen and a cooperative system outside is an interesting way of managing release and what happens in a cell.

## References


Bezprozvanny I, Watras J, Ehrlich BE 1991 Bell-shaped calcium-response curves of Ins(1,4,5)$P_3$- and calcium-gated channels from endoplasmic reticulum of cerebellum. Nature 351:751–754

Etter EF, Kuhn MA, Fay FS 1994 Detection of changes in near-membrane $Ca^{2+}$ concentration using a novel membrane-associated $Ca^{2+}$ indicator. J Biol Chem 269:10141–10149

Evans MG, Marty A 1986 Calcium-dependent chloride currents in isolated cells from rat lacrimal glands. J Physiol 378:437–460

Györke S, Fill M 1993 Ryanodine receptor adaptation: control mechanism of $Ca^{2+}$-induced $Ca^{2+}$ release in heart. Science 260:807–809

Hanson PI, Meyer T, Stryer L, Schulman H 1994 Dual role of calmodulin in autophosphorylation of multifunctional CaM kinase may underlie decoding of calcium signals. Neuron 12:943–956

Ikemoto N, Antonin B, Kang JJ, Meszaros LG, Ronjat M 1991 Intravesicular calcium transient during calcium release from sarcoplasmic reticulum. Biochemistry 30:5230–5237

Meyer T, Hanson PI, Stryer L, Schulman H 1992 Calmodulin trapping by calcium–calmodulin-dependent protein kinase. Science 256:1190–1202

Missiaen L, Taylor CW, Berridge MJ 1991 Spontaneous calcium release from inositol trisphosphate-sensitive calcium stores. Nature 352:241–244
Petersen OH, Petersen CCH, Kasai H 1994 Calcium and hormone action. Annu Rev Physiol 56:297–319
Taylor CW, Richardson A 1991 Structure and function of inositol trisphosphate receptors. Pharmacol & Ther 51:97–137
Whitaker M, Patel R 1990 Calcium and cell-cycle control. Development 108:525–542

# Summing up

*Berridge:* At the conclusion of such an exciting conference it is not appropriate to try to summarize all the talks, so I will highlight those aspects that I found particularly interesting. One of the major problems that remains to be solved in $Ca^{2+}$ signalling is the nature of the initiation process. I believe that the description of $Ca^{2+}$ sparks in cardiac cells and $Ca^{2+}$ puffs in *Xenopus* oocytes represents a major breakthrough in our understanding of how $Ca^{2+}$ signals are initiated. These sparks and puffs represent the unitary events whose summation or coordination is responsible for initiating a typical response. The picture that emerges is that a single or a small group of ryanodine and inositol 1,4,5-trisphosphate ($InsP_3$) receptors can open stochastically to give a brief pulse of $Ca^{2+}$ (a spark or puff), which can either dissipate or, if conditions are right, trigger a regenerative process whereby neighbouring receptors, recruited by the initial event, act to spread a $Ca^{2+}$ wave throughout the rest of the cell. Future studies will have to address how this sequence of events is controlled to give regular $Ca^{2+}$ spikes.

One aspect that has to be accommodated in any future explanation of $Ca^{2+}$ signalling concerns the way in which such $Ca^{2+}$ signals can be graded, even though they seem to depend upon the regenerative process of $Ca^{2+}$-induced $Ca^{2+}$ release. One way in which this paradox might be resolved emerged from the recent work on cardiac cells described by Dr Wier. The voltage-operated $Ca^{2+}$ channels, which gate the small pulse of trigger $Ca^{2+}$, appear to be tightly coupled to the underlying ryanodine receptors to form a 'calcium synapse'. By restricting the process of $Ca^{2+}$-induced $Ca^{2+}$ release to these microdomains, the amount of $Ca^{2+}$ released from the sarcoplasmic reticulum can be graded with membrane potential. Just how $Ca^{2+}$ signals are graded in non-muscle cells remains to be worked out. As many cells contain both ryanodine and $InsP_3$ receptors, attention is being focused on how these two intracellular receptors might be interacting with each other. The picture is complicated by the finding that these receptors have variable subtypes with markedly different properties. For example, the ryanodine receptor (RYR) has three forms; all three forms bind ryanodine but only RYR1 and RYR2 seem to be sensitive to caffeine. Perhaps the most interesting feature of these receptors is that they display a use-dependence in response to ryanodine. This was clearly demonstrated in the case of the sympathetic neurons which become quiescent one spike after the addition of ryanodine. Ryanodine was also seen to inhibit the spiking induced by cholecystokinin in pancreatic acinar cells. What is even more interesting is that the action of cyclic ADP-ribose, the putative second messenger of ryanodine receptors, is inhibited by heparin, which is known to block $InsP_3$ receptors. Observations of this kind have led Dr Petersen and his colleagues to propose that ryanodine receptors are intimately connected to the action of $InsP_3$ receptors during the onset of $Ca^{2+}$ signals. Interactions between these two receptors may be central to the generation of $Ca^{2+}$ signals in many other cell types.

I should like to finish by considering the functionality of such spatially organized $Ca^{2+}$ signals, beginning with early development. It is important to realize that the immature *Xenopus* oocyte is surrounded by pronounced electrical fields. There is an increased sensitivity of the $InsP_3$ system in the animal pole, so one wonders whether the gradient of activity of the enzymes associated with $InsP_3$ production and sensitivity varies from the animal to the vegetal pole and contributes to the establishment of this field, which may have a role in setting up some of the axes during early development. At the time of fertilization, the sperm activates the egg to generate a wave, which sweeps through the cell and is responsible for initiating the developmental programme.

After development comes differentiation and there is evidence for a role of $Ca^{2+}$ waves and gradients in several functional systems. A push–pull model has been proposed for $Ca^{2+}$ waves initiating in the apical region of the pancreatic acinar cell before propagating to the basal side. Ions leave at the apical side and are then sucked in across the basal membrane. Eosinophils undergoing chemotaxis can detect a concentration gradient from the front to the back of only 2%. Somehow this small spatial gradient is translated into a steep gradient of $Ca^{2+}$ within the cell. It seems quite magical that the cell can detect such a small difference in concentration between its front and its back as it moves around in its murky world.

In the nervous system, there are fascinating interactions operating between neurons and glia. It has been apparent for some time that glutamate from the nerve terminals can activate surrounding glial cells. Now we have to consider the possibility that the glia might be able to talk back to the neurons, as suggested by Dr Smith. This is potentially very interesting, especially with regard to the mammalian suprachiasmatic nucleus, which contains the biological clock. Somehow the activity of this population of clock neurons must be synchronized. The suprachiasmatic nucleus is stuffed full of glial cells. The glial cells might provide a matrix which enables the neurons to communicate with each other so that they fire in synchrony. Such synchronization might be achieved by means of the intercellular waves that have been characterized by Dr Sanderson.

In closing, I would like to thank you all for coming and for being a super group for me to chair. I have not had to think of a single question, which is a real testament to a good meeting.

# Index of contributors

*Non-participating co-authors are indicated by asterisks. Entries in bold type indicate papers; other entries refer to discussion contributions.*

*Indexes compiled by Liza Weinkove*

# Subject index